Who You Are

Who You Are

The Science of Connectedness

Michael J. Spivey

The MIT Press
Cambridge, Massachusetts
London, England

© 2020 The Massachusetts Institute of Technology.

All rights reserved. No part of this book may be reproduced in any form by any electronic or mechanical means (including photocopying, recording, or information storage and retrieval) without permission in writing from the publisher.

This book was set in Stone Serif and Stone Sans by Westchester Publishing Services. Printed and bound in the United States of America.

Library of Congress Cataloging-in-Publication Data is available.

ISBN: 978-0-262-04395-3

10 9 8 7 6 5 4 3 2 1

For Daddy Ossie, Grampa Bill, Uncle Nat, Dion, Jennie Mama, Sean, Glynn, Gramma Donna, Jon, Chris, Aunt Sarah, Dick, Guy, Michael, Isamu, Bruce, Jeff, and Dennis

Contents

Acknowledgments ix
Prologue: Who You Might Think You Are Now xi

1 Let Go of Your Self 1
2 From Your Soul to Your Prefrontal Cortex 19
3 From Your Frontal Cortex to Your Whole Brain 47
4 From Your Brain to Your Whole Body 73
5 From Your Body to Your Environment 113
6 From Your Environment to Other Humans 147
7 From Other Humans to All Life 177
8 From All Life to Everything 207
9 Who Are You Now? 239

Notes 257
References 295
Index 353

Acknowledgments

As with any major project, there are countless intellectual influences that contributed to my writing of this book. My parents, my sister, and Steve helped turn me into the kind of person who would write a book like this. Specific suggested topics, feedback, and revisions were offered by three anonymous reviewers retained by MIT Press, and also by my friends and colleagues Ramesh Balasubramaniam, Mahzarin Banaji, Ben Bergen, Barbara Finlay, Chris Kello, Paul Maglio, Teenie Matlock, Dave Noelle, Paul Smaldino, and Georg Theiner, among others. Elizabeth Stark was especially helpful at helping me strike the right tone. Over the years, I have drawn inspiration for this work from my conversations with my students and with experts, including Larry Barsalou, Claudia Carello, Tony Chemero, Andy Clark, Rick Dale, Shimon Edelman, Jeff Elman, Riccardo Fusaroli, Ray Gibbs, Scott Jordan, Günther Knoblich, Jay McClelland, Ken McRae, Claire Michaels, Daniel Richardson, Mike Tanenhaus, Michael Turvey, Guy Van Orden, Jeff Yoshimi, and also my son Samuel Rex Spivey (find him on soundcloud.com). My wife, Cynthia, was extremely insightful at the metatheoretical level. Without her encouragement, this book would have been completely unreadable by a general audience, and it wouldn't have the "directions for use" sections. The guidance from Phil Laughlin, my acquisitions editor at MIT Press, was crucial in allowing me to see a light at the end of the tunnel. And finally, I am grateful to Melody Negron at Westchester Publishing Services, and Hal Henglein, who helped me avoid numerous embarrassing mistakes (including reducing the number of sentences that start with the word "And.")

Prologue: Who You Might Think You Are Now

Do you know who you are?
—Keaton Henson, "You Don't Know How Lucky You Are"

The phrase "who you are" could refer to multiple things. For some people, it might refer to their conscious awareness. For others, it might refer to their self-image or perhaps to their personality. Still others might conceive of "who they are" as referring to their personal reputation among their family or their community. "Who you are" can be many things, and I encourage you to think of it as *all these things* and more. In this book, we will go on a journey through an ever-expanding definition of the physical material that makes up "who you are." Like a Venn diagram that starts out small, as a circle that contains the stuff that makes you you, initially it might contain only a particular region of your brain. Then, as we progress through the chapters, that circle cannot help but expand to include your whole brain, then your whole body, then more and more. For any learning exercise, starting out small is good.

Many people operate under an unspoken assumption that "who they really are" is something deep inside them. We develop this impression, in part, because sometimes we have to put on a "social mask" that is somewhat different from how we really feel or how we really want to behave. A popular metaphor here is that of peeling an onion. Each surface that looks as if it might be who you are can actually be peeled away to reveal a more "core" version of who you are, and it, too, can be peeled away to reveal something even more central. If you start with that assumed sense of who you are, something deep inside, then the chapters in this book are actually designed to reverse that curious metaphor incrementally. By carefully analyzing the content and makeup of each of those layers, we will in fact see that they are all part of who you are. After all, if that social mask were not part of who you are, then you wouldn't be able to use it so effortlessly, would you? And, believe it or not, there are quite a few more layers to your onion than you might think.

Knowing who you are, and embracing all those layers, is important because an inaccurate conception of who you are can lead to decisions and actions that hinder or harm you. Imagine that an onion somehow thought that only its inner bulb was "who it was." Then it might neglect its outer layers, not feeding those cells as it grew. What a sick and sad onion that would be! When you know in your heart that something outside your body is actually "part of you," such as a child, a pet, or even a favorite book, you take good care of that something because by doing so you are taking care of your self. If you take good care of all those different parts of your expanded self, then you will live a better and longer life.

This book is specifically designed to shake up those narrow assumptions about who you are, what stuff makes you you. It does not do it with poetic pronouncements that sound, on their own, like they might be true, although there might be a few flowery turns of phrase here and there. It does not do it with philosophical "thought experiments" that pump your intuitions up out of the ground like subterranean water, although you will encounter a couple of those. The primary tool of persuasion in this book is scientific study. Science is the best method by which we humans have been able to determine reliable facts about how this world works. So many other methods have relied too much on poetry and intuition, and have a bad habit of making failed predictions or being proven wrong in other ways—usually by science. But you do not have to be a scientist to enjoy what is being delivered in this book. The scientific studies are described in just enough detail for you to understand and trust the results that were obtained, and if you want to look up and read the actual scientific studies that I describe here, each chapter subheading has its own section in the Notes at the end of the book, with some more in-depth discussion and lots of references.

Here is what this book has in store for you. As each chapter moves along, it will stretch and expand one particular definition of who you are, or what the self is. Each chapter's minor stretching or expansion is designed to be small and gentle enough that it should not be too painful, mind-boggling, or unbelievable. Chapter 1 is designed to help you let go of your biases and assumptions about who you are, so you can approach the chapters that follow with an open mind. As you work your way from chapter 2 to chapter 3 to chapter 4, the examples of scientific experiments will give you compelling evidence for expanding your definition of your mind to include all your brain and all your body. If that sounds obvious to you right now, then you are exactly the kind of person who should be reading this book. As you work your way from chapter 5 to chapter 6, you will encounter powerful scientific evidence for treating the tools and people in your environment as additional parts of who you are. If that sounds ridiculous to you right now, then you are exactly the kind of person who should be reading this book. You just might find yourself convinced to change your mind about what your mind is. Then, if you make it agreeably through chapters 7 and 8—and, to be honest, many readers will have objections there—you will see scientific evidence for nonhuman life and even nonliving matter being *intelligent* like you and even *part* of you. You do not have to accept the claims being made in chapters 7 and 8, but you will find it worthwhile to peruse the scientific evidence. Let the evidence wash over you like a light rain. Put away your mental umbrella. Some of that rain will soak right into you, and some might bead up and slide right off. Then, when you move on to chapter 9, the final chapter, you can see whether my summary will help you tie it all up and put a bow on what you've learned. That is what this book has in store for you.

I certainly do not expect every reader to come along with me happily all the way through all eight chapters of mind-boggling scientific findings, but I do expect you to give those scientific findings a fair shot at figuratively "expanding your mind" and literally expanding your definition of who you are. That's why you cracked open this book in the first place. Irrespective of how far you follow along with me in this book, my hope is that I can pull you along further than you would have gone while relying solely on your intuitions. Importantly, each chapter will not only get you to *learn something new* but also get you to *do something new* with that fresh knowledge. At the end of each chapter is a "directions for use" section, aimed at having you try out some of your newly acquired wisdom in the real world. When you get into the habit of *applying* your expanded mind and not just reveling in it, you will find that you can help introduce change into the world that makes it a better place for everyone and everything.

As you will see in this book, the cognitive and neural sciences are showing us that who you are is much more than an immaterial soul. It necessarily includes the reasoning parts of your brain, and those reasoning parts are so intimately dependent on the perception and action parts of your brain that who you are has to include them as well. Further scientific findings are showing that your peripheral nervous system, your muscles and joints, and even the gristle between your muscles are all transmitting information patterns related to forces that are acting on your body. Your entire body is doing some of your thinking for you. It is part of your mind. If you are willing to allow those non-nerve-based transmissions of information and data (across muscles and connective tissue) to be part of who you are, then you should be open to seeing *other* non-nerve-based information patterns as part of your mind when they directly impact your thinking, such as when your hand catches a ball or when your foot presses the accelerator on your car. Those external objects become part of you, and amid those toys and tools that you rely on so much, there are also other humans that you routinely rely on. Think of a conversation where you and the other person completed each other's sentences and communicated entire ideas with just a nod. In a good conversation, that other person can become part of who you are. Extend that idea to the communication you achieve with your favorite pet. Is he or she part of who you are? What about the plants in your garden or the soil that feeds them? Wherever you draw the line and stop your expansion of self, that spot might seem a bit arbitrary to an unbiased onlooker. But don't just take my word for it. Please don't fall for those cutesy examples just cited. Trust the results of carefully designed scientific experimentation and analysis. This book is littered with them.

Throughout this book, I will continue to address the reader as "you," as if we were just hanging out and chatting. I hope you don't mind. I will continue to address you that way even after the content of a chapter has encouraged you to expand your sense of self to include people and things outside your body. That expanded sense of self will still be using the eyes in *your* head to read this text, so when I address you directly, you can let that word "you" refer to as wide a scope of your *self* as you like. When we get to chapter 8, my use of the word "you" will actually be referring to everything on Earth and more. Will you be ready for that?

Who You Are is not quite a choose-your-own-adventure book, but it's close. You are encouraged to choose your own conclusion as you work your way through the chapters. Whatever station you get off at on this train ride, my hope is that the scientific evidence was sufficient to take you further along the track than you had been. All aboard!

1 Let Go of Your Self

> There is no pilot. You are not alone.
> —Laurie Anderson, "From the Air"

Breathe

Take a deep breath and let it out slowly. Go ahead, humor me. Close your eyes for several seconds, breathe in deeply through your nose until you can't anymore, and then let the air out slowly through your mouth. Do it now.

Now we're back. Well done. Breathing is important. When you pay attention to your breathing, you realize that the air around you isn't just some empty vacuum in which your body moves around. We are all constantly submerged in a chemical gas cocktail that our bodies need to survive. It is made of nitrogen, oxygen, hydrogen, and several other elements. As you breathe in that chemical gas, give some thought to the fact that those chemicals are now becoming part of who you are. Biologist Curt Stager has his readers take a breath as well in his book *Your Atomic Self*. He then goes into beautiful detail about the trillions of molecules that get sucked into your lungs. Our brains and the rest of our bodies are made of these molecules, and so is everything else around us. One of the oxygen molecules in the breath you just took could wind up fueling a neuron just in time to receive a waterfall of glutamate neurotransmitter molecules in a synapse that helps your brain process this very sentence as your eyes pass over it, and that's why you comprehend it. You just helped an oxygen atom understand itself!

For decades, health scientist Andrew Weil has been saying that the number one piece of health advice he would give everyone is that they need to learn how to breathe properly. You may think you already know how to breathe, but you might not. Breathing is something we all do all the time, so we can all improve our general health simply by improving our breathing. When we get stressed or overly focused, we often literally forget to breathe for a few seconds. Those few seconds rob your brain of essential oxygen. If you notice yourself doing it, then force yourself to take a deep breath. Breathing exercises are also an important part of many meditation techniques. Proper breathing helps the brain stay attentive and calm at the same time. Since we're talking about it, why not take another deep breath and let it out slowly, right now. You're going to need your brain to be well oxygenated while you read this book. There are some challenges ahead; or, as the Hunt-Lenox globe of Earth from the year 1510 warns in its uncharted waters, "Here be dragons."

You Are Not Who You Think You Are

One of the challenges you will face in this book is the idea that the you that you *think* you are is not the you that you *really* are. But this book will not tell you *who you are*. Each person must decide that for herself or himself. This book will give you example after example of scientific evidence that will encourage you to expand your definition of who you are, but first you have to let go of your old definition of who you are. How far you come along with me in this gradually expanding sense of self is up to you—whoever you are.

It actually shouldn't be surprising that you are wrong about who you really are. We are all routinely wrong about all kinds of things in our everyday experience. In fact, cultural critic Chuck Klosterman wrote a bestselling book about it, *But What If We're Wrong?* He explores examples of when our society was wrong about things in the past and it took decades to figure it out. He then asks what we might be wrong about today. I know I am frequently wrong about all kinds of things. It's part of being human. For example, we momentarily misunderstand each other in conversations. Usually, we detect it and query each other until we get back on the same page. While reading a sentence, we occasionally misread a word, and when we get to the end of the sentence, we realize it didn't make sense. Rather than conclude that the author is illiterate and throw away the book, we simply reread the sentence and find the right wording, which was there all along. Sometimes while writing, we produce a typo and fall to notice it. And once in a while, we see a fleeting shadow or reflection in the corner of our eye that looks as if it might be an unexpected object or person in the room with us. Then we point our eyes directly at the location of the visual effect and see nothing unusual there. Some people conclude that this momentary misperception is proof of a spiritual presence, but most of us understand that it was just a trick of the light. It is important to *have some humility about your interpretation of the world around you.*

Don't Always Trust Your Perception

Good old-fashioned magic tricks are some of the best examples of why we should not trust our initial perception of the world around us. A good magician will tell you that he is not performing magic; that it's really just a trick and that your perception of the event is actually a misperception. A bad magician will perform the trick in so clumsy a fashion that you will be able to see *both* the intended misperception and the actual state of affairs, and a rebellious magician will tell you *how* the trick is performed!

James Randi ("The Amazing Randi") is one of those rebellious magicians who teaches people how magic tricks work. When you know how those magic tricks work, you can better identify the sleights of hand that happen all around us all the time—sometimes with catastrophic consequences. For example, a skilled psychic who pretends to communicate with the dead could perform some magic tricks to convince you that your deceased father wants you to invest all your inheritance money in a company that you've never heard of. Later, you might find out that the psychic owns a large interest in that company! These things happen. Randi devoted his career to debunking exactly those kinds of confidence schemes.

After many appearances on *The Tonight Show* with Johnny Carson, as well as Penn and Teller's TV show, James Randi became particularly famous for his Million Dollar Paranormal Challenge. For decades, he and his foundation offered a prize of one million dollars to anyone who could convincingly demonstrate psychic powers under controlled test conditions. No one ever succeeded.

When you know what to look for, you can easily detect how a skilled con man will try to fool you by misdirecting your attention at just the right time or picking up on a microexpression that your face reveals—all in about half a second. Not long ago, The Amazing Randi teamed up with my old college lab mate Steve Macknik to write a review of the brain science behind magic tricks. The brain of the observer plays a crucial role in most magic tricks. A key point Randi and Macknik make is that your eyes and your brain are typically focused on just one thing at a time, and therefore things that are outside your focus at a given moment can be exchanged with other things and you simply will not notice.

One version of this phenomenon that Randi and Macknik write about is called "change blindness." When two images that are almost identical are alternated back and forth—with a very brief blank period between them—the difference between the two images can be very difficult to identify. It could be the engine on the wing of a plane that is present in one image and edited out in the other, but it will take viewers over a dozen viewings back and forth to finally find that very important difference. You wouldn't mind flying in one of those planes, but flying in the other one would be a very bad idea. Yet people are initially blind to the change in these two photos. The internet has a variety of change blindness demonstrations that you can easily find for yourself so you can see just how hard it is to detect the changes. Change blindness even works for gradual changes, where an object slowly changes its color while the person is viewing an otherwise static picture. It works with movies, too. In fact, on a Hollywood movie set, there is usually a "continuity director," whose job is to pay attention to little details in the surroundings to make sure that when the scene is performed multiple times, and different parts from different performances are spliced together, it still looks like one coherent scene. Next time you watch a movie where some people are talking while seated and there are drinks on the table, pay close attention to the drinks. As the camera switches back and forth between speakers, one of the drinks might be half-full at one moment and then three-quarters full two seconds later. When you look for these continuity errors, you can sometimes catch them, but usually we take for granted the consistency of our surroundings. This is why change blindness happens, why continuity errors escape us, and why magic tricks work.

For cognitive scientists and neuroscientists, however, change blindness is much more than just a cute trick. It is a powerful demonstration that, despite our everyday intuition that we are fully aware of our surroundings, our brains actually encode and store surprisingly little about our immediate environment. Change blindness can even happen in real-life face-to-face human interaction. Twenty years ago, while I was a professor in the Psychology Department at Cornell University, two of our graduate students, Dan Simons and Dan Levin, conducted a live-conversation change blindness experiment right outside the window of my third-floor office. In their experiment, another graduate student, Leon Rozenblit, would accost pedestrians on campus and ask for help with finding directions on a paper map that he had in his hands. After about 15 seconds of conversation, Leon and his unwitting experiment participant would be interrupted by two people (the Dans) carrying a door lengthwise. They would walk right between Leon and his conversation partner. Then, just before the back end of the door finished obstructing the two people talking, Dan Simons would trade places with Leon and take the map from his hands. Simons would then resume the conversation with the pedestrian as if nothing strange had just happened.

Imagine this happening to you. You're walking along the sidewalk, minding your own business, and someone with a map in his hands asks you for directions. Deciding to be charitable, you agree to spend a couple of minutes helping this lost soul find his way among the streets and buildings surrounding both of you. After you chat a bit, suddenly some rude people carrying a door walk right between the two of you, and when they finally get past you, the person you've been talking to has changed into a different guy. His voice is different, he's wearing a different coat, he's about two inches shorter, his hair has changed from black to brown, and he's not as handsome as he was a second ago. (Just kidding, Dan.) Of course you would notice the change, right?

Wrong. Odds are you wouldn't notice the change. Less than half the pedestrians reported noticing anything unusual happening during the conversation (aside from the rudeness of the two guys carrying the door). Most of them just continued happily assisting Simons with finding his directions on the map, not even realizing that he wasn't the same person with whom they had started the conversation. I guess it really is important to *have some humility about your interpretation of the world around you.*

Don't Always Trust Your Memory

If your perception of the world around you is unreliable over the course of just a couple of minutes, then it shouldn't be surprising that your memory of those perceptions years later can be even more unreliable. But sometimes we feel as if particular important memories are pristine and unaltered. In the 1970s, cognitive psychologists identified a form of memory in which it feels as if the event being remembered was "burned" into your brain in its raw original form, a bit like how the flashbulb from an old-fashioned camera "burns" an afterimage of the scene onto your eyes. They called these memories "flashbulb memories." Most people have one or two flashbulb memories from intense experiences in their lifetime, where it feels as if the memory can be invoked in your imagination with all the tiny details of the original experience. Many Americans who were adults in fall 1963 reported having flashbulb memories of what they were doing when they found out that President John F. Kennedy had been shot. For example, they might be able to report exactly which seat at the bar they were sitting in, where on the wall the TV was, what drink they were drinking, and who was sitting next to them—even though it happened decades ago. And, of course, many people around the world have flashbulb memories of what they were doing when they found out about the September 11, 2001, terrorist attack that killed thousands of people and brought down the Twin Towers in New York City. However, some flashbulb memories are more specific to a subset of the population, such as sports fans who see their team win a championship for the first time ever, residents of a neighborhood that gets destroyed by a tornado, or guests at a children's birthday party watching the clown pull out a gun and rob all the parents. These things happen. When people have an intense emotional experience, more fine-grain details of that experience seem to be stored in memory than get stored for normal, everyday events.

Almost a decade before 9/11, I acquired a flashbulb memory of the day I found out that musician Kurt Cobain had killed himself. I was a graduate student at the University of Rochester in spring 1994 and a devoted fan of Cobain's rock band, Nirvana. I was in an office that I shared with two other graduate students, and the phone rang. Upon learning of the sad news, which was tentative and uncertain at the time, I went to a campus restaurant at the Rochester Institute of Technology and drank a pitcher of Killian's Red beer by myself while watching MTV News cover the unfolding story on a large-screen TV. At least, that's what my flashbulb memory tells me.

Another decade before *that*, on January 28, 1986, the space shuttle *Challenger* exploded shortly after liftoff, and many people acquired a flashbulb memory of that event. At that time, psychologist Ulric Neisser was at Emory University in Atlanta, and he realized that this was an opportunity to actually test the theory of flashbulb memories scientifically. Neisser was the father of cognitive psychology. After all, he wrote the book on it—literally. In the mid-1960s, when behaviorism still dominated psychology and most scientific psychologists were busy training rats and pigeons to do slightly interesting things, Neisser wrote a hugely influential book that combined psychology, computation, and linguistics to study the human mind. He titled the book *Cognitive Psychology*. Two decades later, Neisser's research program was focusing on memory. He knew that the only way to really test the theory of flashbulb memories was to have an original record of the "ground truth" for a person's remembered event and then find a way to compare that to the "flashbulb memory" years later. So, on the morning after the *Challenger* accident, Neisser's research team asked a group of college students to write down the detailed circumstances of how they found out about the explosion of the *Challenger*. Many of them wrote down a full page of fine-grain details. Some had heard from a friend in their dorm. Some had seen it on the TV news. Others heard it over the phone.

Nearly three years later, Neisser tracked down almost all those students and brought them into his cognitive psychology lab again. The students were asked whether they had vivid, detailed ("flashbulb") memories of how they found out about the *Challenger* explosion. Many of them did, so the experimenter asked them to write down these flashbulb memories. When each student was done writing down their flashbulb memory, the experimenter would walk over to the file cabinet, pull out the original description from that student, handwritten on the morning after the event three years earlier, and hand it to them. The students could clearly see that the two descriptions (current and original) were in their own handwriting, but about half of them had wildly different stories when comparing their current description to the original one. Locations were different. Informants were different. The interesting thing is that, upon being confronted with their own handwritten original description, several of the students found themselves unable to revive the original accurate memory in their minds and replace the inaccurate one with it. They felt tempted to distrust their own original handwritten story. In one case, a student's flashbulb memory was that she was eating lunch at the campus cafeteria when a student came running in and announced the tragedy to everyone. However, her original description stated that she was by herself in her dorm and saw it on TV. She had trouble accepting that her flashbulb memory was wrong.

It should be noted that I am drawing from memory myself when I provide that particular example student story. Ulric Neisser and I were at Cornell together for a dozen years, and I have vivid, detailed memories of chatting about flashbulb memories with him, watching Super Bowl XXXV with him, and going to our first demolition derby together. At least, I think those memories are accurate. As I said, it's important to *have some humility about your interpretation of the world around you.*

When an everyday memory is inaccurate, it can be pretty embarrassing, but when the memory of a *criminal act* is inaccurate, the consequences can be devastating. For example, the Innocence Project in the United States has documented hundreds of criminal cases where eyewitness testimony was later proven inaccurate with DNA evidence. It is all too frequent that witnesses of terrible crimes accidentally pick out the wrong person in a police lineup. Later, after replaying the event in their mind with that person's face mentally superimposed on the memory, the witness shows up in court and feels even more confident than before that the defendant sitting at that table was indeed the perpetrator. The innocent defendant then gets convicted and sentenced to multiple years in prison.

In his Cornell laboratory, David Dunning conducted experimental tests of eyewitness testimony and found evidence that a careful deliberation of facial features and a detailed discussion of selection procedures can actually be a sign of an *inaccurate* identification. It's when people find themselves unable to explain why they recognize the person, saying things like "his face just popped out at me," that they tend to be accurate more often. Sometimes our first, immediate, automatic reaction to a situation is the truest rendition of what our mind is really doing. That very first impression can also be more accurate about the world than the deliberative, reasoned self-narrative can be. In his book *Blink*, Malcolm Gladwell describes a variety of studies in psychology and behavioral economics that demonstrate the superior performance of relatively unconscious first guesses compared to logical step-by-step justifications for a decision.

Almost three-quarters of the criminal cases that the Innocence Project has overturned with DNA evidence involved highly confident, but apparently inaccurate, eyewitness testimony. What might be even more surprising is that more than a quarter of the cases overturned by DNA evidence involved a *false confession* by the accused during interrogation. Why in the world would someone confess to a crime they didn't commit?

As it turns out, police interrogation procedures can occasionally induce false confessions, in much the same way that a psychiatrist might inadvertently implant false "recovered" childhood memories in a patient or an investigator might accidentally induce a false memory in a child. These things happen. Anyone who is suggestible enough to be hypnotized by a stage magician is suggestible enough to crack under the pressure of a police interrogation. After several hours of questioning, a suggestible adult can begin to imagine that anything would be better than withstanding further questioning. Confessing would at least bring the interrogation to an end for now. And some of the very same mind games that good investigators use to trick a guilty person into revealing knowledge that only the criminal would have are exactly the kinds of mind games that can accidentally trick a suggestible innocent person into falsely confessing—such as pretending there was a witness that saw you do it.

Saul Kassin at Williams College conducted a laboratory experiment in which he successfully induced about half the participants into signing a false confession that they had damaged a laboratory computer's software by pressing a key that they were instructed not to touch. When there was a false witness claiming that she saw the participant press the forbidden button, the rate of confessions nearly doubled! A few of the participants even provided detailed descriptions of how they accidentally pressed the key that they never actually touched.

Mistakenly accepting a false characterization of our selves and our actions—and even going so far as to justify and explain that inaccurate characterization—can happen in a wide variety of contexts. For example, Lars Hall of Lund University in Sweden converted the change blindness experimental paradigm described earlier into a "choice blindness" experimental paradigm. He and his team used a good old-fashioned sleight-of-hand card trick to fool Swedish voters into agreeing with political positions to which they had just reported they were opposed. Sound like a crazy magic trick? That's exactly what it is.

Imagine there's an important vote coming up and a pollster on the street asks you to answer some questions. Not in a hurry, and feeling generous, you agree to offer your opinions and fill out a short questionnaire on a clipboard. The questions address 12 different political issues, and they require you to place a mark on a left-right scale that reports the magnitude of your political leaning toward the liberal (left) or conservative (right) sides of that scale. Unbeknownst to you, while you fill out your questionnaire, the "pollster" watches your responses and completes an identical questionnaire using the same style of check marks or x's that you used, and in the same color ink, but he fills in responses that lean in a slightly different political direction than your responses did. When you return the questionnaire, the pollster places his clipboard on top of yours, at which point it secretly glues his questionnaire on top of your questionnaire. Then the pollster and you look at your (altered) questionnaire together, and he asks you to logically justify the responses on it. But you're no dummy. You'd interrupt him and say, "Hey, those aren't the responses I just gave you." Wouldn't you? Probably not. Only 22 percent of the people noticed the changed answers. The other 78 percent made no complaint and went ahead and provided justifications for the responses on the sham questionnaire as if those responses were their own! After discussing each response, the pollster then asked the person to examine their overall view and determine which political party they would vote for. After that experience of being tricked into thinking their political opinions were different than they originally supposed, and finding themselves justifying them, almost half the people said they were now open to considering a left or right shift to another political party.

Don't Always Trust Your Judgment

Getting someone to embrace an idea that they just said they were opposed to a few minutes ago is a pretty good trick, but what if you could get them to embrace your preferred idea *during* their initial response in the first place? That's what a salesman tries to do when selling you a car. In that case, you *know* that's what he's doing, but what if you didn't know it was happening?

For someone's thought process to be visible enough for a person, or perhaps a computer system, to manipulate it, you would need a way to "see into their mind" while they were ruminating over a decision. Well, as it turns out, when you waffle between two options, your eyes and hands often reveal this waffling from one fraction of a second to the next. The evidence that your mental waffling between response options is revealed by your momentary eye and hand movements comes from a variety of sources, not the least of which is a good poker player who knows how to identify "tells." There have been several books written on poker "tells," to teach the reader how to detect a facial expression or a hand gesture as evidence that your opponent has a good or bad poker hand, but even professional poker players still make a lot of mistakes because—just like you and me—they can fall prey to change blindness, memory failure, and poor judgment. By contrast, when you record a person's eye movements and reaching movements with scientific equipment, you have all the data on file. You don't miss a thing.

One example of this is research that my former student Rick Dale did with Chris McKinstry. Chris was an amateur artificial intelligence aficionado who "crowdsourced" a huge database of true and false statements by inviting anyone on the internet to contribute to its knowledge. Early in the new millennium, people would go to his Mindpixel website and rate as true or false 20 arbitrary statements. Once they completed that, they were allowed to contribute one new statement to the database. This way, each statement in the database eventually collected a percentage of truthfulness ranging from 0.0 (false) to 1.0 (true). McKinstry somewhat grandiosely referred to the database as a Generic Artificial Consciousness (GAC), which he pronounced as "Jack." According to GAC, the statement "TV is bad for you" is 0.52 true, "Life is cyclical" is 0.61 true, and "All politicians are liars" is 0.62 true.

I think it is safe to say that Chris McKinstry was a mad genius, and by that phrase I don't mean he was a genius who was slightly mad, such as Ernest Hemingway. Nor do I mean he was a madman who was slightly a genius, such as Hunter S. Thompson. He was perfectly balanced between both of those descriptors. That's exactly why so many people loved him and why so many people hated him. In fact, in the hard-to-find documentary about him, *The Man behind the Curtain*, it seems clear that some people did both. And just like Hemingway and Thompson, Chris McKinstry took his own life.

What McKinstry left behind was a handful of friends, a handful of critics, and a still-living inspiration to evolve a commonsense knowledge base for artificial intelligence by sampling the knowledge bases of thousands of willing participants—rather than having a few graduate students program it from scratch in a laboratory basement. The sample of GAC that I have on my computer contains 80,000 relatively random statements with crowd-sourced truth values that range from 0.0 (false) to 1.0 (true). It's all I have left of Chris, but if you think about it, that's a lot more raw information than most of us have about our long-lost friends.

In 2005, Rick Dale and I were collaborating with McKinstry to extract 11 carefully chosen statements from GAC that had truth ratings of 0.0, 0.1, 0.2, 0.3, 0.4, 0.5, 0.6, 0.7, 0.8, 0.9, and 1.0. We presented them on a computer screen as questions to experiment participants who were instructed to respond to them by mouse clicking a "Yes" box or a "No" box. For example, the question "Are humans logical?" was perfectly balanced between those answers, according to GAC. If 0.0 is false and 1.0 is true, then the statement "Humans are logical" is what you might call 0.5 true, and sure enough, when asked "Are humans logical?," about half our experiment participants clicked the "Yes" box and the rest clicked the "No" box. But which box they clicked wasn't really what we were measuring. Like the professional poker player, we were looking for "tells" that the person might give away during the one-second process of giving that response, so we recorded the x, y pixel positions of the mouse cursor as it moved from the bottom of the screen to a top corner, where the response boxes were.

With the questions that were at the extremes of the continuum, 0.0 (totally false) and 1.0 (totally true), the computer-mouse trajectory was an almost exactly straight line to the "No" box at one top corner of the screen or the "Yes" box at the other top corner. For the 0.0 true question "Is a thousand more than a billion?," everybody made a quick, straight mouse movement up to the "No" box at the top left corner of the screen. Thank goodness. For the 1.0 true question "Should you brush your teeth every day?," everybody made a quick, straight mouse movement from the bottom of the screen up to the "Yes" box at the top right corner. Whew. But something interesting happened with the questions whose truth values were somewhere between true and false. When asked "Are humans logical?" or "Is murder ever justifiable?," most of the experiment's participants began their mouse movement upward toward the middle of the screen. Then the mouse trajectory curved midflight to settle into their chosen response, about half settling on "Yes" and about half on "No." Thus, when a person was responding to a question that everybody happened to agree on, they produced a straight, confident movement in clicking on their chosen response, but when a person was responding to a question that people tend to disagree on, they produced a movement that revealed their uncertainty and wavered a bit between the two possible responses. While they may have finally clicked the "Yes" button, the computer-mouse movement revealed that they were briefly (to some degree) considering clicking the "No" button. They gave away a "tell," and our computer was able to detect it.

Now, once you have a real-time measure of a person's thought process, either by recording their hand movements or eye movements or by other methods, you can take steps to influence that process before it finishes. You can get them to embrace your preferred idea *during* their initial response in the first place—even better than a car salesman can. In 2007, Daniel Richardson and I developed the germ of an experiment idea that would do exactly that. Daniel was a student of mine and is now on the faculty at University College London. We were in Toronto together "talking shop" and having our respective Samuels meet each other. His was about three years old at the time and mine was about two. Over a bottle of scotch and into the wee hours that night, we came up with an idea for an experiment that involved recording people's eye movements to response options on the computer while they answered questions that it asked them. Your eyes jump from object to object about three times per second, and they are usually looking at objects that you are thinking about, so Daniel and I imagined an experiment in which the computer software could "catch" the person's mind while its waffling process was leaning toward a particular prechosen response option—as revealed by where their eyes were briefly looking. Then the experimenter might be able to trigger an early decision that went in that prechosen direction. That is to say, the millisecond timing of the environment interacting with the human could influence that person's choice.

Daniel Richardson and I spent years revising and replicating our eye-tracking experiment, and it took a lot of help to make it work. Finally, a student of Lars Hall (the choice blindness guy discussed earlier), Philip Pärnamets, came along and perfected it with us. Philip's genius was to focus on moral statements, such as "Murder is sometimes justifiable," with response options such as "sometimes justifiable" and "never justifiable," or "It is important to be a team player, even if that means censoring oneself at times," with response options such as "team comes first" and "I come first." By tailoring the response options to each statement and pretesting them with people, each moral statement used in the experiment can be made pretty close to an even 50/50 split for the two response options. This way, each person reading the response options for a given moral statement would wind up looking back and forth at them for a few seconds while struggling with the moral decision. And if we hadn't been tracking their eye movements, then each response option would have been chosen 50 percent of the time. But we *were* tracking their eye movements. By tracking those eye movements, the computer could "tell" when a person was leaning toward a particular response option (even if only temporarily). In the Pärnamets experiment, the computer had already prechosen a preferred response, and it would prompt a decision from the person when they just happened to be looking at that prechosen response—not when they happened to be looking at the other response. Because of the timing of that decision prompt, the computer's preferred response option was chosen by participants 58 percent of the time instead of 50 percent. If this were a close election between two candidates, that would be a 16-point lead you just finagled!

But this isn't just a parlor trick to show off how manipulable your mind is. It's not some terrible scientific secret that salesmen and policymakers are now going to misuse to control your mind. Outside the laboratory, no one is going to be tracking your eye movements accurately, without your knowledge, anytime soon. Advertisers learned long ago that blatant in-your-face advertising is far more effective than any sneaky tricks that might hypnotize you. For example, subliminal messages have occasionally been inserted into radio, television, and movies for almost a hundred years now, but when controlled experiments test these effects, they come up short. Very briefly flashed images or words have a rather small effect on your decisions. By contrast, simply and bluntly telling your audience "buy our product" is far more effective and transparent, and, as an added bonus, the Federal Trade Commission won't come after your business for deceptive practices when you do it that way. No, the take-home message of the Pärnamets experiment is not that your choices will soon be manipulated by evil cyborgs that track your eye movements. Rather, the take-home message is that the *you* that you think you are—even your moral sense of right and wrong—doesn't originate purely from inside your brain. Instead, this you (and its moral compass) emerges organically in the moment as a result of the environment interacting with your brain and body in real time. There are numerous real-world examples of this kind of process happening every day. For example, imagine a woman who has recently decided never to eat veal again—even though veal parmigiana is her favorite dish—because of the moral issues surrounding how the baby cows are raised. After months of abstinence, she sits down at an Italian restaurant and looks at the menu. She knows she should stick to her principles and choose the chicken parmigiana, but the veal parmigiana is "calling her." As her eyes flit back and forth several times between the veal and the chicken on the menu, the waiter comes up and asks for her order. The moment at which he finishes his question, "What can I get for you?," the woman's eyes just happen to be looking at the veal parmigiana on the menu. If only he'd walked up to the table half a second earlier or later, or spoken a little faster or slower. But just as her mind is briefly (what would have been temporarily) leaning toward the veal option, she feels pressed to make her choice, and she orders the veal. These things happen. The precise timing of the waiter's query is not intentional. No one is being nefariously manipulated on purpose. Nonetheless, this woman's moral compass has just shifted. This is a lot like the Pärnamets experiment, brought into a real-life scenario of a type that happens all the time in all kinds of situations. And I'm sure it was delicious.

Letting Go of Your Self

Letting go of your self and "letting yourself go" are not the same thing. I don't want to get accused of telling someone they don't need to exercise and eat right. In fact, *not* "letting yourself go" will probably help you in letting go of your self. A healthy body helps promote a healthy brain, and it takes a healthy brain to nurture the healthy skepticism about your sense of self that the preceding experimental findings encourage.

Once more, with feeling: it is important to *have some humility about your interpretation of the world around you*, because the world around you is not just "around you." It has a way of getting inside you, too. Your environment has such a constant moment-to-moment influence on your deepest definitions of self, including your morals, that the environment is essentially part of who you are. Therefore, whenever possible, you should take great care in selecting the environments in which you place yourself.

If you cannot fully trust your perception and memory of the world, and the world is routinely seeping into who you are, then logically you cannot fully trust your own sense of who you are. And realizing that fact is a healthy realization to embrace. Someone who thinks he is *always* right about himself, and about the people and things around him, is obviously lying to himself. If he has ever changed his opinion about something or someone, then clearly his original opinion was wrong. He was wrong. It is healthy for a person to acknowledge this and also to learn from it.

If you are reading this book, it is probably because you were already not quite 100 percent sure about who you are and were curious about whether this book could inform you on that topic. As I noted, this book will not *tell you* who you are. That's not my job. But don't forget, as the quotation from musician Laurie Anderson states in the epigraph that begins this chapter, while there may be no actual "pilot," *you are not alone* in this predicament. Be thankful. You can share this struggle with your friends and family because they don't fully know who they are either. Only together do we truly have a chance to figure it out.

Instead of telling you who you are, what this book will do is provide concrete demonstrations of scientific evidence that should encourage you to consider expanding your definition of what makes you *you*. Each chapter details a field of neuroscience, cognitive, social, or ecological research, where a handful of findings are showcased that compellingly support one more tiny expansion of your definition of "self," from what was reached in the previous chapter to what is warranted by the next chapter. From chapter to chapter, you will be progressively invited to expand your sense of self from the soul to the prefrontal cortex; to your whole brain; to your brain-and-body; to your brain-body-environment; to the people and things around you; to the species of humans; to a class of mammals; to a kingdom of animals; to the domain of life; and beyond even that.

Now that you have a little humility in your perception, your memory, your judgment, and even your personal sense of self, you should be somewhat ready for what the chapters ahead are going to try to do to you. Most of these tiny incremental expansions of selfhood from one chapter to the next should feel relatively easy to accept, but at some point you may find one of those tiny expansions to be too much for you. Be ready for that moment. It is going to come. When it happens, ask yourself why such a tiny shift goes beyond your personal definition of self. Is the science really unconvincing, or are you just resisting for personal reasons? That's a decision for you to make—whoever you are.

It comes down to this. Your perception of the world around you is not accurate about what's really there. Your memory of your past is not accurate about what really happened. Even your sense of self is not accurate about what really motivates your decisions. And, to top it all off, the upcoming chapters in this book are going to tell you to include things outside you as *part of you*. This is a lot to process, but don't worry. Everything's going to be okay. Just take a deep breath and let it out slowly.

Directions for Use

At the end of each chapter, this book will provide concrete instructions for how to take the information just provided and use it in your daily life. It could be a specific meditation, a contribution to the public good, or some other assignment. It could be an experimental exercise with a friend, or, in the case of this chapter, it is an innocent little prank that you play on someone. To get a real-world understanding of how powerful change blindness is, how unaware we usually are about our physical surroundings, your "directions for use" in this chapter are to exchange a pair of pictures or decorative objects in your home or the home of someone you love. For example, when no one else is around, you might trade the locations of two vases in the living room. See how long it takes your spouse, children, or parents to notice. If you live alone, then next time you are visiting a close friend or family member, secretly find a pair of similar-sized pictures in their bathroom or a pair of decorative objects in a hidden hallway and carefully, quietly exchange their locations with one another. After a few days, if no one has said anything, ask if they've noticed a change in that room. Then tell them what you did—and maybe start a conversation about how easy it is to take our environment for granted and our selves, too.

2 From Your Soul to Your Prefrontal Cortex

Who do you think you are?
Bless your soul.
You really think you're in control?
—Gnarls Barkley, "Crazy"

Imagine that your left hand has a will of its own. While chatting with friends over coffee, your left hand reaches up to your chest without your intending it to and starts unbuttoning your top. Your friends look confused. You look down at your hand doing this without your intent, and you feel mortally embarrassed. Before things get too weird, you have to put down your coffee cup and use your right hand to restrain your left hand. These things happen. It is called alien hand syndrome, because people often report that the errant hand seems to be controlled by some alien force outside their body. In some cases of alien hand syndrome, the person's left hand actually hits them in the face. Of course, it is not actually an alien that is controlling the hand. The wayward limb is actually being controlled mostly by damaged networks in the brain. When brain damage leaves the motor movement network improperly connected to the rest of the brain, it sometimes drives one of the hands in a fashion that is not under the conscious control of the person who owns that hand. How does one portion of the brain generate a motor plan to unbutton your top without your permission? In this chapter, we will examine where intent comes from and how decisions are formed.

Now that you've made it through chapter 1, you should be ready for chapter 2. If chapter 1 did to you what it was designed to do, then you've become a little less trusting of your intuitions about your self. This means

Chapter 2

to look at the science of "who you are" with a clear head. e just might surprise you.

In this chapter, I'm going to show you scientific evidence indicating that your sense of self, and authorship of your destiny, doesn't happen as a quantum trick at the subatomic scale, from mysterious ethereal signals in a gland in the center of your brain, or from the flick of a switch in a single neuron. The reasoned decisions that you make in everyday circumstances, which you often treat as defining *who you are*, emerge chiefly from billions of neurons cooperating together in your prefrontal cortex—the frontmost quarter of your brain.

Feel Free

When Cee-Lo Green, of the music duo Gnarls Barkley, laughingly asks, "You really think you're in control?," he might be suggesting that God is actually in control or that the Universe is actually in control. Or maybe those two options aren't that different. Maybe it's as physician Deepak Chopra says: "God did not create the Universe. God became the Universe." Who do *you* think is in control?

When you make a decision about something, you often feel as if you exert control from somewhere deep inside your self. Is that inside your brain? Inside your soul? Sometimes it feels like a force of *will* that comes *free* of any external causes, hence the term "free will." No person or circumstance caused you to make that choice; you own it all on your own. Is there some part, or aspect, of your brain that carries out this free will, that functions as your soul?

To begin to answer that question, let's look at the basics of how your brain works. For starters, it is not the case that you only use 10 percent of your brain. That, my friend, is a myth. You have many billions of neurons in your brain, and they all conduct important work sending electrochemical signals to one another, and generating electromagnetic fields, to do all kinds of things. They regulate your heart rate; pump your lungs; move your limbs, eyes, and mouth; understand language; recognize faces; calculate math; dream; and fall in love. You do not have a bunch of neurons sitting quietly in your brain waiting to be given a job. You do have billions of glial cells in your brain, and their important job is to hold in place the neurons and their connections; they also help with the delivery of oxygen and calcium to those hungry neurons.

A typical neuron in your cortex is around 10 micrometers (one hundredth of a millimeter) wide at its cell body. It is usually connected to almost ten thousand other neurons, so, with billions of neurons in your brain, and each one connected to almost ten thousand other neurons, that means there are hundreds of trillions of synaptic connections in your brain. When neuron 1 sends a signal to neuron 2, it often does so via an electrochemical impulse that travels along the length of the neuron's axon—which looks kind of like a tail—and causes the release of chemicals (neurotransmitters such as glutamate, acetylcholine, and serotonin) at the end of the axon, which are then received by the dendrites of neuron 2. However, sometimes the signal travels between dendrites of the two neurons, or even between a dendrite of one and the cell body of the other. To carry out the synapse (the signal transmission from the first neuron to the second neuron), those chemicals travel across a 20 nanometer (20 millionths of a millimeter) gap between the two neurons. Each individual synapse carries a very simple signal. The neurotransmitter molecules reach the membrane of the second neuron and can influence it only by increasing or decreasing the likelihood that it will also send an electrochemical signal down its own axon. The synapse is either excitatory, telling the second neuron to "get active," or it is inhibitory, telling it to "be quiet."

These minutiae are important because it has been suggested that your free will, your ability to make choices that are all your own, just might originate in those tiny details of the synapse. Award-winning mathematician Sir Roger Penrose of Oxford University proposed that at the tiny spatial scale of that synapse, quantum randomness may be at play. So the reason we feel that our choices and decisions are not predetermined by our genes and our circumstances, and that our decisions even surprise ourselves on occasion, is because random neural events at the subatomic scale sometimes alter the results that those genetic and circumstantial forces tried to bring about. With help from anesthesiologist Stuart Hameroff, the Penrose-Hameroff theory of consciousness came into being as follows. An individual molecule of a neurotransmitter needs to land perfectly in the receptor site of the second neuron's membrane in order to carry out its synaptic influence. If it misses by a fraction of a nanometer, then it doesn't lock into place in the receptor site, and it fails to influence the second neuron. According to this account, because of quantum physics, the atoms that make up the microtubules that form that receptor site may randomly fluctuate in their structure just enough for this randomness to sometimes be responsible for a neurotransmitter molecule *missing* its receptor site when it should have locked in or *binding* to its receptor site when it should have missed. Therefore, the unpredictability of quantum physics may play a role in making our decisions. Quantum randomness may be responsible for our consciousness and our sense of free will!

Well, not so fast. There are many thousands of those molecules being released in the synapse, so the uncertain fate of one particular molecule, or microtubule structure, may not be so consequential. And the quantum uncertainty of their atoms' positions happens on a timescale of quintillionths of a millisecond (femtoseconds), whereas the synapse happens on a timescale of about one millisecond. So, as with most everyday quantum events, the randomness is likely to average out to about the same number of quantumly *disrupted* receptor bindings as quantumly *induced* ones. Besides, a single synapse does not a human decision make. In fact, a single synapse doesn't even decide whether that second neuron will become active or remain quiet. At any given millisecond, that second neuron is receiving anywhere from 5 to 50 synaptic signals at the same time from its various connections. It has to add up all those "get active" and "be quiet" signals to determine whether it will send a signal of its own along its axon. Clearly, a human decision is not determined by whether a single receptor binding site receives its neurotransmitter molecule or not. Nor is it determined by one synapse, or even one neuron. It takes a group of such neurons cooperating in concert to reach a decision, comprehend a word, or even just make an eye movement. A human decision to choose the chicken or the veal is something that emerges from millions of neurons interacting over several seconds to drive the eye movements on the menu, to comprehend the text on the menu, to mull over the pros and cons of the options, to understand the waiter's spoken words, and finally to drive the mouth, lungs, and tongue to utter the speech sounds that finalize that decision. Given the complexity of causes involved, the vast spatial range of cortical real estate involved, and the lengthy duration of quadrillions of femtoseconds involved, it seems rather unlikely that any one atom's tiny and brief quantum uncertainty could, on a regular basis, be what determines our conscious decisions, or our sense of free will.

One Neuron to Rule Them All

If our ability to make choices that seem to be all our own doesn't come from the randomness inherent in quantum physics, then where does it come from? If our thoughts and choices are not determined at the *nano*scopic scale of an atom, are they perhaps instead determined at the *micro*scopic scale of a neuron?

For more than a hundred years, the majority of researchers in the field of neuroscience have embraced one version or another of "the neuron doctrine." The origins of this doctrine come from anatomical methods developed in the nineteenth century by Camillo Golgi and Santiago Ramón y Cajál that allowed biologists to see under a microscope that brains were indeed made of cells in a fashion somewhat similar to other biological tissues. However, unlike the spherical cells in most biological tissues, it was discovered that neurons have branching dendrites and axons that connect them to one another in complex ways. In the 1950s, a great grandson of Charles Darwin, Horace Barlow of Cambridge University, recorded the electrical impulses of neurons in a frog's retina to develop an understanding of how those neurons are connected to one another. Barlow helped develop the notion of a "receptive field," which refers to a region of the sensory surface (in this case, the retina) that, when it receives input (in this case, light), causes a particular neuron connected downstream in the network to become active. Barlow discovered that a neuron in the frog's retina that sends signals to the rest of the frog's brain has a receptive field that includes at least a thousand light-sensitive cells on the retina. If a neuron in the retina responds to spots of light that land on a thousand receptor cells, then it stands to reason that a neuron in the brain might in turn respond to a thousand of those retinal neurons. The receptive field for that neuron could combine inputs from retinal neurons that are arranged in a pattern, such as a contour or a shape. Such a neuron could be the computational element for recognizing a tilted line, a circle, or maybe (particularly relevant for the frog) a fly. In a simplified version of the neuron doctrine, the frog recognizes a housefly because its "housefly neuron" is active. In fact, the activity of that particular neuron *constitutes* the frog's recognition of the housefly. That housefly neuron does not have to signal some other entity in the frog's mind for the frog to realize that a housefly is present. The activation of that particular neuron is itself the frog's recognition of the fly. For the one second that the frog is in a state of recognizing that housefly, that one neuron is *who the frog is*. It alone is very briefly operating as the frog's "self."

But that oversimplified and idealized version of the neuron doctrine has a number of flaws. These flaws are well characterized by the hypothetical example of the "grandmother cell." While a frog may not be able to recognize his grandmother, you can recognize yours. You can see her when you visit her or can look at old photos of her. However, it would be very difficult for a single grandmother cell, with its one specific receptive field on your retina, to accommodate the wildly varying patterns of light that bounce off your grandma and land on your retina when she's smiling, when she's frowning, when she's wearing a hat, when she's not, when she's with a fox, in a box, in a house, or with a mouse. I do not like this theory. Do you? The patterns of light that hit your retina when grandma is wearing her gardening overalls versus a dress, or sitting versus standing, are so incredibly different that it would be impossible for the receptive field of *one neuron* to be flexible enough to accommodate those variations.

Rather than a single "grandmother cell," a sparse distributed code (of a smallish number of neurons tightly coupled in their activity) allows for easy learning of new concepts and also rich contextualization. For learning a new concept or the name of a new acquaintance, a sparse code allows the brain to combine neurons that already code for tiny features of visual images or speech sounds. As an example, imagine you are seven years old and this is your first time meeting a person with blond hair and a beard, and the first time you've heard the name Cody. You've met people with blond hair, and other people with beards, and you've met someone named Corey and someone else named Jodie. If your brain was trying to remember Cody with a single neuron, there's a good chance that at age seven you wouldn't have any unemployed neurons sitting around doing nothing and waiting for a job like this. By age seven, pretty much all your neurons are involved to some degree in constituting the meanings of very important things, such as remembering the names of cartoon characters and the lyrics to your favorite children's songs. The "grandmother cell" style of coding just won't work for learning this new name and face. With a sparse distributed coding scheme, however, your seven-year-old brain can learn this new person's face and name by combining the activations of the different neurons that have already learned to code for blond hair in some people and beards on other people, the first syllable "Co" (from Corey), and the second syllable "dy" (from Jodie). When you recognize Cody again, that sparse code of neurons doesn't have to be a symbol or representation that some other part of your brain observes and interprets. That sparse code, with its cobbled-together collection of information, *is* your recognition of Cody, just as with the frog and the fly. For that one brief second that you are in a state of recognizing Cody but not yet doing anything else, this sparse code of active neurons functions as *who you are*. It alone is very briefly operating as your "self." Now your little seven-year-old brain has a brand new name and face it is learning, simply by connecting groups of neurons that already had jobs coding for the relevant features. So rather than a "grandmother cell" for Cody, you probably have a sparse population code for Cody.

In addition to their use in learning new concepts, sparse neural codes are also good for contextualization. Contextualization is what causes you to see grandma slightly differently every time you look at her. With a simple "grandmother cell" code, this would be impossible. Every time you saw grandma, she would seem the same to you, because your brain state that constitutes your recognition of grandma would be the same every time. With sparse population coding, one routinely has the capacity for activating additional neurons that accompany the sparse code and provide context for each particular recognition event. You see grandma a little differently every time not just because she is wearing different things at different times but also because you interpret her and understand her differently when you look at her at different times. For example, imagine that you just heard some old stories from your mother about how hard your grandma worked to raise her children in tough times. Now when you look at that same photo for the hundredth time, you see her a little differently this time. These things happen. The reason you have that different perception of her is because a slightly different collection of neurons are active this time. That sparse code for recognizing her face is active, of course, but so are many other neurons. Those additional active neurons often provide the context for this "new view" of grandma. If you contemplate this new view of grandma while looking at her, you could train those additional context neurons to become part of the core sparse code. Then your basic concept of grandma will have changed forever—until the next "new view" of her comes along.

Of Ghosts and Grandmas

A sparse code of a handful of linked active neurons might be an important part of how the visual areas of the brain recognize a face or how the language areas of the brain recognize a word, but where in the brain are the sparse codes that determine our decisions and our choices? Where in the brain is the *self*, with its precious free will?

In the seventeenth century, before Golgi and Ramon y Cajal revealed the cellular structure of brain tissue, philosopher René Descartes developed a rather speculative answer for that question. Descartes suggested that, deep in the center of your brain, the pineal gland might be the organ that connects your soul to your body and allows your will to be converted into physical action. The pineal gland is an anatomically separate collection of glial cells, right between your brain stem and your cortex, and it releases a variety of neurotransmitters, including serotonin, melatonin, and even endogenous dimethyltryptamine (the powerful hallucinogen DMT). But it doesn't have neurons, so it doesn't do sparse neuronal coding. When the pineal gland and other brain regions release neurotransmitters, those chemicals wash over the neural networks of the brain and can dramatically alter the connectivity pattern, strengthening some synapses and weakening others. Changes in the chemistry of the brain can make the very same neural network produce very different behaviors.

But if Descartes was right and the pineal gland not only bathes your brain's networks in chemicals that alter their function but also provides the conduit between the spiritual world of the soul and the physical world of the brain-and-body, then we would have a lot of explaining to do. For starters, exactly how could the pineal gland convert nonphysical intentions of a soul into physical activation patterns of neurons that drive human behavior? And does a rat have a soul? After all, it does have a pineal gland. If your "self" is located in a soul that belongs in a nonphysical realm, does that mean it persists after your brain and body die? Can that soul produce and hear speech sounds? Can it perceive the light that bounces off physical objects such as grandmas and houseflies? If so, how?

Perhaps we should briefly examine some of the logical quandaries that go along with Descartes's notion of a nonphysical soul, a form of consciousness that is not brought about by the physical processes of the brain. If the brain is a kind of biological machine, then Descartes's soul is a "ghost in the machine." Can your ghost get out of its machine and have an out-of-body experience? Is that what happens when you die or when you have a near-death experience? How can a disembodied soul perceive the physical world when it doesn't have biological eyes to uniquely filter the electromagnetic radiation we experience as light, or ears to uniquely filter the air vibrations we experience as sound? Does the disembodied soul simply not filter light and sound input at all and thus perceives all frequencies of light (from radio waves, to "visible" light, to gamma rays) and all frequencies of sound (from infrasonic whale sounds to ultrasonic sonogram sounds)? That would be a radically different sensory experience than we humans are accustomed to perceiving. Or does a disembodied soul somehow filter light and sound in the physical world the same way its body did? Does that mean that when a color-blind man dies, his soul still has trouble distinguishing green from red, and when a hearing-impaired person dies, that their soul is hard of hearing? What about the 3 percent of women who are tetrachromats, with an extra type of color receptor in their retinas that allows them to see finer distinctions of color than the rest of us? Will her ghost continue to see more colors in the rainbow than you and I do? That's a lot of questions, isn't it? Luckily, there have actually been some controlled scientific experiments that point to some answers.

Near-death experiences in the emergency room are occasionally accompanied by reports of out-of-body experiences. For example, almost 5 percent of people who recover from cardiac arrest later report some form of out-of-body experience, such as a memory of seeing their body, and the emergency room technicians, from above—as if their soul had departed their body and was levitating over the scene and looking down at it. In chapter 1, you saw how unreliable memory can be, so you know to take self-reports like that with a grain of salt. Well done. What is needed is an unbiased test of those claims, right? Well, in the mid-1980s, Janice Minor Holden designed that test. She was the first to conduct a controlled experiment to test for evidence of paranormal visual perception during a spontaneous near-death experience. She placed written messages on the tops of shelves and cabinets in the emergency room. If a floating soul were elevated in the room and looking down, it should be able to see those random unpredictable messages in the same way that it is seeing the people in the room. Then, when interviewed afterward, the patient who reported having an out-of-body experience should be able to answer questions about those hidden messages. In 2007, Keith Augustine, editor of the *Internet Infidels*, wrote a review of Holden's study and several others that took place over a 20-year period. Across hundreds of emergency room cases and dozens of reported out-of-body experiences, there was not a single instance of a patient being able to report the content of those hidden messages. More recently, in 2014, Sam Parnia and his colleagues reported a study where they interviewed 101 survivors of brain death during cardiac arrest in an emergency room, and while one of them reported a conscious memory of an event that may have taken place during brain death, nobody was able to report the information content of the hidden messages in the room. If there is a nonphysical soul, the evidence clearly suggests that your consciousness cannot travel in it separately from your living body. Besides, the sensory experience that it would provide is sure to be vastly different from what your body is used to. After all, the unique filtering of light and sound that your eyes and ears conduct simply wouldn't be there for your disembodied soul. Thus, while it is possible that you perhaps have an immaterial soul, I'm afraid that we must conclude that Descartes's pineal gland theory just leaves too much unexplained regarding the mechanics of how nonphysical intentions interface with physical forces.

But whether the soul is physical or nonphysical is not at the crux of *who you are* anyway. The scientific argument against the existence of a supernatural soul and an afterlife provided by a God that listens to human prayers has been laid out in many books and movies, such as those by Richard Dawkins, Daniel Dennett, Christopher Hitchens, Bill Maher, and others. Likewise, there are many books that argue strongly against atheism, such as those by John Danforth, John Haught, Ian Markham, and others. My message here does not require anyone to alter her or his opinion about the soul, the afterlife, or God. Rather than asking you to revise your definition of what is at the core of your self, this book simply asks you to consider the scientific evidence on the ever-expanding role of the physical world in constituting *who you are*. What is offered here is incremental encouragement to slightly expand the scope of the definition of your soul from chapter to chapter (irrespective of whether its core is physical or nonphysical).

Let us return from this brief diversion now and catch up with the science of where the brain might house its self, its will, or its soul. Descartes's theory of the pineal gland being the source of will in the brain has fallen into disfavor over the centuries mostly because the anatomy and the neuroscience just don't back it up. But if the self/soul isn't in the pineal gland, then where in the motherlovin' brain is it?

One Brain Region to Rule Them All

Some of the first inklings of scientific evidence for what specific part of the brain might contain the self came from medical observations of head injuries and the behavioral changes they caused. The most famous of those is perhaps that of Phineas Gage, an American railroad construction foreman in the mid-nineteenth century. When a drilling explosion drove a 3-foot-long, inch-thick tamping iron through his cheekbone and out the top of his head, he was walking and talking in just a few minutes. I'm not kidding. And when he vomited an hour later, his doctor did report that a chunk of brain tissue was squeezed out of the hole in his skull and splattered on the floor. Still not kidding. Gage spent about two months recovering in the hospital and then a few months more at his parents' home, but he recovered. Try not to let this story interfere with your enjoyment of the next zombie movie you watch, where a simple knife through the skull seems to instantly neutralize them.

The remarkable thing about Phineas Gage is that, after about six months of rest, he no longer experienced any serious cognitive or movement deficits as a result of his losing a substantial portion of his left frontal lobe. He was able to work again. However, his railroad co-workers reported that his personality had changed and that he was more impulsive and less trustworthy than he used to be. They said he was "no longer Gage." Upon invitation, Gage moved to Chile to be a long-distance stagecoach driver, and he kept that job for several years. Given that such a job definitely required not only motor skills but also social skills, planning skills, and a strong work ethic, psychologist Malcolm Macmillan suggests that the "personality change" resulting from Gage's frontal lobe damage may have been relatively temporary, as was his life. About a decade after the accident, Phineas Gage began having epileptic seizures, and a couple of years later, he died from convulsions.

In the century and a half since then, neurology and neuroscience have made great advances in understanding what the frontal lobe of the brain actually does. It affects a lot more than just personality, impulsiveness, and trustworthiness. The anatomically separate anterior portion (front) of the frontal lobes, the prefrontal cortex, combines neural signals from a wide variety of sensory brain regions and motor movement brain regions to perform something like "executive decisions" on how to interpret, plan for, and act in different situations. Various studies have suggested that the prefrontal cortex inhibits reactive urges, maintains planned actions over time, or selects among simultaneously competing ideas. Rather than treating these different functions as separate abilities that might have their own devoted piece of real estate inside the prefrontal cortex, Joaquin Fuster suggests that we think of the prefrontal cortex as an important component in a broad network of brain areas that cooperate together to carry out the "temporal organization of behavior." In order to organize a sequence of behaviors over multiple timescales (e.g., seconds for reaching, minutes for talking, hours for working, days and weeks for studying, months and years for building a family and a career), that same network would frequently have to select among competing ideas, maintain plans over time, and inhibit instinctual outbursts.

People with damage to the prefrontal cortex can recognize the disadvantage of a choice that benefits them now but hurts them even more later. Nonetheless, those same people then select that very choice that they know will hurt them later. People with damage to the prefrontal cortex can easily learn the rules of sorting images by shape or by color, but then they have trouble switching from one of those rules to the other. You can even induce temporary "damage" to the prefrontal cortex in healthy people. In a laboratory, cognitive neuroscientists can temporarily deactivate the frontal lobe by using transcranial direct current stimulation. Sharon Thompson-Schill and her students have induced temporary disruption of the left prefrontal cortex by noninvasively directing weak electric current (about 1 milliampere) through the scalp near the targeted brain region. While the functioning of the left prefrontal cortex is being disrupted in this way, normal participants have some trouble categorizing objects by their color or shape. Interestingly, this kind of mild categorization impairment might not always be a *bug* in the code of the brain. In the right circumstances, it can be a *feature*. Thompson-Schill's lab has shown that when you mildly inhibit the left prefrontal cortex with a little electric current, you can make people more creative in their speculative ideas regarding how to use a mysterious unrecognized tool. By slightly reducing the cognitive control that the prefrontal cortex normally exerts on your mind, new and interesting ideas are allowed to blossom.

The Libet Experiment

Do all ideas (creative and otherwise) have to make it through the gateway of the prefrontal cortex in order to eventuate in action? Is that where the decision for chicken or veal gets made? Psychologist Benjamin Libet asked where and when the brain achieves simple decisions like these. In the 1980s, he used electroencephalography (EEG) to record the millisecond changes in electrical fields that a sparse population of neurons might produce when they become active in synchrony during perception, cognition, and action. You see, not only can electric current travel *into* the skull to modify brain activity, as in Thompson-Schill's experiments, but normal brain activity naturally produces electric currents, and those electric fields, though faint, extend *outside* the skull. They can then be detected by surface electrodes that are pressed harmlessly against the scalp. By recording the changes in those electric fields over time, cognitive neuroscientists can see precisely *when* those neural events are associated with a particular sensory input or motor output. Libet recorded these electrical scalp potentials while people watched a dot circle its way around a clock face and made freely chosen decisions of when to lift their finger.

It's a simple decision. Put yourself in the shoes of one of the experiment's participants. Imagine wearing the EEG net on your head, resting your hand on the table (with an extra surface electrode on your index finger), and watching the dot go around in a circle. Your job is nothing more than (1) lift your finger *whenever you feel like it* and (2) keep track of where the dot is on the clock face when you make that decision to lift your finger. Each time a participant lifted a finger, there were three time measurements that Libet had at his disposal: (A) the time at which the finger actually moved, as recorded by the surface electrode on the finger, (B) the time at which the dot was in the clock location that the person said they saw it in when they felt they had decided to move that finger, and (C) the time at which the EEG signal provided evidence that the brain was indeed planning a movement of that finger. Naturally, time point A was the latest, because it makes sense that the decision to move and the neural preparation of that movement are going to precede the actual movement. On average, time point B was a few hundred milliseconds before time point A. Again, this makes sense, because you should expect it to take a few hundred milliseconds for your decision to lift your finger to be converted into the right sparse neural codes in the motor cortex, which then sends neural signals to the subcortex, which then sends messages to the spinal cord to contract the appropriate muscles in the hand. In fact, it's moderately impressive that all these things can happen in a fraction of a second. The surprising bit in Libet's experiment is that time point C happened, on average, a few hundred milliseconds before time point B. That's right. A person's brain produces a significant change in electrical activity (in the supplementary motor area, right next door to the prefrontal cortex) *before* the person is aware that they've made a decision to move.

Put yourself in that lab again, and imagine what's going on. You watch a dot zipping its way around a clock face, you feel the urge to lift your finger on your own volition just when you happen to see the dot at 3 o'clock, and by the time your finger actually lifts, the dot is right between 3 and 4 o'clock. What you don't realize while you are in the experiment is that the EEG net on your head detected a change in neural activity in your brain when the dot was between 2 and 3 o'clock, before you were aware that you had made your "freely willed" decision to lift your finger. So *you* didn't make that decision. Your brain made it for you, and a fraction of a second later, you became aware of it. A fraction of a second after that, your finger actually moved. And then *you* blithely took credit for it all.

Libet's result set off a firestorm of debate in philosophy, psychology, and cognitive neuroscience that continues today. A wide variety of theorists of the mind were notably rankled at the idea that free will might be nothing more than an illusion, and they set about desperately hunting for logical flaws in the interpretation of the experiment. In fact, Libet himself even tried to find an exception to his own result that might give room for free will to exist in some special circumstances. He speculated that, while an initial decision to act might be preceded by neural activity that is actually responsible for the decision, perhaps a second decision to "veto" that first decision could be something that relies on actual free will—free will of a sort that is not the causal result of neural activation patterns.

Patrick Haggard and his colleagues tested Libet's idea that a "veto" of a previously settled decision might not be so predictable by brain activity that happens before the veto. However, Haggard's team showed that even a veto of a previous decision is still presaged by neural activation patterns that precede the person's awareness of their veto decision. According to Haggard's study, you lack not only "free will" but also "free won't."

From Your Soul to Your Prefrontal Cortex

Since Libet's original experiment in the 1980s using EEG, neuroimaging techniques that are more sensitive have been developed that allow an even richer insight into what the brain is doing while it performs perception, cognition, and action. Functional magnetic resonance imaging (fMRI) detects the change in magnetic properties of blood when it feeds the metabolic process of neurons that are producing their electrochemical impulses. When many neurons are active in a region of the brain, because that region is working harder than usual, those neurons draw oxygen from the blood supply to replenish their energy resources. By detecting the magnetic changes in this deoxygenation process of the blood, fMRI can reveal what regions of the brain are working harder than others. John-Dylan Haynes and his colleagues used fMRI in a Libet-type task that involved people freely deciding to press a left-side button or a right-side button. Haynes and his team were able to use activation patterns in the frontopolar cortex to predict which button a person would press with 60 percent accuracy (reliably better than the 50 percent chance level). The frontopolar cortex is the most anterior portion of the prefrontal cortex, just above your eyeballs. Can you guess how far in advance these neural activation patterns were able to make this above-chance prediction? A lot more than a fraction of a second. It was a full eight or nine seconds before the experiment participant reported being aware of their choice that neural activation patterns in this portion of the prefrontal cortex were already developing that choice.

This is a lot to absorb, so feel free to take a minute. The science here shows that there's a statistical pattern among the activation of neurons in the prefrontal cortex that carries in it the relevant information for biasing whether a person is going to choose, say, chicken or veal. Importantly, that information is present in a person's brain *several seconds before* they know whether they've decided between chicken or veal. To put it in a nutshell, your brain knows what you're going to do before *you* know what you're going to do. Your neural activation patterns determine your will, not the other way around. The way that your prefrontal cortex is wired, by your genetics and social learning, is what determines your actions. Your conscious will actually shows up quite late to that party. Accepting these scientific findings is an important step in expanding your definition of your self. *Who you are* is not so far removed from the physical world as to be capable of generating a decision that has absolutely no physical precursors. You are more than a soul. *Who you are* includes your brain as well—at least your prefrontal cortex.

Your Brain Knows More Than You Do

In fact, there are many examples of a brain containing information that its owner is unaware of. Not only does your brain know what you're going to do before you do, but it can also perceive and respond to things that you didn't think you even noticed. Remember the change blindness experiments in chapter 1? Have you carried out your assignment from the "directions for use" section in that chapter? You better get on that. Anyway, one of my PhD advisers, Mary Hayhoe, recorded people's eye movements while they looked at images on a computer screen that had changes in them from one moment to another. In situations where people failed to notice the changes in the images, their eyes nonetheless lingered on those locations for a little longer than for other locations. Essentially, part of the person's brain suspected the change and directed the eyes to linger on that object. A number of brain regions contribute to driving eye movements, and the frontal eye fields (inside the prefrontal cortex) are particularly important, but the rest of the brain is not able to collect and recollect enough evidence to report a change in the display. This is just one of many examples of how your reportable impression of your situation (e.g., your "introspection") can often differ substantially from your brain's actual processing of that situation. It is frequently the case that what you *think* your mind is doing is not at all what it's *actually* doing.

Blindsight is a particularly striking example of the brain outperforming the self. When a person suffers damage to their primary visual cortex, in the occipital lobe (at the very back of the brain), they typically lose some visual perception. The patient is clinically established as being unable to see visual information in a certain portion of their field of view. When asked to choose between an "X" or an "O" that is presented in the blind portion of their visual field, they protest that they have no idea because they are blind in that part of their field of view. However, when forced to make a guess anyway, some blind patients perform reliably above chance again and again. It's a little bit like the sighted person who misses a change in Hayhoe's change blindness experiment but their eyes linger on the right spot for that extra fraction of a second. Somewhere in the brain, the visual information is present. It just doesn't make its way to conscious awareness.

In fact, one patient with well-established blindsight was shown to be able to grasp objects in her blind field almost normally. Cognitive neuroscientist

Mel Goodale and his colleagues brought a patient with blindsight to his lab and carefully recorded her hand movements while she was reaching for and grasping everyday objects. When objects were in her blind field, she reported being unable to see them at all. When asked to reach for them anyway, her hand shape adjusted to the right size for grasping each particular object *before* coming into contact with it. Clearly, that visual information is indeed somewhere in her brain. Perhaps it is in visuomotor coordination areas such as the parietal cortex (near the top of the brain), but it's definitely not in brain areas that support visual awareness.

A particularly impressive example of a brain containing information that its user cannot seem to access comes from psycholinguists Lee Osterhout, Judith McLaughlin, and Al Kim. Students just beginning to learn French were brought into the lab, where they wore an EEG cap to record the electrical fields that their neurons produce when they become active. These students were shown a sequence of trials, each presenting two words in a row, and were asked to decide whether both words were actual French words (such as *maison—soif*) or one was a nonword (as in *mot—nasier*). Since they were just barely beginning their training in French, their performance was at chance. They were wrong as frequently as they were right. But the EEG signal of their brain activity told a different story. The difference in the brain's response to French words like *mot* compared to nonwords like *nasier* was already detectable in the EEG signal of these students who had only just started learning French. In fact, that difference was greater for students with a few more hours of exposure than for other students, yet none of them were able to demonstrate the knowledge about these words and nonwords that their brains were clearly starting to pick up on.

Apparently, your brain is doing all kinds of things that you simply aren't aware of. It is subconsciously noticing details about your surroundings, quietly learning words, deciding to lift your finger, and maybe even choosing veal over chicken, *without your even knowing about it*. The idea that your perceptions, thoughts, and actions all find their most immediate origins in the tiny electrical and chemical interactions that microscopic neurons conduct with each other can be a little discombobulating at first. If sensory inputs cause neural activation patterns, and the network's connectivity (acquired through genetics and learning) then causes decisions and actions, who is accountable for those actions? Can *you* be held responsible for your behavior, whoever you are?

Not Feeling So Free Anymore?

Actually, this conundrum shouldn't be surprising at all when you think about it. If a freely willed decision of yours doesn't come from your brain, then where in the world could it come from? Every effect has a cause, right? Or perhaps a complex of causes. If a decision is an effect, then it must have had some preceding causes. Neural activation patterns are the most immediately preceding causes. And those causes are themselves effects that have their own preceding causes. And so on. Your current sensory input, other recent actions that helped produce the current sensory input, cultural context, parental input, and genetic history are some of those preceding causes. Even a decision to merely lift your finger couldn't really just come out of nowhere. We're not ready to rewrite the law of cause and effect, are we? (Not yet, anyway. That happens in chapter 8, which is going to be a challenge. I don't envy you, but don't worry about that for now.)

Psychologist Daniel Wegner said that our *experience* of free will is an illusion. Wegner marshaled evidence of psychological deviations from our usual experience of free will (such as schizophrenia or hypnosis), along with laboratory findings such as the Libet-type experiments, to show that when we become aware of an intention and witness ourselves carrying it out, that sense of ownership/authorship of that intention/action is simply not true. After all, if a scientist like John-Dylan Haynes can record the activity in your prefrontal cortex and predict (reliably above chance) whether you will press a left button or a right button several seconds before you are even aware of which button you intend to freely choose, then your *awareness* of that choice is clearly not the *origin* of that choice. Your sense of free will during that choice can be nothing else but an illusion, right?

But you have to be careful what you do with that conclusion. Whenever someone suggests that free will is not real, the constant refrain is that we can no longer hold people accountable for their actions. It is often suggested that free will *has to be real* because if it isn't, then people would be allowed to commit crimes and not be held responsible for them. Before evaluating the truth of this argument, perhaps we should first evaluate whether it is even logical in the first place. It might be "placing the cart before the horse." Point 1: Before societies were formed, the then unknown natural laws of biology, chemistry, and physics were already determining whether a human brain could produce freely willed actions that have no causal precursors. Point 2: Contemporary humans often feel strongly that, in most circumstances, society needs to assign blame to a person for his or her antisocial behavior. Thus, long before anyone could have suggested that free will "has to be real" for social responsibility reasons, the brain of *Homo sapiens* had already evolved to either have free will or not have it. Since point 1 preceded point 2 by many thousands of years, how can a strongly felt point 2 have any influence whatsoever on the results of point 1? The fact that your definition of civil society demands a certain theory about the brain to be true doesn't make that theory true.

The Libet-type experiments show that your 50/50 arbitrary decisions can be predicted with 60/40 accuracy several seconds in advance by someone who is recording the activity of your prefrontal cortex, and the Pärnamets experiment from chapter 1 shows that your 50/50 moral decisions can even be manipulated into a 60/40 decision predetermined by the computer interface software. So the scientific evidence points against the existence of genuine free will inside a human brain. If your definition of civil society relies heavily on the existence of free will in the brain, then rather than contorting your interpretation of the science about the brain, perhaps you can just slightly modify your definition of civil society instead.

Philosopher Derk Pereboom suggests that, in the absence of free will, we should not assign sole blame to an individual person for their bad acts, nor should we assign sole credit for their good acts. For example, imagine finding a wallet on the sidewalk. Because you know you're a good person, you consider taking it to the closest shop or trying to find its owner's phone number yourself. But then it occurs to you that it is possible someone else already took some money out of it, and you don't want to be the one accused of doing that; some of us have learned the hard way that "no good deed goes unpunished." So you decide to find a police station and anonymously drop it off. While walking there, your curiosity gets the best of you, and you open the wallet to check the ID. After all, you might know the person and could then return it yourself, right? In this little thought experiment, let's say that you don't recognize the face on the driver's license but you do recognize the face on all those $100 bills. There's 13 Benjamin Franklins in this wallet! Before reading any further, give yourself some time to think about what would go through your mind in this situation and what you would do with this wallet.

Okay, let's continue with the thought experiment. Let's say that suddenly the thought crosses your mind to just pocket the whole stash because it would be an awesome windfall for your finances right about now, but you quickly discard that thought because you know you're a good person. Then it occurs to you that if you took just one of those Benjamins, it might not really even be missed. It would leave an even dozen in there, and maybe you've *earned* it as a reward because after all you are returning everything else in the wallet. This thought doesn't get discarded as quickly as the previous one. It lingers at about 50/50 for several seconds while you mull it over. Then the neural activation patterns in your prefrontal cortex finally settle on a decision *not* to take any of the money. Several seconds later, your consciousness becomes aware of this firm, upstanding decision, made by the neurons in your prefrontal cortex, and you blithely assume sole personal responsibility for it. As you hand over the wallet and its full contents at the police station's front desk, you think, "Damn, I'm a really good person!"

Should you really pat only *yourself* on the back for that decision? Derk Pereboom doesn't think so. Don't you think your mother and/or father played a role in teaching you to be a good person, such that they deserve a pat on the back, too? How about other mentors you've had throughout life? Did they contribute to what kind of person you are? What about the kinds of books you've read and the kinds of TV shows you've seen? Did those stories play a role in shaping your brain, such that you were predisposed not to steal? In a world without free will, maybe even your good deeds shouldn't be credited solely to you as an individual.

The same goes for blame and punishment in the case of bad deeds. Derk Pereboom argues convincingly that, when it comes to criminal acts, simple humane imprisonment and incentives for rehabilitation are more logically appropriate than punishment and more effective at preventing a relapse into criminal behavior. (In fact, the Norwegians and the Dutch have proven him right, with their prison rehabilitation methods that produce about 20 percent recidivism, compared to the 75 percent prison recidivism rate in the United States.) Punishment has a bad habit of inducing a return to bad behavior even more than it serves as a deterrent. For example, Mahatma Gandhi is reported to have said, "An eye for an eye leaves the whole world blind." This aphorism is actually a kind of riff on another famous quotation. In the Christian Bible's book of Matthew (5:38–39), Jesus of Nazareth is quoted as saying, "You have heard it said, 'An eye for an eye and a tooth for a tooth,' but I say to you, Offer no resistance to one who is evil. If anyone slaps you on the right cheek, turn to them the other cheek also." The point being made by these three very different wise men is that when a crime is met with punishment, the punishment is often perceived by others as another crime worthy of its own retribution. It sets off a vicious cycle of unstoppable violence. These things happen. Indeed, they continue to happen in many places around the world. The only way to stop it might be to let go of punishment and personal vengeance and turn the other cheek.

Applying blame and punishment, on the assumption that a person's free will is the source of their bad actions, is a decidedly imperfect method of keeping society civil. We often treat the idea of free will as something that gives each of us individuality, specialness, and power, but it might be the opposite. Ironically, in certain circumstances, our faith in free will actually limits us. It makes us essentially helpless when we chalk up a person's decision to commit a crime to some mysterious process of free will that has no causal determinants. If we tell ourselves that there is no cause, then we make ourselves unable to root out the cause and fix it. If instead we acknowledge that problematic antisocial behaviors emerge out of a complex combination of many different causal forces, the nexus of which resides inside the person producing those behaviors, then we empower ourselves to at least try to track down some of those causal forces.

In the hackneyed hypothetical scenario of a person holding a gun to your head and forcing you to steal something, it is easy to track down that causal force and apply corrective measures to the person holding the gun rather than to you. However, in real life, things are much more complicated. For example, what goes into a young person's decision to rob a convenience store at gunpoint? If you could use a futuristic statistical software package to determine why a particular youth robbed a particular convenience store, you would find a myriad of causal forces that interact with one another and feed back on themselves in very complex ways to result in that fateful decision. In a terribly simplified set of pretend initial results of this futuristic statistical analysis, you might find something like this: (a) neighborhood poverty is 15 percent responsible for the behavior, (b) the youth's previous decisions to behave in milder antisocial ways are 12 percent responsible, (c) a taste for revenge that he built up in juvenile hall is 11 percent responsible, (d) poor education is 10 percent responsible, (e) poor parenting is 9 percent responsible, (f) poor job prospects are 8 percent responsible, (g) easy access to handguns is 7 percent responsible, and (h) the remaining 28 percent of what went into that decision is simply unaccounted for. That's right; the causal complexity of a decision like this is likely to be so knotted up that even a *futuristic* statistics software package probably couldn't find out everything.

This combination of causal forces makes up what psychologist Craig Haney refers to as the criminogenic circumstances of an environment, the conditions that give birth to crime. For example, the conditions in most American prisons (and those of many other countries as well) are not really conducive to rehabilitation. Rather than preventing further criminal behavior, the kind of punishment that happens, directly or inadvertently, in American prisons is likely to induce *more* criminal behavior, not less. But don't forget that each of these criminogenic circumstances also has its own causal precursors that preceded it. The poverty, poor job prospects, and poor education in a neighborhood are in part the result of a wide variety of public policy decisions made by government officials. And how did those officials get into office? We the people voted them in, based on what they told us they would do. The parents of that convenience store robber might have poor parenting skills because they were poorly parented and poorly educated themselves. The easy access to handguns could be the result of a growing black market for them, coupled with the specific gun laws in that state and nearby states. Who made those gun laws? Once identified, these kinds of causal forces are certainly not easy to fix, but at least they can—in principle—be improved for the next generation. By contrast, when you blithely attribute the young man's actions to "free will," all you can do is imprison him for a while and hope he doesn't do it again. Good luck with that. Clinging to free will doesn't give you power. It is exactly our clinging to free will that *disempowers* us. In fact, gently relaxing your grip on free will as the evidence for a person's autonomy and self-actualization is empowering for all society.

Rather than telling yourself in vain, for the hundredth time, that your "will power" alone will be what allows you to improve your choices with regard to unhealthy foods, exercise, substance abuse, criminal behavior, or poor life decisions, maybe you should instead try to rearrange those "criminogenic circumstances" in your life. Accept that your free will, by itself, isn't going to make things happen, good or bad. Make adjustments to your environment that will change the context of those decisions and change the options that are immediately available to you. Don't buy the unhealthy food in the first place. Don't hang out with the people who are doing the things you don't want to do. Give yourself personal incentives for the good choices and personal disincentives for the bad ones. Have a close friend help you impose those disincentives. You've heard that it "takes a village to raise a child," right? Lots of other things rely on that village as well. It takes a village to have your prefrontal cortex make you return a wallet unmolested. It takes a village to choose the chicken over the veal. It takes a village to do the right thing every day.

The Will Emerges

Based on the findings described in this chapter, when your prefrontal cortex makes these good and bad decisions for you, it functions a little bit like a funnel. It takes all those different causal influences, on all those different timescales, and swirls them together into a chaotic, turbulent stream of output that turns into choices and actions. As a result of the complex combination of all those multifarious subtle biases, you choose among your available options.

All your intentions emerge out of a complex churning soup of causal forces, and the nexus of that churning, the bottleneck on that flow, just might be your prefrontal cortex. The reason why even a "futuristic statistical software package" wouldn't be able to track and disentangle all the causal influences is because they form a complex dynamical system. In a complex dynamical system, not only do causes produce effects, but some effects can reinforce their own causes. A system with that kind of internal feedback is said to "self-organize," to be able to organize itself. This makes it close to impossible to trace the chain of cause and effect backward in its timeline in a linear fashion. Philosopher Alicia Juarrero describes the new and surprising behaviors that can emerge from the bubbling boil of a complex mind as a *phase transition* in a dynamical system.

A dynamical system is something that carries out its work not by being in a certain state (or condition) that carries meaning but instead by constantly changing from state to state. It is the change in state that carries meaning, not any one state on its own. Take your savings account, for example. If your savings account has less than a certain amount, then your bank may pay you a 0 percent interest rate. The amount of money inside that savings account is what the account *means* to you, and the only time that changes is when you deposit or withdraw money. If you go a few weeks without doing that, the account's state does not change, so its meaning does not change. As a system, it is not very dynamic. A system like that has its meaning, its content, attached to its state. By contrast, if you have money invested in a volatile stock in the stock market, then your money is in a complex dynamical system. Since you need to watch for changes in that stock (and changes in the world market and world events) pretty closely, it is not really the current value of that stock but rather the *change* in its value that carries meaning for you. Most natural systems are dynamical systems. Hundreds of millions of investors interacting with one another make the US stock market a complex dynamical system. Hundreds of millions of neurons interacting with one another make your prefrontal cortex a complex dynamical system.

When you hold a stranger's lost wallet in your hands and try to decide what to do with it, there are innumerable causal influences that go into your brain's dynamical computations. You spend some time in an uncertain state, considering the options. According to Juarrero, that uncertain state is an unstable phase in the dynamical system and cannot last long. The system naturally transitions into a more stable phase: a choice. The choice is not random, but it is also not determined in a simple additive fashion of 15 percent caused by your parents, 10 percent because of your education, and so on. Many of those causal precursors (such as parenting and education) are interdependent with one another, and many others are damn near unidentifiable. Moreover, their influences are often nonlinear. That is, a steady increase in one factor does not necessarily produce a steady increase in the result. For example, the more guidance and protection your parents bestowed on you might tend to increase the chances that you would return the wallet intact—but only up to a point. Overprotective, micromanaged parenting can sometimes induce severe rebellion in a child and thus could contribute to the wallet not getting returned at all. In complex dynamical systems, this chaotic intertwining of nonlinear causal forces induces new phases of behavior to emerge unpredictably, so it should be no surprise that it often looks and feels like "free will."

If your brain is a complex dynamical system capable of self-organization, then we can expect some interesting statistical signatures to show up in its data stream, almost like fingerprints at a crime scene. Cognitive scientist Guy Van Orden identified one of those statistical signatures as evidence for *intentionality* in human cognition. Van Orden's intellectual influence on cognitive science is changing the way people build models of cognition. Guy was quite a character. He could drink you *under* the table, look you *straight* in the eyes, and theorize *over* your head, all at the same time. He will be missed.

According to Van Orden, a traditional linear-additive system exhibits "component-dominant dynamics," such that each separate part of the system carries its meaning, and does its work, by itself. Then the outputs of those parts are added together to produce the system's overall behavior. A calculator does that. A calculator doesn't have intentions. When cognitive scientists use a linear-additive system to model human cognition, they typically add "white noise" (e.g., completely random variation) to the simulation in order to produce the kind of variety in behaviors that human cognition exhibits. But the human mind is not like a calculator, either with or without white noise added. The variation observed in human cognition does not have a white noise signature over time.

The variance in human cognition has a "pink noise" signature over time. Pink noise is when the change in a measurement from one point in time to the next does not have an equal likelihood of all values within some range (i.e., as white noise does), but instead the measurements have a slow up-and-down wave to them over the course of hundreds of measurements. There are long-term correlations in the time series of measurements. The noise tends to be somewhat high for a while, then it tends to be a bit low for a while, and then it goes a bit high again, almost like breathing, in a slightly irregular fashion. A linear-additive system doesn't do that naturally, but a complex dynamical system does. According to Van Orden, a complex dynamical system exhibits "interaction-dominant dynamics," such that each separate part of the system produces system-level meaning, and does system-level work, not by itself but instead via the complex interactions themselves. The meaning in the system emerges *between* its parts, not inside them. An intention is given birth in the interactions among the components, not inside any one component. The nonlinear feedback loops in a system like this, where effects reinforce their own causes, produce pink noise in the system's behavior. Van Orden argued that since human cognition exhibits pink noise, the human mind must be a complex dynamical system. When a human develops a new intention, when they phase transition from one state to another, it is a self-organization process that emerges in a complex dynamical system. As such, any attempt to trace the various causal precursors in the many looped chains of cause and effect is very difficult. Therefore, under typical everyday circumstances, this emergent intention partly *belongs to the human individual*, at least to a greater degree than it does to any one of those causal precursors. It's almost as if this emergent intention is the 28 percent that the futuristic statistical software package was unable to account for.

One of Van Orden's best friends, Chris Kello, tested this idea that pink noise would naturally emerge from a complex dynamical system. He developed a complex dynamical network simulation of more than a thousand neurons interacting with one another. Under certain constraints of connectivity and signal transmission, this network naturally generated pink noise and other statistical signatures of self-organization. Kello's computational simulations identify the neural parameters required to produce the pink noise that neuroscientists now know brains produce. When it comes to intentionality emerging from a system with interaction-dominant dynamics, perhaps these simulations provide a mathematical framework in which to prove that Guy Van Orden was right.

What all this means is that, even though Libet's experiments make you feel like you might not have genuine free will, this shouldn't leave you despondent and feeling like you have no autonomy, self-governance, or uniqueness. You, your brain, and perhaps especially your prefrontal cortex unmistakably have a unique way of combining the thousands of social, genetic, educational, nutritional, and accidental causal forces that go into determining your decisions and actions. In our thought experiment where you found a lost wallet, no one else in the world would have done precisely what you did, not even your genetically identical twin if you have one. Only *you* spent that exact amount of time mulling over your various options and scenarios, in your particular order, with those specific nuances and evaluated consequences, to finally arrive at the choice that you selected. Your will may not be entirely free, but it is yours. You should allow yourself to assume some responsibility for that decision. Just don't give yourself *too much* credit, or blame. Maybe 28 percent?

However Improbable

Many of the findings in this chapter don't point to *who you are* but instead help you rule out *who you aren't*. From an investigative standpoint, that's important progress. In the immortal words of Sir Arthur Conan Doyle's famous sleuth Sherlock Holmes, "When you have eliminated the impossible, whatever remains, however improbable, must be the truth." Based on the scientific evidence, it seems unlikely that *who you are* is nothing more than a nonphysical soul that mysteriously talks to its pineal gland to find out what the brain perceives and then tells it what to do. It also seems unlikely that one atom can regularly determine *who you are*. Perhaps if you were perfectly balanced in all your mental biases for choice A at 50.0000000000 percent and choice B at 50.0000000000 percent, then one atom's subatomic randomness might catch fire in a quantum resonance that spreads across hundreds of neuronal membranes and become the "straw that broke the camel's back." But how often are you that perfectly balanced between your choices? Likewise, it seems unlikely that one neuron can determine *who you are*. You have billions of them in your brain, and they produce your thoughts with their *coordinated population activity*, not through their individual activations.

So if we rule out the impossibly small (nonphysical causal influences), the subatomically small (quantum-mechanical influences), and the microscopically small (individual neuronal forces), then *who you are* must mean something larger than that. Perhaps your prefrontal cortex is who you are. It seems to be the place where seemingly arbitrary decisions (e.g., choosing between a left button or a right button for no reason) appear to get formulated. There are in fact quite a few cognitive scientists who would be happy with characterizing the prefrontal cortex as the *seat of your selfhood*, because it carries out so many "executive functions." The prefrontal cortex is important for reasoning, problem solving, planning, selecting among competing options, inhibiting urges, personality, and many other things.

For the time being, try it out. Try it on like a hat or a pair of sunglasses. Imagine that who you are—the you that's talking and listening when you silently think to yourself—is that frontmost one-quarter of your brain. When you rationalize to yourself why you chose the veal instead of the chicken this time, it's your prefrontal cortex at work, a network of billions of neurons, with a complex pattern of connectivity, sending electrochemical signals to one another. Is that who you are?

Directions for Use

Your assignment for this chapter is intended to make you challenge your own sense of *why* your prefrontal cortex has you do the things you do. I want you to look through your house or apartment and find an unwelcome insect. That's right, a bug. It could be a spider, an ant, or something similar. Check the corners of your closets, for example. Most of us have some kind of informal policy, or standard practice, for what we do when we find an unwelcome insect in our home. Some of us squish them. Some of us capture them and release them outside. (In my case, it varies depending on the type of insect. With most types of bugs, I usually just squish them, but spiders are different. Technically speaking, spiders aren't a member of the insect family. I prefer to capture spiders and release them outside, except for black widows. If there's a black widow in the house, then she's going to have to "take one for the team.") But here's the rub: Whatever your personal policy is for the particular bug you find during your assignment, I want you to "veto" that decision and carry out the opposite policy. If you normally squish this kind of bug, then capture it instead and release it outside (unless it is a venomous bug, in which case please kill it safely and start your assignment over again). If you normally capture and release this kind of bug, then squish it instead. Go do that right now. When you are finished with this first half of your assignment, go to the end of the notes for this chapter and read the second half of your assignment.

3 From Your Frontal Cortex to Your Whole Brain

I love who ya are. I love who ya ain't.
—Outkast, "So Fresh, So Clean"

Imagine if every time you saw the letter "A" it had a reddish glow about it—even when it was actually in black print on a white page. That is exactly what is experienced by people with a certain form of synesthesia (where different perceptual/cognitive channels for vision, hearing, touch, space, and numerosity seem to arbitrarily interact with one another). These things actually happen relatively commonly. The parts of the visual cortex that process written letters can sometimes wind up building strong neural connections to the parts of the visual cortex that process color perception, and a permanent visual illusion is born. In fact, certain songs or spoken words can cause some synesthetes to feel an illusory touch at particular parts of their body. There are even cases of synesthesia where a person may experience numbers, days of the week, or months of the year as being laid out in a spatial sequence that forms a robust imagery that is sometimes overlaid on their visual field. For instance, my father has a curvy number line in his mind's eye that goes up and down like a roller coaster, and he credits his use of that mental roller coaster as the reason he is so quick with arithmetic. Examples of synesthesia are powerful reminders of how brain regions get richly connected to each other throughout development. Therefore, pretending that any single brain region—such as the prefrontal cortex—is somehow so independent from the rest of the brain that it can perform its functions in a vacuum is clearly going to lead one to an incorrect understanding of how the mind works.

Now that you've made it through chapter 2, you should be ready for chapter 3. I realize that some parts of chapter 2 were hard work. I'm genuinely proud of you for making it this far, but I trust you'll forgive me if I refrain from lavishing your individual person with too much credit and accolades for your "will power" in doing so. Maybe you should thank your parents for me.

If chapter 2 did to you what it was supposed to do, then you've embraced the idea that your prefrontal cortex is a very important part of "who you are." In this chapter, I'm going to show you a handful of scientific findings indicating that the functions of your prefrontal cortex are so deeply intertwined with the functions of other parts of your brain that you really ought to expand your definition of who you are to include your entire brain—not just the frontmost one-quarter of it. As chapter 2 showed, you ain't just a soul, you ain't just a molecule, and you ain't just a neuron. And maybe you ain't just a prefrontal cortex either.

The Homunculus and Its Modules

A few decades ago, philosopher Jerry Fodor famously described the mind as being made of "modules." In his account, it was just about everything *outside* the prefrontal cortex that functioned in a modular fashion. He suggested that visual perception was carried out by an "input system" that was dedicated solely to processing information from your eyes. Therefore, knowing the name of an object, its smell, or its history cannot change the way you see it. The way you see that object or person is determined entirely by the pattern of light that lands on your retinas, nothing else. Similarly, Fodor suggested that language comprehension likewise was carried out by an input system whose computational processes were specific to the domain of language. According to this modular account, whether reading or hearing, the only information that the language input system used was linguistic information. Your expectations based on context for what you might read or hear next in a sentence are not part of what that linguistic input system does. Knowing the history of someone whose face you're looking at or having an expectation for where a partial sentence might go next are things that a central executive—like the prefrontal cortex—does. In Fodor's modular theory of the mind, those central executive processes (where "the self" lives) are unable to influence the modular processes of those input systems.

In this kind of theory about the mind, the central executive living inside the prefrontal cortex is almost like a wizard's homunculus. When I was a kid playing Dungeons and Dragons, my friends and I would read about a small, creepy humanoid creature that a wizard in the game could construct from clay and ashes to be his servant and his spy. Whatever the homunculus saw and heard, the wizard would see and hear as well. If the prefrontal cortex functions a bit like a homunculus in a mental "control room," this means it can see the visual information delivered by the visual input system, can hear and read the linguistic information delivered by the linguistic input system, can smell the olfactory information delivered by the olfactory input system, and so on. The homunculus, your central executive, then gets to decide which levers to pull in the control room to drive your limbs, your eyes, and your mouth. I guess it's no wonder that you find out seconds later what choices you've actually made if those decisions are being made by a creepy little monster hiding inside your prefrontal cortex!

But seriously, what the homunculus description really points to is the fact that the most important computation in this modular theory of the mind, the *conversion of sensory inputs into motor outputs*, is completely neglected and left to the mysteries of an unexplained central executive. It may as well be a wizard's homunculus, for all the theory cares. However, as chapter 2 clearly demonstrated with Libet's experiments, the executive functions of the prefrontal cortex are not really as mysterious as one might think. They rely not on magic or quantum randomness but simply the electrochemical interactions of billions of interconnected neurons. These neurons produce electric fields and magnetic fields, and they change blood flow. Science can measure these things.

However, Fodor's modular theory of the mind specifically predicted that central executive processes would never be explainable because they were just too complex. According to his account, central executive processes involved feedback loops that provided context effects and thus were not simple and linear enough for science to ever figure out. By contrast, the input systems were not influenced by context effects, because they did not involve any feedback loops, and thus they were sufficiently modular for science to be able to "divide and conquer" them. The double irony for what eventually happened to this modular theory of mind is that over the course of the past few decades (1) the cognitive and neural sciences have discovered dozens of feedback loops and context effects in those supposedly modular input systems of vision, language, olfaction, and so on, and (2) a number of cognitive scientists have twisted Fodor's suggestion into inspiration for proposing specific modules *within* the prefrontal cortex, each devoted to executive processes such as inhibition, task switching, emotional control, arithmetic, lie detection, and simulating what others know, to name just a few. The perceptual input systems that Fodor said were modular are turning out not to be, and the one place that Fodor said we wouldn't find modules is exactly where his followers are glibly proposing them. Isn't it ironic?

So much for Fodor's modularity of mind. A wide range of context effects have been discovered in vision, language, and other areas of cognition. What this chapter will focus on is how the prefrontal cortex—which (according to chapter 2) plays such an important role in determining "who you are"—is intimately connected to many other brain regions to carry out those context effects. The connections are so dense that we probably should not draw a line around the prefrontal cortex and treat it as an independent component. Despite the theory's errors about what it called modules, Fodor's modularity of mind was right about one thing: the central executive functions of the prefrontal cortex should not be treated as a context-free module that houses the self. Wherever the self is, whoever you are, this chapter will show you that it is something that emerges among the neural interactions of many different brain regions. The prefrontal cortex is one of the more important hubs in that vast network, but it is still just a part in the whole that determines *who you are*.

A Paradigm Drift

Soon after the modularity of mind became a popular idea in the cognitive sciences, James McClelland and David Rumelhart pooled together the Parallel Distributed Processing (PDP) Group at the University of California, San Diego, to explore how the inner workings of interactive nonmodular networks of neurons might actually be explainable after all. The PDP Group and its progeny became generally known as "connectionists." It almost sounds like a bad new wave band from the early 80s, doesn't it? The Connectionists. McClelland, Rumelhart, and their crew designed hundreds of computer simulations of neural networks that converted simplified sensory inputs (such as letters) into simplified outputs (such as recognized words)—without needing to stop in the middle to consult a central executive, or a creepy little monster for that matter. These computer models provided approximate neural simulations of the inputs, internal computations, and outputs related to visual object recognition, word recognition, categorization of plants and animals, and other things without ever requiring a step in the middle that might correspond to a mysterious free-will-wielding entity. Most importantly, their simulation results matched human data from hundreds and hundreds of laboratory experiments.

To this day, as the simulations continue to get larger and more biologically realistic, they do become more difficult to understand and explain, but they are still far more transparent and manipulable than the actual target of the simulation: the human brain. Central to the idea is the *interactions* between neurons and neural subsystems. By building simulations of neural-like systems that perform perceptual and cognitive tasks, the connectionists (and other neural network researchers) have demonstrated that—contra Fodor's admonition—nonmodular systems with feedback loops can be explainable and understandable. It's not easy. You have to be fearless in the face of nonlinear mathematics, so not everybody is ready to do it yet. But you can learn.

The result has been a detectable drift within the larger paradigm shift that cognitive science is currently in the middle of. Call it a paradigm *drift* if you want. Whether a contemporary cognitive scientist admits it or not, his or her theoretical approach to how different parts of the mind produce intelligent behavior has been deeply influenced by the *interactionism* that McClelland, Rumelhart, and their connectionists advanced over the past few decades. Even when a small handful of current philosophers of mind still theorize that cognition is made of "modules," they now acknowledge that those supposed modules somehow interact with one another to exert and receive context effects. It is thanks to the connectionists that they have made that incremental improvement on their theory.

Interactionism in Vision

In the case of visual perception, the scientific evidence for interactionism has been uniquely compelling and unmistakable. This is in part because we can invade the brains of animals that are pretty similar to us and look at how their visual cortex works. I say "invade" because the methods are, after all, referred to as *invasive techniques*. Many monkeys and cats have unwillingly given their lives for this kind of research, just like many chickens and calves have unwillingly given their lives so people can eat delicious poultry and veal—instead of beans and tofu—and many rabbits and mice have given their lives for research that tests for rashes and illnesses caused by cosmetic products. Neuroscience truly is a more noble cause than that. It may not be pretty, but the medical implications of neuroscience research are improving human lives worldwide today. Without the sacrifice of these animals' lives to science, we wouldn't have the medical understanding that we now have of Alzheimer's disease, Parkinson's disease, and vision disorders.

Thanks to the efforts of visual neuroscience researchers, we now know that the human brain has dozens of anatomically separate regions that roughly specialize for different aspects of visual information processing. The retinas respond to light input and send electrochemical signals primarily to a subcortical region called the lateral geniculate nucleus, which then sends signals to the visual cortex, at the back of the brain. In this region of the brain, there are billions of neurons that have receptive fields like those Horace Barlow found in the frog brain (see chapter 2). Each of these neurons receives signals from a field of hundreds or thousands of light receptors on the retina. However, that receptive field structure is now referred to as the "classical receptive field." You know what that means, right? Whenever anything gets called "classical," it means it's out-of-date, less interesting, and only old people listen to it. In visual neuroscience, the *"nonclassical* receptive field" structure refers to the wide range of neural signals that apparently find their way to a visual neuron but did not recently originate from light landing on the retina. Research has been discovering that these visual cortex neurons also receive signals from other parts of the brain: feedback signals coming from other visual, visuomotor, memory, auditory, or emotional brain regions, and even—you guessed it—the prefrontal cortex.

Vision researcher Moshe Bar developed a theory that claims one of the most important feedback signals to the visual cortex comes from the orbitofrontal cortex (aka the ventromedial prefrontal cortex), the underside of the frontmost portion of the prefrontal cortex, hanging just over the tops of your eyeballs. Bar suggests that noisy visual signals quickly reach the prefrontal cortex, and rough imagistic expectations are generated there and sent as feedback to the visual cortex. This feedback biases the way the visual cortex processes its incoming signals from the retina. Most of the time, these expectations and biases are generally on target, and they give you a head start in interpreting your visual environment, but occasionally these biases are incorrect. That is why you sometimes "see what you want to see" rather than what's really there. Moshe Bar and his colleagues combined spatially precise fMRI neuroimaging (the same method John-Dylan Haynes used to show that the prefrontal cortex knows what you're doing before *you* do; see chapter 2) and fast-recording magnetoencephalography neuroimaging (MEG) to identify what brain areas are active, and at what times, when people recognize visual objects. MEG immediately detects the changes in magnetic fields that happen when the brain's neural activity changes, millisecond by millisecond. By contrast, fMRI has finer spatial resolution to identify very specific brain regions, but it has to average its data over about one whole second, and its signal is delayed by a couple of seconds, depending on how long it takes the blood to reach that particular brain area. By combining these two methods, Bar and his team were able to confidently determine that this region of the left prefrontal cortex became active at least 50 milliseconds *before* visual object recognition regions in the temporal cortex became active.

This is a little different from the prefrontal cortex "knowing what you're doing before *you* do," but it's related. Essentially, visual neural signals eventually travel from the retina to the primary visual cortex (in the back of the brain), and then those partially processed signals are quickly sent all the way to the prefrontal cortex, where an "initial guess" is made as to what the object might be. Then, while the temporal visual cortices (on the sides of the brain) are just barely beginning their fine-grain analysis of precisely how to recognize this object, the prefrontal cortex is using its feedback connections to send its "initial guess" back to the temporal visual cortices to bias them a little in how they do their job. Keep in mind that these feedback connections are there all the time, so the bias that the prefrontal cortex sends to the visual cortex doesn't have to be limited to its noisy guess about the current image. The prefrontal cortex is probably always sending biases, based on what it saw *a few seconds ago* or more, to place a thumb on the scale when the visual cortex is trying to figure out what it's looking at *now*. For example, when the visual cortex is trying to resolve the competition between "is it a vase or faces?," "is it a duck or a rabbit?," or "is that a gun or his wallet?," the prefrontal cortex is busy using its biases about the current context and situation to influence that perceptual competition process even before it starts. These things happen all the time.

By combining these prefrontal *biases* with this perceptual *competition* process, you might imagine you could develop a theory of vision that could be called *biased competition*. Well, you're too late. Neuroscientist Robert Desimone and psychologist John Duncan already did that a couple of decades ago. Desimone and Duncan's biased competition theory has essentially revolutionized the way vision researchers think about how the primate brain processes the patterns of light that land on the retina. For example, in 1995, when I was a graduate student at the University of Rochester, I tried to publish a cute little experimental result showing that when tilted lines compete against one another in your visual field, where you apply your attention (separate from your eye position) can influence the strength of the interference effect, but I was shot down by expert reviewers who felt it was too outlandish for me to suggest that signals from the frontal brain regions might be able to alter the functioning of neurons in the primary visual cortex. I felt like the hackneyed "small town detective" who has a hunch about whodunit but just can't quite prove it. That same year, Desimone and Duncan's biased competition theory was published, with a masterful, comprehensive review of the visual attention and neuroscience literature, and everything changed in the field of vision research. They didn't just have a hunch that the frontal cortices could exert a top-down influence on visual perception. They proved it. Soon after that, I resubmitted my cute little experimental result for publication, and rather than reviewers telling me that my suggestion was outlandish, they asked me to tone down how new and surprising the manuscript made it sound, because these kinds of findings were already well accepted at that point.

The past two decades have seen an incredible collection of compelling demonstrations of visual perception being influenced by nonvisual information. Here are just a few examples. In 1997, Allison Sekuler and her colleagues presented to experiment participants a little animation of two circles moving toward each other and passing through one another. But when a simple little click sound was added upon the circles touching each other, the exact same visual event was instead perceived as two circles *bouncing* off each other and reversing their directions. In 1999, Geoffrey Boynton and his colleagues used fMRI to show that activity in the primary visual cortex is altered by instructions to apply your attention (separate from your eye position) to one side of the visual field or the other during a visual discrimination task. In 2000, Ladan Shams and her colleagues presented to participants a *single* brief flash of light but accompanied it with *two* very quick beeps of sound. When the participants were asked whether they saw one or two flashes of light, they frequently said they saw two. This is basically a visual illusion that is caused by auditory input.

In 2010, cognitive scientist Gary Lupyan used a "signal detection task" to show that spoken linguistic input can influence visual perception at the sensory level. In each of 200 experimental trials, participants were presented with a visual letter (e.g., a capital letter M) for a mere 53 milliseconds, and it was then masked by a 700-millisecond scribble of lines. The mask ensures that your visual system has the 53-millisecond signal quickly erased by the scribbled input, so your brain really gets only 53 milliseconds of input from this visual signal. However, on half the trials, there was actually nothing presented during those 53 milliseconds. So after the scribbled mask is gone, the participant has to guess whether there was a visual letter (of any kind) presented in that trial or nothing was presented. People sometimes felt as if they were guessing in this task, but their performance was still better than chance. And here's where Lupyan inserted some nonvisual information that seems to influence visual processing. On some trials, the 53-millisecond presentation window was preceded by a spoken letter cue (e.g., a voice saying "emm"). But remember, some of those trials with the spoken letter cue did *not* have a visual signal presented during the 53-millisecond window. That is how the signal detection task keeps your reports of sensory perception honest. If the participant were to respond positively in a trial like that, it would count as an incorrect "false alarm" trial and be subtracted from the measure of their performance. In Lupyan's experiment, detection of targets was improved by 10 percent when the verbal cue preceded the visual stimulus, and false alarms (saying "present" when there was no target) were unaffected by the verbal cue.

Let's imagine being in this experiment. You're looking at the center of a computer screen, over headphones you hear a recorded voice say "emm," about a second later the screen flashes something that might be letter-like and then some brief display of squiggled lines, and then you're asked whether you think there was a letter presented right before those squiggled lines. You're not entirely sure, but you think maybe there was a letter briefly flashed there, so you say "yes." It's as if your language brain areas told your visual brain areas what to look for, so they were better prepared to see it. Then, in another trial, you might hear nothing over the headphones, then see a quick flash of squiggled lines, and conclude that there wasn't a letter presented. But in fact, in that trial there was a letter M presented for 53 milliseconds right before the squiggled lines were displayed. You missed it because your visual system didn't get that boost of preparedness from other brain areas. When you pool together the data from your 200 trials like that, along with 200 trials from each of a few dozen other participants, you get statistically reliable scientific evidence that spoken verbal cues can influence visual perception at the early stage of simply detecting the presence of a visual stimulus.

These kinds of effects on visual perception and visual cortical processing could only happen if brain areas such as the prefrontal cortex were constantly sending synaptic signals to the visual cortex, biasing how you see the world. It is precisely because we know so much about the cortical "wiring diagram" of the visual cortex and the rest of the brain that we are able to develop a deep functional understanding of how the frontal brain regions are connected, bidirectionally, to areas of the brain that process visual information. The visual cortex is constantly telling the prefrontal cortex what it *thinks* it is seeing, and the prefrontal cortex is constantly telling the visual cortex what it *expects* to see. With this continuous immediate sharing of information back and forth between two sprawling brain areas, it makes sense to treat them as *one* large network. It is amid this vast expanse of interconnected neural tissue, stretching from the front of the brain to the back of the brain, that our experience of visually perceiving the world emerges.

Interactionism in Language

In the case of language processing, it has been a little more difficult to develop as precise a cortical "wiring diagram" as we have for visual perception. We can't dissect the brains of nonhuman animals to figure out the blueprints of the language cortex, because we don't have any nonhuman animals that speak a full-fledged language, and if we did, we probably wouldn't feel too good about dissecting their brains. But we can examine the auditory cortex of animals that hear, and that's a start.

As a powerful demonstration of how opportunistic they are with their interactions, neurons in the primary auditory cortex of the ferret don't seem to give a heck what kind of information they receive from the sensory system. Neuroscientist Sarah Pallas and her colleagues performed brain surgery on a newborn ferret and redirected its optic nerve to send *visual* information to the part of the thalamus that sends information to the primary *auditory* cortex. Thus, visual input signals were now reaching the auditory cortex instead of the visual cortex. The amazing thing is that those neurons in the auditory cortex were more than happy to develop *visual* receptive fields. That ferret was able to see the world with its auditory cortex instead of its visual cortex. Evidently, there is nothing innate about the auditory cortex that forces it to process only auditory information.

There are also more naturalistic explorations of the neuroplasticity of the auditory cortex that have been conducted in the neuroscience laboratory. For example, neuroscientist Jon Kaas and his colleagues carefully mapped out the receptive fields of the primary auditory cortex in a live adult monkey, so they knew exactly what tiny regions of the exposed auditory cortex responded to low-pitch, medium-pitch, and high-pitch sounds. Then they lesioned the cochlea (in the inner ear) at the high-pitch sensitivity region, so that the auditory nerve no longer transmitted any high-frequency sound information to the primary auditory cortex. As a result, this monkey was no longer able to hear high-pitch sounds in that ear—kind of like my dad. High-frequency hearing loss is a common problem we primates experience as we age. It will probably happen to me, too. The interesting thing is what takes place inside the primary auditory cortex when this happens.

When Jon Kaas and his team examined the auditory cortex of this monkey three months later, the map of receptive fields had substantially reorganized itself. Those neurons that used to be selective for high-pitch sounds (but hadn't been receiving any such signals for the past three months) had now become selective for medium-pitch sounds. Basically, the strong synaptic connections with high-frequency inputs that they used to have were no longer transmitting signals, so the weak synaptic connections they had with medium-frequency inputs had grown stronger. Neurons that fire together wire together, right? And neurons that don't fire together anymore will tend to unravel that wiring. This kind of neural reorganization surely happened in my dad's primary auditory cortex. Now that he has more cortical real estate devoted to that medium-pitch range than he used to, I wonder if it might make him better at discriminating among those medium-pitch sounds.

While mild hearing loss is a relatively natural occurrence, we tend to frown on performing disruptive cortical rewiring surgery in humans. In order to perform those kinds of experiments, we usually have to settle for injecting strange conflicting sensory signals into their ears and eyes instead, to test for interactions between perceptual systems. Recall earlier how vision researcher Ladan Shams produced a *visual* illusion that was caused by *auditory* input. A single flash of light was misperceived as *two* flashes of light, because two quick beeps accompanied it. Well, in the 1970s, Harry McGurk was able to similarly produce an *auditory* illusion that was caused by *visual* input. Imagine you're in an experiment where you are supposed to watch a video of a person repeating a spoken syllable over and over. For example, it might be a close-up of a person's face as they say "ba-ba-ba-ba…" for a dozen seconds. Close your eyes for a minute and generate a rich mental image of what it would look like and sound like. Go on, do it. Please? I'm not going to continue this story until you do.

Okay, thank you. What did your mental image look like? Did the person's lips close each time the /b/ sound was made with their mouth? Good. That's because the only way to make the /b/ sound is for the lips to touch. For about 50 milliseconds, the *two lips* are *stopping* the air from coming out of your mouth, so linguists call that /b/ sound a *bilabial stop* consonant. Aren't you glad we have names for these things? Now that you have a sense of what people in McGurk's experiment saw and heard, I can tell you about the strange conflicting signals he then subjected them to.

In one particularly interesting experimental condition, McGurk dubbed and synchronized the "ba-ba-ba-ba" audio track onto a video of a person who was actually saying "ga-ga-ga-ga" at the same tempo. Now that you have some linguistics training from the previous paragraph, you can probably figure out that the /g/ sound is *not* a bilabial consonant. (That is, your lips do not close when you say "Lady *Gaga*." Nor should they.) So, when participants watched this altered video, it didn't *look* like the person was saying "ba-ba-ba-ba," because the lips weren't touching. The amazing thing was that it didn't *sound* like it either. Instead of hearing that audio track as the "ba-ba-ba-ba" that it really was, most people heard it as "da-da-da-da." This is the next closest speech sound, because the /d/ sound is a stop consonant just like /b/, and the closure happens near the lips (as the tongue touches the roof of the mouth), but it does not involve the visible closing of the lips, so it is consistent with the face in the video. Importantly, people did not perceive this altered video as a confusing, conflicting event. They did not find themselves reasoning about what might be a sensible reinterpretation of the speech sound into a "da-da-da-da" event. When they looked at the video, they instantly felt they were hearing "da-da-da-da." And when they closed their eyes, eliminating the visual context, they could suddenly hear the actual "ba-ba-ba-ba" of the audio track. In fact, cognitive neuroscientists Tony Shahin and Kristina Backer have evidence from brain imaging studies suggesting that this very effect of vision influencing speech perception is happening inside the auditory cortex—not in some decision-making part of the brain.

Findings like these start to make sense when we get a glimpse into the wiring diagram of cortical language areas. Riki Matsumoto and his team developed an invasive electrode technique that can be safely used with epilepsy patients who are already undergoing brain surgery anyway. During their brain surgery to help identify and alleviate their epilepsy symptoms, Matsumoto used implanted electrodes to stimulate one brain region and record which other brain regions exhibit the spread of neuronal activation. When he stimulated language and speech areas in the temporal lobe, near the auditory cortex, the activation spread to language areas in the prefrontal cortex. This part isn't surprising, because it makes sense that speech areas next to the auditory cortex are accustomed to receiving spoken speech input and then transmitting their information to frontal language areas for a more complete understanding of the sentence. What was surprising at the time was that when Matsumoto stimulated language areas in the prefrontal cortex, the neuronal activation spread backward to language and speech areas in the temporal cortex. The result shows that the neuroanatomy of language involves bidirectional connections between temporal brain areas that process speech sounds and frontal brain areas that process the meaning, grammar, and context of sentences.

This kind of neural architecture is exactly what Jay McClelland and Jeff Elman envisioned when they developed their connectionist neural network model of spoken word recognition in the mid-1980s. In this framework, there is a continuous two-way sharing of information back and forth between linguistic subsystems. For instance, the subsystem that recognizes 40-millisecond speech sounds, such as the phonemes /v/, /o/, and /t/ in the word "vote," naturally feeds that sequence of phonemes into the subsystem that recognizes whole words. But that subsystem that recognizes whole words is also simultaneously sending feedback to the speech-sound subsystem to help it do its job. McClelland and Elman's network simulation, with its bidirectional connections between word neurons and speech-sound neurons, predicted that the recognition of a *previous* word could help resolve uncertainty in the speech input of the *current* speech sound. They, and a number of psycholinguists, have since proven that prediction correct by demonstrating a wide variety of context effects where the recognition of a current spoken word is influenced by the meaning, grammar, and statistical properties of previous words in the sentence.

Take, for example, a spoken word that is ambiguous between "dusk" and "tusk." Imagine the speaker had a piece of gum in her mouth when she said it, or maybe some guy nearby coughed just as she uttered the word. In isolation, you might find yourself completely uncertain whether the spoken word was "dusk" or "tusk." Luckily, it is rare for a word to be delivered in isolation. There is almost always some preceding context for it, either linguistic or situational. (Even when someone yells the single word "fire!," we know it means different things in different situations, depending on whether you're in a crowded theater, playing a first-person shooter video game, or watching reruns of The Apprentice.) In the case of hearing the noisy spoken word "d/t-usk," if the preceding sentence fragment had been "The moon rises at...," research shows that you would seamlessly hear it as "dusk," and if the preceding sentence had been "The walrus was missing a...," then you would seamlessly hear it as "tusk." Just as with the McGurk effect, you wouldn't feel as if you were reasoning and debating about how to resolve the uncertainty in the word. Thanks to the context narrowing your interpretation *while the ambiguous word was entering your brain*, you wouldn't even notice that its spoken delivery had been noisy in the first place.

The behavioral experiments show that this context effect works, but what goes on *inside the brain* when it happens? This is difficult to measure because, as you will recall, fMRI neuroimaging has to average over a whole second or more, and it only takes about one-third of a second to say "dusk" or "tusk," so cognitive neuroscientists David Gow and Bruna Olson had to bring out the big guns to study this linguistic context effect. Gow and Olson recorded both MEG and EEG brain patterns at the same time, taking more than a thousand samples per second, to get as much data as possible from the magnetic and electric fields that emanate out past the skull when large groups of neurons transmit their electrochemical signals to one another. Participants listened to prerecorded sentences such as, "After four beers, he was t/d-runk." Data were recorded during the recognition of that last word, which on its own would sound ambiguous between "trunk" and "drunk." However, in the context of the rest of that sentence, people nearly always reported hearing it as "drunk." Since MEG and EEG do not have very precise spatial localization, Gow and Olson also performed structure magnetic resonance imaging (MRI) on every participant to help identify each person's specific brain regions that are roughly specialized for speech perception, word recognition, meaning comprehension, and grammatical processing. With quantitatively identified brain regions, and millions of data points from the MEG and EEG time series, they were able to perform a statistical analysis that revealed which brain regions were influencing which other brain regions and when. The Granger causality analysis they used is able to track what changes in a time series of events (e.g., activity in one brain area) are traceable back in time to similar changes in the time series of a different set of events (e.g., activity in a different brain area). It's almost as powerful as the "futuristic statistical software package" that I joked about in chapter 2. Gow and Olson's results clearly established that, in those trials where the sentence context altered the person's comprehension of the final ambiguous spoken word, word recognition areas and meaning comprehension areas of the cortex were causally influencing the activation going on in a speech perception area of the cortex. People's brains really were *hearing* that ambiguous word differently in the two sentence contexts, not just *interpreting* it differently.

Just as we saw in the relationship between the prefrontal cortex and visual cortex, the prefrontal cortex also forms an expansive network with language areas, stretching all the way back to the temporal lobe, at the side of the brain. The temporal cortex is constantly telling the prefrontal cortex what it *thinks* people are saying, and the prefrontal cortex is constantly telling the speech area what it *expects* them to say. In this complex interconnected network of multiple brain regions, areas that roughly specialize in meaning, grammar, speech sounds, or written letters are all sending signals back and forth to one another to help you understand all the words you hear or read as they stream into your ears or eyes at three or four per second.

Interactionism in Concepts

Imagine you're at the zoo with your nephew, and he starts saying "Giraffe! Giraffe! Giraffe!" because he's so excited about the two of you seeing this amazing creature up close and in person. When you recognize a spoken word and perceive a visual object associated with that word, those two input streams spread practically all over the cortex, interact with each other in multiple places, and feed back on themselves, but one place where they definitely have to meet up and "get on the same page" is in the prefrontal cortex. Some researchers think that the prefrontal cortex is where your concepts are stored. What is your stored concept of a giraffe made of?

Several decades ago, cognitive psychologists seemed to think that you could access the information in a person's stored conceptual representation just by asking them about it. They developed an experimental method that involved simply asking participants to list the features of a concept (e.g., "Please list all the features of an eagle"). Participants would then dutifully write down the first several things they could think of, such as "wings, feathers, talons, flies fast, catches rodents and snakes," and so on. By pooling together data across many participants, cognitive psychologists got it in their heads that with this method they were finding out what was inside people's heads. The assumption was that the process elicited by that experimental method essentially involved the person accessing their stored concept for *eagle* and providing a readout of the content listed under that mental entry, in a manner similar to finding a word in a dictionary and reading its definition. In this dictionary metaphor, whenever and however you access the concept, you should get the same information every time, because that content is simply waiting there for you, generally unchanged over time.

Some useful observations regarding how people think about their concepts and categories were obtained by this method, such as a remarkable consistency between what different people wrote down for the same concept. However, this is probably not because the method actually provided a readout of the content in a mental entry that was already sitting there waiting to be accessed. In fact, Larry Barsalou showed that this very same experimental method could provide rich coherent information about concepts that couldn't possibly have been already sitting there in some "concepts module," waiting to be inspected. He asked participants to list the properties of categories made on the fly, such as "things you'd sell at a garage sale" or "things you'd remove from a burning house." As with more standard concepts such as "giraffe," there was again a remarkable similarity in what people wrote down. Therefore, the consistency across people in how they respond in this task should not be taken as evidence that the information being generated is coming from a stored mental entry.

Over the past few decades, Barsalou has developed a more sophisticated account of how concepts are encoded in the brain that includes a vast network of different perceptual components. Rather than storing abstract logical symbols in a "concepts module" inside the prefrontal cortex, Barsalou's perceptual symbol systems theory suggests that every concept you entertain is made of learned patterns of neural activation that are spread across many different sensory, motor, and cognitive areas of the brain. To develop this scientific theory, he draws from findings in cognitive neuroscience, cognitive psychology, linguistics, and philosophy. For example, he and his student Kyle Simmons collaborated with neuroscientist Alex Martin to discover that merely *looking* at pictures of tasty foods produces activation in the gustatory cortex. Thus, with no sensory input to the taste buds at all, simply seeing and thinking about a tasty food gets the taste region of your brain revved up.

Larry Barsalou, with his student Ling-Ling Wu, demonstrated how the "mental access of a stored concept" is really better described as a "perceptual simulation" of that concept. They gave people two types of instructions to elicit the list of features for a concept. With one group, they said, "Describe the characteristics that are typically true of the concept [conceptname]." With another group, they said, "Construct a mental image of a [conceptname] and describe the contents of your image." People's responses were nearly identical for these two types of queries, suggesting that when people "access" their "conceptual representations," they're not really looking up an entry in a dictionary and reporting back the information listed in that entry. Rather, they are generating what Barsalou calls a "perceptual simulation" of that concept on the fly. Importantly, subtle adjustments in the way you generate that perceptual simulation can produce large systematic changes in the features that get listed by the participant. Small tweaks of the concept name can cause big changes in the features that people list. For example, with the concept "watermelon," people tend to list things such as "green" and "round." However, for the same basic concept "half a watermelon," the conceptual features people list most are "red" and "seeds." Now, you probably wouldn't want to list a separate entry in your conceptual dictionary for every object and also for every *half* of an object. And conceptual combination won't help you here either. The concept *watermelon* on its own doesn't have the features "red" and "seeds" figuring prominently, and the concept *half* certainly doesn't have "red" and "seeds" stored inside it. So, how could those features of "red" and "seeds" become so prominent in the concept for "half a watermelon?" The conclusion is it must be that a perceptual simulation of a watermelon that's been cut in half is being mentally generated when a person thinks of that concept. Thus, rather than the prefrontal cortex working all by itself to activate a logical symbol (or dictionary entry) for that concept, it is instead cooperating with sensory brain areas to generate a global pattern of neural activation that is slightly similar to what the brain would do if it were actually perceiving that object.

The real-time process of thinking about a concept and then thinking about its features is clearly sensitive to some mental timing constraints. For example, if you feel that the relationship between a given concept (e.g., bird) and one of its features (e.g., has feathers) came to your mind really quickly, you might feel that this particular feature must be a very important part of that concept. By contrast, if you feel that it took you a while to see the link between a given concept (e.g., bird) and one of its features (e.g., has kidneys), you might feel that it's not a prominent part of that concept. Importantly, it might simply be your own perception of how readily or how slowly that relationship was able to come to mind that determines your guess of how prominent that feature is for that concept. That's exactly what Danny Oppenheimer and Mike Frank found out when they showed people concepts and tested their ratings of how typical they thought certain features were. For example, on a scale from 1 to 10, with 10 being the most typical, they might ask, "How typical is it for a dog to chase cats?" However, Oppenheimer and Frank visually tricked people into thinking that the relationship between the concept and the feature was slow to come to mind. When the features were presented in a cursive typeface or as a faded grayscale, making people a tiny bit slower at reading them, the averaged typicality ratings for the very same features were slightly lower than when those features were presented in a standard print typeface. Basically people were inadvertently interpreting their slightly slowed fluency in processing a feature such as "chases cats" in a cursive handwritten typeface as mental evidence that it must not be quite that strongly linked to the concept *dog*.

You can also manipulate the typeface of the concept itself to alter the way it gets perceptually simulated in the first place. For example, Chelsea Gordon and Sarah Anderson gave people a questionnaire that asked them to "list the features of a *diamond*," and compared the results to when it asked them to "list the features of a `diamond`." People more frequently produce "necklace" for the fancy Edwardian typeface than they do for the typewriter-style Courier typeface. By contrast, they more frequently produce "hard" for the Courier typeface than they do for the Edwardian typeface. It works with abstract concepts, too. You can ask them to "list the features of *justice*" versus "list the features of `justice`." People more frequently produce "judge" for the Edwardian typeface, and they more frequently produce "police" for the Courier typeface. Subtle differences in the way the sensory input triggers the activation of a concept can substantially alter the way you think about the concept at that moment. Clearly, you are not accessing some fixed dictionary entry located solely in a logical reasoning part of your brain.

Instead of two different typefaces for the same concept, imagine having two very different *words* for the same concept. If you are fluently bilingual, you don't have to imagine it; you experience it every day. Is it possible that the way a language *refers* to a concept could affect the way you *think* about that concept? Lera Boroditsky and her colleagues took a group of native Spanish speakers (who were also fluent in English) and a group of native German speakers (who were also fluent in English) and asked them to write down in English the first three adjectives that came to mind for a set of two dozen concepts that were all presented in English. For half the concepts, the German translation had masculine grammatical gender and the Spanish translation had feminine grammatical gender. For the other dozen concepts, the reverse was true. The grammatical gender of nouns in Romance languages is generally considered to be an arbitrary grammatical assignment that says nothing intrinsic about the meaning of the noun itself. Take, for example, the word "bridge" in English, which in Spanish is "el puente." The determiner "el" reveals that in Spanish the grammatical gender for this word is masculine, but the very same object is referred to in German with feminine gender, "die Brücke." Boroditsky's results showed that the grammatical gender wasn't so arbitrary after all, because these Spanish speakers and German speakers produced descriptions of their concepts that were significantly affected by the gender marking. For example, Spanish bilinguals described the concept *bridge* as "big," "strong," and "towering," clearly placing emphasis on masculine-like qualities of bridges. By contrast, German bilinguals tended to describe the very same concept as "beautiful," "fragile," and "elegant." The reverse happened with concepts such as *key*. In Spanish, it is feminine, "la llave," and in German it is masculine, "der Schlüssel." So, for the concept *key*, Spanish bilinguals listed

English adjectives such as "lovely," "little," and "intricate," while German bilinguals listed English adjectives such as "jagged," "heavy," and "hard."

It seems clear that giving someone a word and asking them to access their internal mental entry for that concept is not as straightforward a process as traditional cognitive psychologists once thought. The precise way that you trigger activation of that concept has dramatic effects on what kind of information comes up. Evidently, the method of asking people to list features of their concepts is not really accessing a stored mental entry in some "concepts module" in the prefrontal cortex. In fact, it might be the case that you simply don't have stored mental entries for your concepts in the first place. When sensory input travels through multiple perceptual systems and activates a shared population code in the prefrontal cortex, the exact set of neurons that becomes active is surely going to be a little different each time. Perhaps you partially reinvent your concept of a giraffe, justice, or your grandmother every time you think of them.

A Self in the Subcortex?

For the purposes of trying to point to a location in the brain that houses *who you are*, it may become especially apparent that the prefrontal cortex is *not* the solitary source of the self when you look at a variety of nonhuman animals. We humans have a whole lot of cortex layered on top of our (evolutionarily older) subcortical brain structures, but the cortex and subcortex are densely interconnected with one another. Cats and dogs do have frontal cortices, but when you compare them to those of human brains, they are noticeably smaller relative to their subcortical brain mass. And yet, with relatively little cortex, our household pets nonetheless seem to behave in ways that often suggest they have a "mental life" that is somewhat similar to ours. Our pets may not ruminate over their choices with complete sentences in their heads the way we humans sometimes do. However, they are definitely paying attention to their environment, deciding what they want to do, and choosing who they like and who they don't like. Years ago, I had a pair of brother and sister cats, Toby and Squeaky. Toby got sick and was away at Cornell University Veterinary Hospital for a couple of weeks before finally passing. Squeaky was noticeably depressed and unusually quiet, and had no idea where her brother had disappeared to. One day, I was looking at a full-screen photo of Toby the cat on my computer. Squeaky jumped up on my lap, saw the image on the screen, stood up straight with her tail and ears perked up and her eyes wide open, and made her trademark squeaking sound. She then moved her head from side to side and finally seemed to realize that it was just a flat picture, not a magic window into a 3-D world that contained her long-lost brother. She then slumped back down on my lap and nuzzled her cheek on my leg. Our pets are thinking about what's going on around them a lot more than we sometimes give them credit for.

Now, imagine an animal that doesn't even have a cortex at all. Could it have a mental life? If you didn't have a cortex, would there be a "who" that "you are?" What we call the mammalian subcortex, resting on top of the brain stem and spinal cord, has been a remarkably well-conserved brain structure for about half a billion years of vertebrate evolution. Therefore, an amazing variety of vertebrate animals have a core brain structure that is very similar to our own subcortex. For example, although dinosaurs, which evolved into birds, separated from our common evolutionary ancestor tetrapods about 300 million years ago, the bird brain's subpallium is surprisingly similar in structure to our human subcortex. Since then, however, birds have evolved a pallium instead of a cortex, and it is substantially different from our cortex in its cellular and anatomical makeup. Nonetheless, most observers would agree that, even without a cortex, birds appear to have some form of mental life. Crows in captivity can solve elaborate puzzles of all kinds. Ravens and jays store food for later consumption, indicating they have basic planning and spatial memory abilities, and when a competing bird comes near their stored cache of food, ravens have been observed to tactically distract the bird away from the food location and then later distinguish between birds that have been fooled and birds that have not. Songbirds provide a social context for their young that shapes their learning of singing behavior, in a manner that is remarkably similar to how human parents tend to shape the babbling behavior of their infants. In a rich, socially interactive context, parrots can learn to produce hundreds of English words and even combine them in novel coherent ways that they have not heard before. Of course, this is all done with a brain that has no cortex. Instead, the parrot has a subpallium (similar to our subcortex) plus a pallium on top of that. What if you only had a subcortex and nothing more?

To test the abilities of a subcortex by itself, neuroscientists have surgically removed the entire cortex in rats and then observed their behavior. The amazing thing is that these rats, who are using only the subcortex to produce their mental life and generate their behaviors, are remarkably similar to normal rats. They groom, feed, and defend themselves almost the same way that rats with a cortex do. If the surgery is performed when they are babies, these cortexless rat pups grow up and play with each other about the same as normal rats. They can learn to avoid a shock when warned with a preceding light or sound, and they can even learn a spatial task one way and then relearn it when it's spatially reversed. However, cortexless rats do generally lose their ability to store food for later consumption. With only a subcortex, perhaps these rats still have a kind of "short-term mental life," just one that does not involve much planning for the future—kind of like some people you may know.

When you look at how various subcortical brain structures work, you can see that they have some basic similarities with how cortical structures work. They combine sensory information with internal goals and send commands for motor movement. Neuroscientist Björn Merker points out that the subcortex can do an awful lot on its own. Even separate from what the cortex is doing, there are structures in the subcortex that integrate sensory information (e.g., "What's around me?") with motivational information (e.g., "What do I want?") for the purposes of selecting actions (e.g., "What do I do next?"). If that core moment-to-moment sense of "who you are" is nothing more than those three things, then the colliculus, the hypothalamus, and the basal ganglia are practically all you need. The colliculus is a set of neural networks forming a topographical map that integrates spatial location information from visual and auditory inputs. The hypothalamus is a mechanism for determining motivational drives. (These are what neuroscientist Karl Pribram called "the four Fs" of animal behavior, the four basic motivational drives that determine survival for evolution. They are feeding, fighting, fleeing, and, um...sex. You generally need all four of those if you are going to contribute to the next generation of the gene pool.) Combine that sensory colliculus and that motivational hypothalamus with the basal ganglia for sending motor commands to the spinal cord, and you have a network of subcortical brain areas that can integrate multiple parallel sensory inputs and then map them onto sequential motor outputs. Merker places emphasis on this "integration for action" process as being central to what generates conscious experience in a human, or any other animal for that matter. For *chess* moves and *career* decisions, the prefrontal cortex in humans may be a key brain area for this "integration for action." However, for *dance* moves and *food* decisions, it may be the subcortex of humans that has the key brain areas for these particular types of "integration for action."

Rather than finding the one region of the brain that contains the "who" of who you are, perhaps what the field of neuroscience has been finding all this time are multiple critical *hubs* in the network of cortical and subcortical brain structures throughout the brain that generate your sense of self. For example, Sir Francis Crick and Christof Koch suggested that the claustrum (a thin sheet of neurons resting along the underside of the cortex and richly connected to many cortical and subcortical regions) might act like a conductor in the orchestra of consciousness. Many brain areas coordinate to form the orchestra from which your mental life arises. If the claustrum is like an orchestra conductor, then perhaps the prefrontal cortex is like the composer, and then maybe the subcortical structures are like the security guards for the concert hall. They determine what gets in and what goes out. The important organizational roles of these particular hublike brain regions basically arise from the fact that they receive a wider variety of incoming neural signals than most brain areas, and they send out a wider variety of outgoing neural signals than most brain areas. These densely interconnected hubs in the network do not *house* the self, but they assist in the coordination of all the different instruments so that a coherent symphony can emerge from them all. *Who you are* is the symphony, not any one particular brain region.

Interactionism in Who You Are

It should be obvious by now that any single brain region is not what makes you who you are. You have literally hundreds of brain regions that cooperate to help generate your sense of who you are. Pretending that one of them is the "seat of the soul" is as foolhardy as pretending that a single neuron can be the engine of a particular thought. If you were to examine the many hundreds of cognitive brain imaging experiments that have been published, you'd find that just about every brain region you could look at has already been implicated in *two or more* cognitive abilities. In fact, cognitive scientist Mike Anderson has done exactly that. He looked at 472 brain imaging experiments from the past couple of decades and found a remarkable amount of overlap between the network of brain areas that appears to process *vision* and the one that appears to process *reasoning*; between the network that processes *language* and the one that processes *memory*; between the network that processes *emotion* and the one that processes *visual imagery*; and between the network that processes *attention* and the one that processes *action*. Anderson performed a statistical analysis to determine numerically how similar these cognitive networks are to one another (the Sørensen-Dice index). On a scale from 0 to 1 (where 0 is not similar at all and 1 is identical), this database of almost five hundred brain imaging experiments shows that those eight cognitive subsystems share brain regions among one another with a Sørensen-Dice similarity index of .81. What this means is that if you threw a dart at a map of the cortex from a dozen feet away, the odds are very good that you'd miss the map entirely and poke a hole in your pristine wall. That's because a dozen feet is way beyond the regulation dart-throwing distance of eight feet. But after a few tries, you'd probably hit the map of the cortex. Importantly, whatever brain region your dart hit would almost certainly be a region that plays a role in *two or more* of the neural networks that are responsible for vision, reasoning, language, memory, emotion, visual imagery, attention, and action. There are no isolated neural modules devoted exclusively to vision, language, reasoning, or *who you are*. They simply don't exist.

Based on the findings described in this chapter, if you think of the prefrontal cortex as a component in your brain, then it is part of a network that is connecting many other components. The bidirectional connections throughout all these various components clearly make your brain produce "interaction-dominant dynamics," in the words of the late, great Guy Van Orden (from chapter 2). According to Van Orden, in an interactive system with feedback loops like that, it would be inaccurate to attribute the self and its intentions to any one of those components. Instead, the self and its intentions—who you are—*emerge* between those components as they interact with one another so fluidly and continuously.

Chapter 2 showed you that the continuous interaction between billions of neurons inside your prefrontal cortex makes it impossible to treat any particular neuron (or "grandmother cell") as the sole determinant of your state of mind at any point in time. In the same way, this chapter shows you that the continuous interaction between hundreds of different brain areas inside your skull makes it impossible to treat any particular brain region as the sole determinant of who you are at any point in time. Whoever the heck you are, you are *not* just your prefrontal cortex. If your prefrontal cortex was removed and placed into someone else's brain, replacing their prefrontal cortex, your relocated prefrontal cortex would have a mind that is very different from what it used to have. It would see differently, hear differently, and act on things differently. *Who you are* would be different.

Rather than just part of a brain, perhaps you are your whole brain: a network of a hundred billion neurons, with a very complex pattern of connectivity, constantly sending electrochemical signals (and perhaps electric fields) back and forth to one another. Is that who you are? In this complex, interconnected network of multiple brain regions, areas that roughly specialize in reasoning, vision, language, concepts, memory, action, and emotion are all sending signals around and around to one another to help you recognize all the objects you see and all the words you take in as they pour into your eyes and ears, a dozen of them every few seconds! It sounds exhausting! It can get you out of breath just thinking about it. In fact, you probably haven't taken a proper breath since chapter 1. Maybe that's why you can barely breathe right now. Let's stop here and just breathe. Sit up straight, and stretch your back a little. Take a deep breath through your nose and let it out through your mouth. Be thankful to the air. Let that oxygen flow through your bloodstream and feed those hungry neurons that we now know are working so hard in the prefrontal cortex, visual cortex, language cortices, and subcortical brain structures, trying their damnedest to make sense of the different types of information they keep sending each other.

Directions for Use

This chapter shows how different brain areas are connected to each other and constantly send electrochemical signals back and forth to one another. This is what allows uncertainty in one brain area to be pretty quickly resolved by the context provided by another brain area. Even though these different brain areas tend to be somewhat specialized for a particular type of information processing (e.g., visual, auditory, or linguistic), they interact with one another frequently enough that each one knows a little bit about what its neighbors do. This is what allows them to "pitch in" and help out when there's focal damage to the brain from stroke, disease, or trauma. Because of the brain damage, a certain set of cognitive abilities will be substantially impaired after the damage takes place. However, to some degree, those lost cognitive abilities can often be recovered. (Recall the story of Phineas Gage in chapter 2 and how his personality changes and impulsivity probably diminished over the years.) This is not because the damaged brain area grows back. It is because all those nearby undamaged brain regions that already knew a little bit about how the damaged area did its job learn over the course of months and years to be a little better at that vacated job. So, I guess what I'm saying is that getting focal brain damage is a little bit like having your business downsized. At first, it's a terribly disruptive shock to everybody involved, and certain functions are not being taken care of. However, over time, the remaining employees can often learn how to add those recently neglected tasks to their own repertoire and do an acceptable job of it.

There is an important lesson that you can extract from this knowledge. The more your brain's "officemates" know about what each of the others does, the better off you will be as you get older and undergo natural risks for brain damage. How do you get your different brain areas to interact enough to know about each other's jobs even more than usual? Games and skills. Not just any game or skill, but particularly ones that challenge your attention and cognitive control. So, your "directions for use" in this chapter are no small feat. I want you to use the knowledge you gained in this chapter to improve your brain by taking up a new game or skill. It could be a new addiction to metal wire puzzles, learn how to solve Rubik's Cube if you never did before, teach yourself to juggle, or learn a new language. Take up a new action- or puzzle-based video game of a type that you haven't played in the past. Take piano or guitar lessons. Learn how to read music. Learn a new computer programming language. There must be something you've always wanted to learn and kept putting off. Start now. The future health of your brain, and therefore of *who you are*, depends on it.

4 From Your Brain to Your Whole Body

Who are you? I really wanna know.
—The Who, "Who Are You"

When your body does some of your thinking for your brain, people sometimes call it "muscle memory." Years ago, I spent a morning mulling over some sentences in my mind, trying to find the best way to write down a complicated theoretical point. The right sequence of sentences just wasn't coalescing for me. Later that day, I was trapped in a lecture by a visiting scientist when suddenly the words fell into place. I grabbed a pen and paper and frantically wrote them down before they could fade in my memory. (The visiting speaker probably thought that he had just said something really moving and that I was feverishly taking notes on the wisdom he had just dropped on me.) After getting the ideas down on paper, I felt a rush of urgency about somehow preserving those scribbled sentences so they wouldn't vanish, but they were right there in front of me in ink on paper. What further preservation did they need? I conducted a kind of mental inventory on my body to see if that rush of urgency was localized anywhere. I felt some residual activity in my left shoulder. It traveled down my arm to my left hand. I wiggled my fingers to see what it was they had wanted to do to help preserve those sentences. That's when I realized that my rush of urgency to preserve those scrawls of ink had been centered on my left thumb and left middle finger—but I am right-handed. Those two fingers had wanted to do something to save my precious handwritten sentences. Suddenly, it occurred to me that those two fingers are the ones I use to press command-s on my Mac keyboard to "save" a document I am working on. Now that's muscle memory.

If chapter 3 did to you what it was supposed to do, then you have successfully embraced the idea that your entire brain is a very important part of "who you are." But is that *all* you are? What about your eyes, ears, mouth, skin, muscles, and bones? Are they just part of some robotic exoskeleton that encases the real you, or are they part of your mind, too? Who the heck are you?

In this chapter, I'm going to show you a truckload of scientific findings indicating that the functions of your brain are so deeply intertwined with the functions of the rest of your body that you really ought to expand your definition of who you are to include your entire brain-and-body. This chapter, and chapters 5 and 6, are especially chock full of scientific laboratory results because these next few steps in expanding your definition of self are sometimes the hardest to accept emotionally. But the evidence is overwhelming. The present chapter should convince you that the information that makes you who you are includes not just the information carried inside your skull but also the information carried by the rest of your body. As it turns out, your body carries a whole lot of information.

What If You Didn't Have a Body?

To get a handle on how your body carries cognitive information that your brain doesn't have, let us start by examining what your mind would be like if your brain did not have a functioning body at all. If your disconnected brain were alive, floating in a vat of nutrient fluid with its blood vessels being fed with oxygen-rich blood, would you still have a mind? Would you think about things the same way you do now? What if sensory and motor signals were delivered and received via electrodes? If your brain was receiving electronic signals wired into the sensory neurons from a powerful computer that generated a convincing virtual reality experience and it also recorded the activity patterns in your motor movement neurons to tell the computer how to move your virtual eyes and virtual limbs in that virtual world, then you would have no way of knowing that you weren't in the real world. This is, of course, the basic premise for the 1999 movie *The Matrix*, where the protagonist, Neo, finds out that his real body is actually in a vat with cables connected to his spinal cord and muscles. There have been all manner of discussions about where the Wachowski brothers borrowed their ideas from for that story line, such as from ingenious science fiction author Philip K. Dick. The notion that your sensory experience of the world might be fake or simulated has been around for a very long time, but one scholarly source who has been quite influential in academic writings on the philosophy of mind was Hilary Putnam.

In 1981, Putnam gave an in-depth treatment to the thought experiment of "What would it be like to be a brain in a vat?" His argument went as follows. If you are a brain in a vat, then when you say "brain" and "vat," you are actually referring to artificial sensory impressions of a brain and of a vat that get programmed on a computer and delivered to your brain via electronic circuits and wires. Your use of those words does not actually refer to the real things out in the real world, since you've never experienced those real things in the real world. Therefore, if you are a brain in a vat, then your very attempt to attain self-knowledge by saying "I am a brain in a vat" cannot actually be true, because your use of those words refers not to the real things but to artificial simulations of them. Basically, you may indeed be trying to proclaim that you know that you "are a brain in a vat," but your proclamation actually amounts to saying, "I am a simulation of a brain in a simulation of a vat," and that's neither true about the real world in that case nor is it what you intended to say. Therefore, the statement "I am a brain in a vat" can never be logically true. Those philosophers are clever, aren't they?

In some ways, though, early versions of artificial intelligence were in fact a lot like brains in a vat. Chess-playing programs, invented as early as the 1950s, were like a very simple mind that only cared about chess, and this mind received electronic "sensory" inputs in the form of a logical description of an opponent's chess move. The chess mind then used that input to update its internal model of its next possible moves and generated what it "thought" would be its own most advantageous next move. Not long after those programs were starting to become competitive (though not at the chess master level yet), the first chatbots were being developed as well. For example, computer scientist Joseph Weizenbaum developed ELIZA, a computer program posing as a therapist that can carry on a text-based conversation with you and get close to sounding intelligent. Here is a conversation that I just had with ELIZA on the internet:

ELIZA: How do you do. Please tell me your problem.

ME: I want the reader to see your strengths and weaknesses.

ELIZA: Why are you concerned over my strengths and weaknesses?

ME: Because you are the first chatbot and the reader should see what you are like.

ELIZA: Could there really be some connection?

ME: Yes. Your strengths and weaknesses are indeed connected to the fact that you are the first chatbot.

ELIZA: You are sure?

ME: Are you demonstrating one of your weaknesses right now?

ELIZA: What is it you really want to know?

ME: I want to know if you feel like a brain in a vat.

ELIZA: What does this speculation lead to?

ME: It leads to understanding how you might be improved if you had a body.

ELIZA: What do you know about I had a body?

Soon after ELIZA was designed, computer scientist Terry Winograd developed SHRDLU, a chatbot that can carry on an intelligent conversation as long it is only about an imaginary world of different-colored blocks and pyramids. ELIZA and SHRDLU are a lot like brains in a vat. They do not have bodies. Their inputs are essentially simulations of a (very simple) world. Numerous chatbots came after that, and today they even compete for prize money in a variety of organized contests. So far, none of them can carry on a fully unconstrained conversation that would fool anybody into thinking it was a human, but in the 1990s, a chess program named Deeper Blue, developed by IBM, actually beat Garry Kasparov, the chess grandmaster at the time. In 2011, IBM's Watson beat two former champions at *Jeopardy!*, a general knowledge game show, and recently Google developed its Duplex app, which can call a restaurant or hair salon for you, speak to a live person, and make a reservation. Each of these computer programs, big and small, impressive and unimpressive, is essentially a brain in a vat. They have no bodies.

Philosopher Hubert Dreyfus famously argued that artificial intelligence programs will never truly match human intelligence until they have humanoid robotic bodies that are seamlessly integrated with their computer programming. Your body does so much thinking for you, and an AI's body would do so much thinking for it, that without it both you and the AI would be severely limited in some very basic understandings of how the world works. Dreyfus argued this almost half a century ago, and it is beginning to look like he might be right. We still do not have artificial intelligence that rivals our own intelligence, or even one that can have a sensible conversation about current events for more than a minute, and it may be mostly because for decades AI and robotics developed their scientific advances independently of one another. The state of the art for building smart artificial intelligence inside a humanoid robot is just not there yet, but it might get there someday. Let's bookmark *that* conversation until the middle of chapter 8. I promise you, we will get into the thick of it then. You know what they say: be careful what you wish for; you just might get it.

Anyway, if Dreyfus is right, then it means not only that computer programs need bodies in order to be humanly intelligent but that we humans also need our bodies in order to be humanly intelligent. Our bodies are essential for making our brains produce humanlike intelligence. So, if all that was left of you was a disconnected living brain floating in a vat, you wouldn't be thinking the same way you are now. You would have a different self. Therefore, *who you are* is not just your brain.

There are a handful of interesting examples of what a human brain does when it is more or less detached from its body. In fact, it happens to you every night when you go to sleep. Your dreams are made of neural activation patterns in your cortex, and your subcortex has to disconnect the brain from (a) sensory inputs that might wake you up and (b) motor movement signals that might cause you to hurt yourself or someone else. Have you ever had a dream where you are trying to run and it feels like your legs are in quicksand? Or maybe you are trying to take a swing at a bad guy and feel like your arm is moving in slow motion, so he easily blocks your punch. This motor movement restriction isn't just because the blankets on top of you are holding you back. It is because your subcortex is preventing the cortical movement signal from reaching your spinal cord—and thank goodness it is, or you might kick or punch someone who is asleep right next to you!

So, while you are dreaming, you are essentially a brain in a vat, and now you can ask yourself, "Am I the same person in my dreams that I am in real life?" When your brain is disconnected from the rest of your body, is *who you are* in your dreams the same as *who you are* in reality? Think about a dream where you did something dramatic or dangerous, or maybe even something morally objectionable. Maybe you did something physically impossible, but found it unsurprising. In your dreams, your reactions to complex situations are often different from what they would be in real life. I once had a dream where instead of letting myself get embroiled in a senseless bar fight I concentrated on levitating to the high ceiling of the saloon, and it worked. But if my waking self was ever in a bar when a large brawl was breaking out for real, I'm pretty sure that concentrating on levitation would *not* be my first course of action. Have you ever had one of those dreams where you are in your underwear in public? Did you immediately leave the situation and find a place where you could get some clothes or at least borrow a coat? That's probably what the real you would do if that happened in real life, but that's probably not quite what you did in the dream. Or what about that dream where your teeth start to fall out? You fiddle with one tooth in the mirror, it starts to come out, and then others start getting loose, too. If that happened in real life, I sure hope you would immediately stop fiddling with your teeth and call a dentist. But that's not what you did in the dream, is it? In your dreams, maybe even the *shape of your body* is different. In your dreams, you occasionally do things that you would not, or could not, ever do in real life. In your dreams, when you are a brain in a vat, you have a slightly different self.

Psychologists and neuroscientists have explored other ways to disconnect the brain from the body as well. In the 1940s, behaviorism was the mainstream theory of psychology, and its definition of the mind depended entirely on the relations between stimuli (input) and responses (output). Therefore, a simple version of behaviorism naturally predicted that if the brain received no sensory stimuli and produced no motor responses, then it would have nothing to do and would basically go quiescent or fall asleep. Donald Hebb was the first to test that theory experimentally. He put students into rooms with no discernible sound, had them wear goggles that permitted no light, and encased their limbs in cardboard so they couldn't move very much or even scratch an itch. Needless to say, most of those students found this experience extremely unpleasant. But, contrary to behaviorism, their minds did not fall asleep. Their minds wandered. Some of them hallucinated, and some of them damn near freaked the heck out. The absence of environmental stimulation, and the inability to act on the environment, does not a normal mind make.

A few years after Hebb's experiments, John Lilly developed the flotation tank as a friendlier, and more effective, means of achieving sensory deprivation. His experiments with the flotation tank similarly induced hallucinatory experiences, but in the context of meditation techniques (and perhaps also the ability to scratch an itch on your nose if you damn well need to) they led to much more pleasant explorations of "inner space." In a flotation tank, the water is constantly filtered and heated to body temperature, and it is chock full of bath salts to keep you buoyant as you lie on your back. When you close the door to your soundproof tank, you see nothing and hear almost nothing (except the breathing of your lungs and the beating of your heart). It's so dark that you may not even be able to tell whether your eyes are open or closed, and if you don't move around, you can lose touch with your body entirely. After a few minutes of boredom, your mind gets used to this unusual situation and takes advantage of its newfound freedom. Flights of fancy are more extravagant, imagination is more vibrant, and meditation is more transformative. The flotation tank is also supposed to be good for your back, according to some chiropractors.

No, when detached from the body, the brain does not go to sleep, as predicted by early behaviorists, but it also does not operate in a normal fashion, as predicted by those who think your brain is all there is to *who you are*. When the chief editor of *Elle* magazine, Jean-Dominique Bauby, was tragically left completely paralyzed by a stroke in his brain stem, his life would never be the same. He was in a coma for a few weeks, and when he awoke, he was unable to move any part of his body except his left eyelid, to blink. In cases like these, it is usually the family—not the doctor—that detects evidence of consciousness in the patient's communicative movements of their eyes or eyelids. This type of "locked-in syndrome" is definitely an imperfect example of a "brain in a vat," because the patient can usually hear and see the things and people in the environment just fine. The input systems are generally intact. It is the output systems that are devastated. Perhaps it is an example of a brain with eyes and ears, in a vat. In Bauby's case, his visual, auditory, and olfactory perception were functioning normally. He communicated by having an attendant verbally list letters of the alphabet, in order of their frequency of use, and when they got to the letter he wanted to use next, he blinked his left eye. It could take minutes to produce just a single word, but it worked. In fact, by blinking his left eye he wrote an entire book, *The Diving Bell and the Butterfly*. In this book, Bauby describes his experience of feeling that his mind was a butterfly trapped inside a diving bell deep underwater. He reminisces, fantasizes, and fumes about his past and present. His mind takes flight, flitting about like the butterfly in the title, fantasizing almost uncontrollably about food, love, and other memories, and then he also comments on how he would happily murder the hospital attendant who was neglectful of him. As cognizant and mentally disciplined as Bauby clearly had to be in order to write this memoir, it seems clear upon reading it that his mind was not operating quite the same as it used to. As chief editor of *Elle* magazine, he was not someone who would normally get lost in mind wandering or fantasize about murdering someone for being lazy. It shouldn't be surprising that being a little bit like a brain in a vat changes *who you are*, at least a little bit.

When a brain loses touch with a subset of its sensorimotor information but not all of it, then one can naturally expect a *partial* alteration of that mind. For example, social cognitive neuroscientist Simone Bosbach studied two patients who had degenerative diseases of the peripheral nervous system, causing them to be unable to sense touch, pressure, or temperature, on their skin. These patients wouldn't know if their hand was resting on a hot stove until they saw or smelled their own cooked meat. These patients have to use a walker and visual feedback just to stand, because their feet cannot tell when the body's center of mass is aligned over the leg. Nonetheless, these patients can watch a video of someone lifting a box, and their unconscious knowledge of motor dynamics allows them to accurately guess whether the box was heavy or light—just like you and I can. However, these patients are not able to detect when the lifter's expectations about the weight of that box are violated. Have you ever lifted a milk carton or juice carton in your refrigerator, thinking it was full, and then found yourself banging it against the inner ceiling of the fridge because the carton was nearly empty? That's because your expectation of its weight was violated. When you and I watch a person lifting a box that's heavier or lighter than they expected, we would be good at detecting the brief imperfection in their lifting style—and we might chuckle a bit, too. These two patients, who can see just fine and make visual inferences about the weight of a box, are terrible at inferring the mismatch between the lifter's *expectation* of the box's weight and the box's *actual* weight. This task involves a *cognitive* inference about the state of mind of the lifter, and a *sensory* disability has impaired it in these two people. Their years of inexperience with the touch feedback that one receives while lifting everyday objects has left their brains unable to guess the mental state of someone else who is applying too much force or too little in lifting an object. Their understanding of how people interact with objects is a little different from what it used to be. *Who they are* is a little different from what it used to be.

From Your Brain to Your Whole Body

It seems clear that your brain, all by itself, is not solely responsible for making you who you are. The "I am a brain in a vat" thought experiment here is reminiscent of the "I am my prefrontal cortex" argument from chapter 2. When you compare the two ideas, the prefrontal cortex is like the "brain," and the sensory and motor cortices are like the virtual reality computer sending and receiving signals to and from that prefrontal cortex. Remember, though, that in chapter 3, mountains of evidence from neuroscience were marshaled to show that the continuous, fluid flow of information back and forth between the prefrontal cortex and the sensory and motor cortices makes it impossible to pretend that the prefrontal cortex is *who you are*, while the sensory and motor cortices are not. So, by the end of chapter 3, the definition of who you are had to be expanded to include your whole brain. Therefore, in the analogy between that story and this "brain in a vat" thought experiment, an outside observer of your brain in a vat would have to conclude that, because of the continuous fluid flow of information back and forth between your brain and the virtual reality computer, *who you are* must be the brain-and-computer, not just the brain. But since Hilary Putnam proved that you cannot truthfully say that "I am a brain in a vat," let's instead say that you are "a brain in a body." This chapter is intended to convince you that an accurate definition of *who you are* must include the brain-and-body, not just the brain.

The Psychology of the Embodied Mind

So, if your body carries cognitive information that is part of who you are, what exactly is that information? If your body does some of your thinking for you, exactly how does that work? Researchers across a dozen fields have collected scientific evidence that makes a convincing case for how a normally functioning human mind is not solely a product of the brain but instead a product of the brain and body working together. Your mind is embodied.

Some of these experimental examples will focus on how neural components of sensory impressions and neural components of motor movement plans play important roles in thinking. This does more than simply show that the prefrontal cortex is linked up with various other cortical brain regions, as done in chapter 3. These sensory impressions and motor movement plans are patterns of neural information that are closely connected to the patterns of environmental information as they impact your body's sensors and effectors (i.e., your senses and your limbs). In principle, these neural associations between cognitive, sensory, and motor regions could take place in a brain in a vat—but only if it had an extremely realistic virtual-reality world with an extremely accurate physics engine. Having a body simply allows the *actual* world to accurately provide that reality and those physics for free. After starting with experimental examples that focus on sensory and motor cortices, we will then transition into examples of sensorimotor information that lives inside the nonneural tissue of the limbs. We will turn to how the morphology of your body does some of your cognitive computation for you.

Larry Barsalou (from chapter 3) has helped the field of psychology come to terms with this idea of embodied cognition. His perceptual symbol systems theory has inspired a variety of experiments that show how people use sensory and motor information to do even abstract thinking. Take mental imagery, for example. Visual imagery is something you can do while being completely closed off from the world. You do a lot of it when you're in a flotation tank, believe me. But Barsalou's perceptual symbol systems theory predicts that you don't simply rev up your frontal lobes when you passively imagine visual objects and move logical symbols around in the "software of your mind." You use some of the same sensory and motor information patterns—some of the same neuronal population codes throughout your brain—that would be active if you were actually *seeing* those imagined objects for real. If I ask you to close your eyes and imagine a female mountain climber wearing a red jacket, your frontal lobes generate the abstract notion of that image first, but pretty soon feedback signals from the prefrontal cortex tell sensory and motor cortices to assist in fleshing out the image in greater detail. As proof of this, cognitive neuroscientist Steve Kosslyn and his colleagues have used brain imaging to discover that, when the eyes are closed, the visual cortex is more active during visual imagery than during a control condition that does not involve imagery. Thus, even when no visual stimulation is entering your brain, your visual cortex gets activated by your prefrontal cortex when you generate visual imagery.

Let's do a little imagery experiment together. In a minute, I'll want you to turn and face a blank patch of wall near you, with your eyes open, and imagine again that mountain climber with the red jacket. Imagine she's at the top of a cliff face, a hundred yards away from you, with ropes attached for rappelling down. Then imagine her rappelling down, making a 10-foot drop, and then resting her feet against the cliff. Then she makes another 10-foot drop and rests. After about eight such drops, she finally reaches the bottom of the cliff. Got that? Read it again if necessary. You'll want the event roughly memorized for this experiment. Now that you know the sequence of events to imagine, set the book aside, look at that patch of blank wall, and imagine the event just described. It should take about 15 seconds. Do it now.

Okay, we're back. While you looked at that blank patch of wall and imagined a mountain climber in red rappelling down a cliff face, your visual cortex was generating partially active neuronal population codes of a human shape, red jacket, ropes, and—most importantly—downward motion. Did you notice what your eyes did while you generated that mental image superimposed on that blank patch of wall? Did your eyes move downward a little? Maybe they even moved downward eight times, once for each 10-foot drop of the mountain climber. Cognitive neuroscientist Joy Geng recorded people's eye movements while they did that task, and she found that most of them were indeed making downward eye movements while they heard a story about a climber rappelling downward. With an upward motion story, most of them made upward eye movements, even though there was nothing to look at on the blank white patch of wall. Basically, when you generate visual imagery in your "mind's eye," that partial activation of the visual cortex (discovered by Kosslyn) naturally leads to partial activation of the oculomotor cortices (brain areas that drive eye movements), and thus the eyes move accordingly.

In fact, as predicted by Barsalou's theory, even just understanding a simple sentence has a tendency to initiate some partial visual imagery that you might not even consciously notice. Cognitive psychologist Rolf Zwaan was able to find evidence for it. He had people read sentences and then verify whether an object in a picture was mentioned in the sentence. For example, he had people read a sentence such as "The carpenter hammered the nail into the wall." Then they were presented a picture of a nail. For half the people, the nail was oriented vertically, pointing downward. For the other half of the people, the nail was oriented horizontally, pointing rightward. Everybody correctly verified that the nail had indeed been mentioned in the sentence, but the people who saw the horizontal nail were faster at doing so. Why would that be? Well, when you read "...nail into the wall," your brain cannot help but generate a little bit of faint visual imagery of that nail going into the wall, and obviously it would be in a relatively horizontal orientation while doing so. Then, a second later, when you are looking at an image of a nail, the horizontal nail image is easier to recognize because your brain has already been thinking about an image like that. The neuronal population code for visually recognizing a horizontal nail had a head start thanks to the sentence, compared to the neuronal population code for visually recognizing a vertical nail. Importantly, Zwaan's data pattern reversed with sentences such as "The carpenter hammered the nail into the floor."

These linguistically activated visual population codes can even alter how you perceive a visual motion stimulus. Psychologist Lotte Meteyard had people listen to verbs that implied downward motion while they viewed a display of randomly moving dots. When they had to guess the direction in which the random dots tended to drift, they guessed downward. With upward verbs, they guessed upward. Meteyard's effect works the other way around, too. When people watched a display of randomly moving dots with a statistical drift among the dots that tended downward, those people were faster to respond to verbs that described downward motion. With a subtle upward drift in the random dots, they were faster with upward verbs. So, not only does perception get influenced by language, but also language in turn gets *embodied* by perceptual information.

These examples barely scratch the surface of the evidence for sensory processes of the body being fundamental to cognition, to what your mind does, and to *who you are*. The embodied cognition movement is exploding with results that indicate the mind is not a separate entity detached from the body. In addition to sensory processes, motor movement processes of the body also play a large role in forming mental content. For example, psychologist Art Glenberg had people read sentences on a computer screen and press a button close to them (a couple of inches) to say it didn't make sense and press a button farther away (about one foot) to say it made sense. (In the second half of the experiment, he reversed this arrangement.) The critical sentences in this experiment described actions that were directed away from them (e.g., "You handed Courtney the notebook") or toward them (e.g., "Courtney handed you the notebook"). When the button far away was for a "makes sense" response, people tended to be faster with sentences describing actions that were directed away from them than with sentences that described actions directed toward them, and when the button nearby was for a "makes sense" response, this pattern generally reversed. Glenberg's student Michael Kaschak has since replicated and extended these findings to explore the time course of this effect. If the person's reach toward the button begins at different points in time in the sentence (while hearing or reading it), the effect can be stronger or weaker depending on the specific content of the sentence at that moment in time, and some of that specific content varies as a result of the grammar of the sentence. Cognitive linguist Ben Bergen found that when the sentence uses progressive aspect, as in "Richard is beating the drum" and "Richard is beating his chest," these effects of cognitive embodiment are especially robust. A sentence like "Richard beat the drum" is not as effective, because it emphasizes the completion of the event; the event is no longer happening. By contrast, Bergen suggests that the progressive aspect in "Richard is beating the drum" emphasizes an ongoingness of the event described by the sentence and thus engages your motor system more because it feels as if this event (which your own limbs could very well carry out, too) is happening right now.

The ownership of that motor movement turns out to be important, too. You can't always just combine the *motion* of an object with the *meaning* of that object and trust that you'll get an effect of embodiment. Sometimes the body has to be *responsible* for the motion. The motion has to be an action that the body is carrying out. For example, developmental cognitive scientist Linda B. Smith showed a novel sample object to two-year-olds, and she moved it horizontally back and forth six times while giving it a novel name, such as "This is a wug. Watch the wug." Then she showed the kids two other novel objects that looked similar to the first object, except that one was stretched out horizontally and the other was stretched out vertically. She then asked the kids to match one of these new objects to the sample presented before; for example, "Which one of these is also a wug?" In this version of the experiment, where the kids simply *watched* the sample object move left and right, they chose either of the two matching objects (horizontally or vertically stretched) with equal likelihood. That is, the horizontal motion of the sample object did not influence their mental category for that kind of object when they later tried to match a new object to that category. But then Smith did a different version of the experiment, where she put the sample object in the hands of the kids and instructed them to move it horizontally in space themselves so now the kids weren't simply *watching* the motion but were *responsible* for it. After doing that for several seconds, the kids were then presented with two objects to match to the previous sample, one that was stretched horizontally and one that was stretched vertically. About two-thirds of the time, these kids chose the horizontally stretched object as a match to the sample object. That is, the horizontal motion that they had imparted to the sample object, via their own actions, was clearly influencing their category of "wug." And, of course, when the motion imparted to the sample object was vertical, up and down for several seconds, the results reversed accordingly.

These psychology experiments are typical examples of laboratory demonstrations that when your brain thinks about meaning, concepts, and categories, it doesn't do so in a vacuum. Your brain does not function like a mind encased in a mechanical robot body that merely follows instructions. Your mind is not *encased*; it is *embodied*. Your body is part of your mind, part of *who you are*.

The Language of the Embodied Mind

Did I say there were a *dozen* research fields that show evidence for the embodiment of mind? Did you think I was exaggerating? Heck, there's *at least* a dozen. In addition to psychology, cognitive psychology, and developmental psychology, there's cognitive science, social psychology, motor movement science, neuroscience, cognitive neuroscience, philosophy of mind, robotics; the list goes on. Let's focus here on linguistics and psycholinguistics.

Our natural everyday use of language is a kind of laboratory all its own. Cognitive linguist George Lakoff and philosopher Mark Johnson have noticed for decades that the way we use language already reveals how we rely on our bodily sensations as metaphors for all sorts of linguistic meanings. It is no accident that when you talk about someone not understanding the information that is coming at them, you say that it went "over his head." The information missed his eyes and ears and was never captured by his sensory apparatus. And when the information is instead imposed on someone who may be unwilling to accept it, we say it was "forced down his throat." We don't have to talk about throats and heads when we refer to someone comprehending or not comprehending some information, but we do anyway. These metaphors show up all over the place in everyday language use, and often we don't even notice that they are metaphorical. In addition to using body metaphors to talk about abstract things such as comprehension, we also use external physical events as metaphors for bodily experiences such as emotions. For example, we routinely talk about anger as if it were hot fluid in a container. We say "he blew his stack," "she flipped her lid," and "he blew a gasket." There are dozens of idioms that fit with this conceptual metaphor of *Anger is heated fluid in a container*. See if you can think of some more body metaphors on your own. (Did you notice how I used the word "see" there, even though this is something you could do with your eyes closed?)

Cognitive scientist Raymond Gibbs has spent most of his career using his psycholinguistics laboratory to study how people comprehend metaphors, idioms, and even proverbs. His body of work shows that figurative language automatically activates these sensorimotor associations in our minds when we read them and hear them. For example, Gibbs had experimental participants write down their detailed mental imagery for various proverbs, such as "Let sleeping dogs lie" or "A rolling stone gathers no moss." What he found was that there is an impressive degree of consistency between different people's mental imagery for these old chestnuts, suggesting that the way they are understood may rely heavily on Lakoffian conceptual metaphors—such as *Anger is heated fluid in a container*, *Emotions are locations*, or *Time is a landscape we move through*—that are common to us all. Even fine-grain details of the proverb imagery turned out to be the same across most of Gibbs's participants. For instance, take a minute to close your eyes and generate some mental imagery for the proverb "A rolling stone gathers no moss." Don't read any further. Close your eyes and do it, and then come back.

Good job. Was the stone rolling down the side of a hill? Was it in a rightward direction? Did the hill have some grass on it? Was there also moss, and maybe some small bushes? As the stone rolled down the hill, did it roll pretty quickly? Did it meander a little, not rolling in a perfectly straight line? Were there perhaps a few moguls that made the stone bounce just a little? And was the stone mostly ball shaped but a little irregular? Yeah, I thought so. That's what Gibbs found as well with his experiment participants.

In addition to figurative language sentences that naturally evoke sensory impressions and motor movement, linguists have also studied how individual verbs and prepositions tend to give themselves over to a two-dimensional visual depiction. This is a way of representing what might be a rather abstract concept in the form of a schematic visuospatial image that is rooted in your body's sensory experience. For example, the meaning of the preposition "over" can be abstractly pictured as an unlabeled target object (e.g., an X) that is resting or moving above a reference object (e.g., a square). Ron Langacker calls the target object a trajector and the reference object a landmark. This image schema of "over" can be used to represent the meaning of sentences such as "The bird flew over the lake," "The lamp hung over the table," or even "He pored over the data."

Len Talmy used these image schemas to describe a wide array of events and spatial relationships that show up in our everyday language use. He developed a vocabulary of basic schematic elements for visually depicting not just spatial relations but also force dynamics: how one thing in a sentence affects another thing. Sentences such as "The door will not open" or "The wind turned the pages of my book" can be depicted entirely visually by using specific combinations of his schematic elements (not unlike Feynman diagrams in quantum physics). Talmy's force dynamics image schema system allows one to represent the generic meaning of a wide variety of sentences in a fashion that relies heavily on a two-dimensional spatial layout along with a temporal component to allow for those force dynamics. What's more, he developed these insights into a theory of how visual pictures can embody the meanings of phrases, all while he was gradually becoming legally blind himself. That's right; cognitive linguist Len Talmy can't see the hallway in front of him, but he can see into your mind just fine. I visited him once at Buffalo State University, and he had me guide him through the crowds of students as we took a winding path through several hallways to the cafeteria for lunch. His visual imagery of the route was absolutely exquisite. If the crowds of students hadn't been there, he could have easily walked that complicated maze unassisted. And when Talmy gives speeches, he has the entire sequence of arguments memorized in incredible detail—obviously not using any visual aids to facilitate his memory. It is remarkable that Talmy has been able to envision these two-dimensional image schemas, with their force dynamics, to such powerful success in accounting for the meanings of so many verbs, prepositions, and phrases.

One common theme among the image schemas that describe so many English verbal phrases is that they tend to treat the trajector (or subject) in a sentence as starting on the left side of visual space and the event (or verb) as taking place in a rightward direction. In fact, neurologist Anjan Chatterjee has carefully documented how English speakers do in fact exhibit a visual preference for that spatial arrangement and that directionality of action when they envision events that are described by verbs. This may in part result from the fact that English speakers *read* from left to right, because there is some evidence suggesting that speakers of languages that read from right to left might have image schemas in which the action moves from right to left. It does seem likely that many of these image schema representations of events and spatial relationships are learned by children in one way or another, because if they were innate, then they wouldn't vary at all across languages and cultures.

Drawing from the work of Lakoff, Langacker, and Talmy, developmental cognitive scientist Jean Mandler provided a detailed account of how an infant's sensorimotor experience can, very early on, provide the basis for learning many of these image-schematic concepts. Long before learning any words or phrases to go along with them, simply by observing events in their environment and interacting with them on occasion, an infant can learn concrete concepts such as *support* and *containment*, and even abstract concepts such as *causality* and *animacy*. Since this knowledge wouldn't be linguistic for an infant, it would need to be in some other format of information. Image schemas make perfect sense. For example, when an infant observes (and participates in) a wide variety of smaller objects being placed inside the concave portions of larger objects—everything from fluid in a bottle, to snacks in a bowl, to toys in a toy box—she can begin to infer some of the spatial and motoric properties that are common to most of those events. That tiny brain, with its exuberant growth of new synaptic connections, condenses those common properties into a nascent concept of *containment*, where the concave portion of just about any object can be imagined to partially or completely surround another smaller object. Can you wrap your head around that idea? This can work for abstract concepts as well. As the infant observes, over time, a wide variety of objects move toward a second object, impact it, and cause that second object to move, this gives that little brain enough statistical information to extract a simple image-schematic understanding of what physicists refer to as deterministic "billiard ball" *causality*—what philosophers call "efficient cause" or what you and I might call basic "cause and effect." Moreover, if either of those objects were to move in a curvy, unpredictable fashion, before or after the impact, this could provide evidence for the infant to develop a basic concept of *animacy*, the key difference between a stuffed animal and a real pet. Mandler's work clearly suggests that a one-year-old can acquire a far richer set of concepts (albeit prelinguistic ones) than famous developmental psychologist Jean Piaget ever imagined.

In an adult, these image-schematic concepts, which rely so heavily on the body's sensory experience, are a bit like the *unconscious partial visual imagery* that explains Rolf Zwaan's findings with sentences such as "The carpenter hammered the nail into the wall." Whenever you read or hear a sentence, part of how you understand the meaning of that sentence involves activating a set of neuronal population codes that correspond to the particular image schemas for that sentence. To more clearly demonstrate this fact, Daniel Richardson teamed up with Larry Barsalou, Ken McRae, and me to conduct some laboratory experiments. First, Richardson tapped into people's intuitions about the two-dimensional shapes that verbs take. To avoid getting any effects from the meaning of the nouns, he gave them rebus sentences, where some words are replaced by shapes, such as "○ chases □" or "○ respects □." Then people were instructed to choose which of four image schemas best matched the meaning of that rebus sentence. It sounds like a weird random task, but people's choices were not random at all. As predicted by Chatterjee, with an action verb like "chased," most people chose the image schema with the circle on the left and the arrow extending from it rightward, and the square on the right, where the tip of the arrow was pointed. The image schema for many action verbs involves a depiction of the action happening from left to right. By contrast, for verbs such as "respected" and "hoped for," most people chose the image schema with the circle at the bottom and the arrow extending upward from it, and the square at the top where the tip of the arrow was pointed. It shouldn't be surprising that the image schema for the verb *respect* involves a spatial layout that depicts the respecter literally "looking up to" the respectee.

Once Richardson knew the orientations of the image schemas for these verbs (some of them horizontal, some of them vertical), he was ready to test the effects of these visuospatial layouts during real-time spoken-sentence comprehension. For example, when you hear a sentence about a person respecting someone else, does your brain really generate some unconscious visual imagery that takes a vertically extended shape? In a simple perception experiment, Richardson discovered that the image schema being activated does in fact "use up" some of your visual processing resources in the corresponding regions of visual space for that image schema. That is, the activated image schema interferes with new visual stimuli arriving in locations that are occupied by the image schema's shape. Richardson adapted the Perky effect to prove this. The Perky effect refers to psychologist Cheves West Perky's finding, a century ago, that while you are engaged in visual imagery, you are somewhat less sensitive to new incoming visual stimulation. While Richardson's participants heard spoken sentences, such as "John respected his father," they were looking at the center of a computer screen and trying to determine whether a briefly flashed object at the edge of the screen was a square or a circle, and they were slower to do so when the shape of the unconsciously activated image schema overlapped with the location where the actual circle or square showed up on the screen. This finding proved that the linguistic understanding of that spoken sentence automatically activated some unconscious partial visual imagery, such as an image schema for "respect" that was extended vertically in mental visual space.

This unconscious partial visual imagery (i.e., image schema) that is activated by language is not limited to static shapes, such as vertical or horizontal arrangements of subject and object. As predicted by Len Talmy, they also happen for sentences that describe dynamic events that change over time. Cognitive neuroscientists Zoe Kourtzi and Nancy Kanwisher observed that the brain area that processes visual motion perception is active when people look at still photos of dynamic events, such as a freeze frame of a dolphin doing a flip above the surface of the ocean. Essentially, your knowledge about what that event would have to look like in real life in order to obtain a still photo like that engenders some partial visual imagery that involves motion. So, when cognitive linguist Teenie Matlock teamed up with Daniel Richardson to test for the existence of dynamic image schemas for *sentences* instead of *photos*, it made perfect sense to predict that even metaphorical motion might also induce this kind of unconscious partial visual imagery. Len Talmy called these sentences *fictive motion sentences*, such as "The fence runs from the corner of the house to the garage." There's nothing actually running, or moving at all, in that sentence, yet the verb "run" is used to describe the spatial extent of a nonmoving object: the fence. These fictive motion sentences are quite common in our everyday language use, so much so that we often don't even notice them as metaphorical. Matlock and Richardson recorded people's eye movements while they viewed static drawings of fences, roads, and tree lines and listened to spoken descriptions of the displays. When the context described the terrain as difficult to travel (i.e., rocky or full of potholes), fictive motion sentences such as "The road runs through the valley" produced eye movements that wandered up and down the image of the road for a long time. By contrast, in the same context, a sentence such as "The road is in the valley" produced eye movements that looked only briefly at the image of the road. Thus, even though the road isn't really supposed to be literally moving when the verb "run" is used, some unconscious partial visual imagery involves some motion, and this is revealed by the eye movements traveling up and down the length of the road. Think of it this way: despite the fact that a certain metaphorical use of language may not require particular sensory properties for you to understand the sentence, your brain activates those sensory properties anyway. That's what the embodiment of language is all about.

The Emotionality of the Embodied Mind

Now that we've seen how your concepts are embodied and how your language is embodied, let us turn to some examples of how your emotions are embodied. Cognitive neuroscientists have discovered that your body's nervous system can develop strong bidirectional associations with your emotional states, so much so that, after an emotional state has activated your nervous system enough times, there comes a time when your nervous system alone can reverse the signal and activate that emotional state all by itself. It's not just that your brain makes your body do things. Sometimes your body makes your brain do things, too.

Cognitive neuroscientist Antonio Damasio has developed a theory that your emotional reactions play a stronger role than you might think in your attitudes, decisions, and actions. In part, this is because emotions sneak into those cognitive processes without your realizing it. Damasio's somatic marker hypothesis suggests that dramatic circumstances in your environment often directly elicit changes in heart rate, muscle tone, and facial expressions, and it is how your brain has learned to interpret those bodily reactions that determines what emotion you actually feel. Then that emotion frequently influences your decisions and actions in response to the dramatic situation. The bodily (somatic) reactions happen first in response to a stimulus or situation, and then the associations that have developed between that somatic marker and some emotional state are what activate the emotional state. For example, one person may have developed an association between an increased heart rate and emotions of happiness (perhaps from being on roller coasters and in fast cars). Then, when almost anything makes their heart rate increase, they feel good—and this can influence their decision-making under those conditions. A different person may have developed an association between increased heart rate and emotions of fear (because of frightening domestic situations). Then, when almost anything increases that person's heart rate (even a mild roller coaster), they feel fear—and that fear affects their general decision-making processes. So, if your girlfriend tends to interpret increased heart rate as a fearful thing, then you probably shouldn't propose to her right after watching an intense action movie. Her already increased heart rate just might make her think that the sensation of fear she is experiencing is related to your proposal rather than the movie. These things happen.

It doesn't take much to have your brain respond to an emotional stimulus in a way that you might not want. For example, Jonathan Freeman briefly flashed pictures of emotionally charged faces to people while they were in an fMRI brain scanner, and he found that the subcortical brain region that processes emotion, the amygdala, was active even when they didn't consciously notice the emotional face image. The emotionally charged faces were flashed for a mere 33 milliseconds (one-thirtieth of a second) and then replaced with a neutral face for one-sixth of a second. People could see the neutral face but were unable to detect that a different face had very briefly preceded it. Freeman digitally manipulated some of the emotionally charged faces to look untrustworthy by adding a slight squint, arched eyebrows, and a subtle frown, and made other faces look trustworthy by widening the eyes, curving the eyebrows, and adding a slight smile. Compared to the trustworthy-looking faces, those untrustworthy-looking faces (which people denied ever seeing) produced greater activation of the amygdala, a part of the brain that supports emotion processing. That's right, the part of the brain that reacts to emotional situations was reacting to these untrustworthy-looking faces, even though people reported being unable to see them.

Decades ago, the amygdala and its related brain structures that form the limbic system were thought to function like a separate module in the brain that carries out emotion processing. For example, people with bilateral (both sides) damage to the amygdala have trouble recognizing emotional facial expressions. In the past, findings like that were often interpreted as evidence for that part of the brain being a module dedicated to processing only that type of information. However, damage to one brain region typically disrupts the processing of a *network* of multiple brain regions, and it is usually that network of brain regions that carries out the mental process in question (and other processes as well). A contemporary understanding of how cortical and subcortical brain structures are richly interconnected with one another reveals just how misleading that modular approach was. If you don't believe me, then you need to reread chapter 3.

Emotion plays a larger role in cognition than many of us would like to admit. If your emotion (or even an undercurrent of emotion that you are unaware of) tells your body to react and then your body's reaction tells your brain what to think, then perhaps we need to develop a better understanding of how emotion works. Current theories of emotion processing in the brain are moving toward an account in which emotions are not processed solely by a separate module but instead interact quite a bit with cognitive processes. Therefore, since your cognition is embodied, then so are your emotions. Psychologist Lisa Feldman Barrett is an emotion researcher who is leading the charge to pull emotion research into the twenty-first century, chiefly by integrating it with cognitive neuroscience and experimental psychology. She suggests that how you experience emotion is influenced by how you choose to think and talk about it, and how you choose to think and talk about it is influenced by your social context. How you cognitively categorize your own emotions has a substantial effect on how you experience them.

For example, do you know what you're supposed to do when a toddler is sitting on the floor and accidentally falls over and bumps her head on the hardwood? No, dummy, you're *not* supposed to freak out and run over to her and try to console her before she begins crying. That will actually *start* her crying. Rather than asking if she is hurt, you are supposed to playfully ask her if she made a dent in the floor—or distract her in some other way. Parents learn these things; usually from each other, but sometimes the hard way. Rather than encouraging the child to focus on the brief pain on her scalp and the emotional embarrassment of clumsiness in front of others, the social context can provide a different set of concepts to think about while the pain subsides, and the result can be little or no crying. That kind of social scaffolding of how to cognitively deal with potentially emotional experiences goes on all the time, and slightly different social/cognitive structures will produce slightly different patterns of emotion representation in the child's brain.

According to Lisa Barrett, your core emotions may not actually be separable into the famous discrete categories of sadness, happiness, love, hate, anger, and so on. Instead, it may be our social learning of those conceptual categories, with those linguistic labels, that trains us to feel those emotions in such sharply defined types. If so, then subtle differences in the way those labels are used might help explain the subtle individual differences that are seen in people's reported emotional experiences *within* a given culture. And larger differences in the way those labels are used when comparing different languages and cultures could help explain the larger differences that are seen in people's reported emotional experiences *across* cultures.

Such a huge part of *who you are* is shaped by the social structures that surround you and how they frame your embodied emotions. It shouldn't be surprising that your body's sensorimotor and emotional training with these social interactions would become part and parcel of your social self. The movements of your body (e.g., eyes, face, mouth, limbs, and fingers) are the primary method of engaging in social interaction. Even if you're talking about electronic social media, you still have to use your fingers to type and swipe. In this millennial generation that has grown up with social media from the beginning, perhaps their finger movements will develop motoric associations with certain social and emotional attitudes. Perhaps the act of "swiping right" with a finger will become associated with liking and acceptance. Their social embodiment will be slightly different from that of us old farts.

The Biology of the Embodied Mind

What makes a mind embodied is not only the way it thinks, talks, and feels, and not only the way it interacts with other minds, but also the way its biological material actually functions. Your mind is embodied not because it is *contained by* a biological body but because it is *made of* a biological body. In the 1980s and 1990s, biologist Francisco Varela and his colleagues developed a wide-reaching theoretical framework in which the evolution and development of any living organism could be seen as a delicate balance between (a) interdependence between the organism and its environment and (b) a form of internal self-organization within the organism, the latter of which Varela called "autopoiesis." *Auto* is Greek for "self," of course, and *poiesis* (pronounced poy-ee'-sis) is Greek for the verb "to make." Say it out loud: otto-poy-ee-sis. That was close. Try again, and make sure you put the accent on "ee." There, that was perfect. When you put those two meanings together, you get *self-making*. What could it possibly mean for something to *make itself*?

We've all heard the myth of the "self-made man" and realize that, as a child, that man was probably taught a strong work ethic from his parents and/or teachers, then probably received important loans from banks (or family members) along the way, and benefited from government investment in various forms of infrastructure that supported his business ventures. So how the heck is he "self-made?" Well, that entire social context is the environmental interdependence part of Varela's theory, and the autopoiesis part is where the organism (e.g., a businessman, bacterium, or Bactrian camel) must have internal processes that maintain that delicate balance between its physical integrity and its permeability. The body of that organism maintains itself as a living being by constantly regulating its interactions with the environment. Cognitive scientist Randy Beer carefully analyzed some idealized computational simulations of artificial life to explore Varela's notion of autopoiesis. In these simulations, he observed that certain interactions between an organism and its environment (i.e., a "glider" in Conway's Game of Life) can be classified as destructive because they lead to the dissolution of the organism, whereas many other interactions are nondestructive because the organism adapts to the perturbations imposed on it by the environment. Typically, when two gliders directly interact in Conway's Game of Life, one or both of them disintegrates. However, in the right context, Randy found that a pair of gliders can sometimes influence each other in a pattern that is equivalent to a form of communication, thus altering the cycle of one of them and leading them into a newly coupled existence. Thus, when a glider (or a bacterium or a person) survives those perturbations, it not only *changes its environment* as it propagates through its surroundings but also *is changed by* that environment. This view of cognition as arising from an organism's dynamic interaction with the environment (not from a passive processing of incoming sensory information) is sometimes called *enactivism*.

Varela's and Beer's treatments of autopoiesis and enactivism can be successfully applied at multiple spatial and temporal scales. For example, whoever you are, you are composed of a collection of trillions of bounded organisms that maintain their autopoietic relationship to their metabolic environment (and are frequently replaced on a regular basis): biological cells. At a larger spatial and temporal scale, this collection of cells maintains a membrane around most of itself: your skin. At a still larger spatial and temporal scale, these skin-bounded humans form collections of people who interact with one another by exchanging information via direct and indirect physical perturbations that are imposed on one another: our social-metabolic environment. Much like the biological cells, these human organisms that form the species of humanity frequently die and get replaced on a regular basis, just on a longer timescale. And a bit like cells do, they have a tendency to form information-based membranes that sometimes cordon off one group of people from other groups of people—unfortunately.

Crucial to allowing these embodied human minds to interact *non*destructively with one another is a biological basis for mutually recognizing their commonality: empathy. The biology of the embodied mind does in fact involve a mechanism that allows the human brain to see itself in others. In the 1990s, neuroscientists Giacomo Rizzolatti and Vittorio Gallese first saw biological evidence for that mechanism in the brains of monkeys. They were electrically recording from a part of the monkey's motor cortex to examine the timing of how neurons activated when the monkey reached for and grasped an object that was resting on a small pedestal. Under the object was a food reward, such as a grape. Monkeys love grapes. Rizzolatti and Gallese found a neuron that rambunctiously produced electrical spikes just as the monkey grasped the object, so they continued recording from that neuron to get as much data as possible. Again and again, they would put a new object on the pedestal, with a grape underneath the object, the monkey would grab the object, the neuron would go nuts, and the monkey would put the grape in its mouth. The monkey probably thought this was awesome. But then something unusual happened. During one of the experimental trials, the grape fell off its pedestal and onto the table before the monkey got the signal to grab the object. Naturally, the experimenter reached into the apparatus and picked up the grape to place it back on the pedestal. And the neuron went nuts. Even though the monkey wasn't moving at all, a neuron in its motor cortex was spiking like crazy when the monkey *saw* someone else's hand grasp the grape. The neuron's activation was reflecting not only its own hand's ability to grasp but also the ability of another person's hand to grasp. So, Rizzolatti and Gallese called that neuron a "mirror neuron."

Since then, a "mirror neuron" mechanism has been identified in humans as well. For example, as long ago as 2000, cognitive neuroscientist Jean Decety and his colleagues recorded brain activity while people viewed illusory apparent motion displays of humans carrying out limb movements. When the limb movements were biomechanically possible, like bending the arm at the elbow in a natural direction, the motor cortex of the observer was active. Even though the observer wasn't moving at all, visually perceiving a limb movement that they were capable of doing caused the motor cortex to become active—perhaps "simulating" how its own motor commands would produce that same kind of movement. By contrast, when similar displays of human limb movements were biomechanically impossible, such as bending the arm at the elbow in an unnatural direction (hyperextending the elbow), the same illusion of apparent motion was visually perceived but the motor cortex in these observers was silent. The motor cortex didn't "mirror" or "simulate" how to make that movement, because it doesn't *know* how to make that movement. It's a good thing, too, because you'd tear a ligament doing that!

Similar evidence for a mirror neuron system in humans comes from work by cognitive neuroscientist Beatriz Calvo-Merino. She and her colleagues put professional dancers in an fMRI brain scanner and had them watch video clips of dance. When female ballet dancers viewed female ballet dance moves, their premotor cortex (right next to the motor cortex) was significantly active—even though they weren't producing any movements. By contrast, when a female ballet dancer viewed male ballet dance moves— moves with which her visual system was very familiar but her motor system was not—her premotor cortex was not active. There was little or no motor simulation of how to carry out those male ballet dance moves because those female brains didn't have practice with them.

How you cognitively understand actions in the world is partly formed by your motor system's understanding of how *you* would (or would not) carry out those actions yourself. Also, how you cognitively understand the things that people say (because speaking is an action after all) is also partly formed by your motor system's understanding of how those words relate to actions in the world. Language and action are encoded together in the embodied brain. For example, cognitive neuroscientist Friedemann Pulvermüller and his colleagues had people perform a "lexical decision task," which simply involves responding "word" or "nonword" to a string of letters, like KICK or GIRP, as quickly and accurately as possible. While these participants carried out this task, Pulvermüller gently activated their motor cortex with transcranial magnetic stimulation (TMS). By mildly activating the subregion of the motor cortex that processes *leg* movement commands, Pulvermüller made people faster at responding to words such as KICK and RUN than to words such as GRAB and THROW. And it worked the other way around, too. By mildly activating the subregion of the motor cortex that processes *arm* movement commands, he made people faster at responding to words like GRAB and THROW than to words like KICK and RUN. These participants were performing a language task, and mild activation of the motor cortex altered the way in which they performed that task.

In fact, you don't even have to use TMS to see this kind of effect. The direct causal role of the body in cognition can be detected by engaging the body in action during a cognitive task. With Zubaida Shebani, Pulvermüller found that rhythmically moving your arms interferes with your memory for arm-action words but not leg-action words, and rhythmically moving your legs interferes with your memory for leg-action words but not arm-action words. While the mild TMS activity in arm or leg regions of the motor cortex primed and improved performance with the corresponding types of action words, overt rhythmic movement of those actual limbs did the opposite. Essentially, maintenance of the robust rhythmic limb movement diverted those motor neurons toward that separate task, and thus they weren't available (when they usually would have been) to assist in the encoding and storing for memory of those corresponding action words.

Not only does motor cortex activation spread into language brain areas and influence language processing, but the reverse happens as well. Language cortex activation can spread into motor brain areas and influence motor movement. Rather than injecting magnetic fields through the skulls of her participants while they looked at words, the way Pulvermüller did, cognitive psychologist Tatiana Nazir instead "injected" written words into their visual fields while they carried out reaching movements with their hands. Even *after* the reaching motion had been initiated, the visual presentation of an action verb significantly altered the acceleration profile of the reaching movement compared to when an object noun was presented on the screen. Action verbs of all kinds, for example JUMP, CRY, and PAINT, went into the visual system, and then neural activation patterns quickly spread through the language cortex and into the motor cortex in time to subtly tweak the execution of the reaching movement while it was being carried out. Your brain's understanding of language about action is intimately linked with its understanding about actual action. The biology of your mind cannot help but represent language in a fashion that embodies it in the context of the perceptual-motor consequences of what those words mean—and also how they are physically produced by the mouth.

For instance, the "motor theory of speech perception" has suggested that the speech motor movement system participates in the process of perceptually recognizing spoken words. Psycholinguist Alvin Liberman argued that part of how your brain recognizes the sounds that make up a word is by partially activating the neural components that would be used if you were to *pronounce* that word yourself. If these neural components of speech production are only partially active while passively listening, then one would not expect them to actually result in motor movement—but maybe they could be encouraged. Neurophysiologist Luciano Fadiga did exactly that with TMS in a collaboration with the mirror neuron master himself, Rizzolatti. They introduced a single pulse of a magnetic field into the left motor cortex of participants while they quietly listened to words, some of which had the Italian double-r sound in the middle, such as "birra" and "terra." If Liberman was right and the recognition of a word that has a strong roll of the tongue in the middle involves partial activation of the motor command to roll the tongue, then encouraging it with a jolt of TMS just might make these silent listeners move their tongue unintentionally. Fadiga recorded tongue muscle activation using surface electrodes attached to the skin of the tongue, and sure enough, Liberman was right. People's tongues moved more (from the TMS jolt) when passively hearing words such as "birra" and "terra" compared to words such as "baffo" and "goffo." Thus, it looks like the neural components of those speech motor movements are partially active (and can be manifested with a little help from TMS) while one is quietly listening to speech.

The speech motor system not only gets recruited for passive listening to speech but also participates in the neural commands of hand movements. Neuroscientist Maurizio Gentilucci instructed people to reach for and grasp large and small objects, and to open their mouth (or read out loud a syllable written on the object) as they began the reaching motion. He and his colleagues found that the opening of the lips was wider (and voice quality was different) when subjects were grasping larger objects than when they were grasping smaller ones. In fact, the millisecond timing of this wider opening of the lips was remarkably similar to the millisecond timing of the wider aperture of the fingers as the hand approached the larger objects. Apparently, the fine-grain details of the neural command to grasp a smaller or larger object are spreading through the network of brain areas to unintentionally influence the oral command to open the mouth and speak.

It's not just that how we think, listen, and speak affects how we use our bodies, but how we use our bodies also affects how we think, listen, and speak. For these relationships between cognition and action to exist, they have to become part of the biology of the brain and the body—not merely part of some abstract notion of the "software" of the mind. That's why these various measures of the biology succeed at finding such clear evidence of embodied cognition. The embodiment of the mind is a physical manifestation of the fact that *who you are* is something that is composed of biological material, and that biological material does not draw crisp boundaries between the mechanisms it uses to carry out quiet ruminative thought and the mechanisms it uses to carry out the action-based consequences of that thought. They are part and parcel of each other.

The Artificial Intelligence of an Embodied Mind

In the science fiction series *The Book of the New Sun*, by Gene Wolfe, one of the characters that you encounter is Jonas, a survivor from a crashed spaceship, who is half man and half robot. Spoiler Alert: The protagonist assumes that Jonas was a human who was repaired with robotic parts after the crash, but he's wrong. It's the other way around. We eventually find out that Jonas was a humanoid robot who sustained damage to various limbs and organs, and the repairs were performed by grafting living human parts onto him. Rather than a human turned cyborg, he is a robot turned cyborg. But then, aren't we all becoming part cyborg as technology advances? Many of us wear visual acuity enhancement devices on our faces (called eyeglasses). Many of us rely on our wristwatches and smartphones to do a great deal of our remembering for us, and some people have electronic hearing aids or even cochlear implants.

As long ago as 1980, consummate cognitive scientist Zenon Pylyshyn described a philosophical thought experiment that might shake up your intuitions about what a mind can be made of. Zenon's thought experiment went as follows. Imagine that a computerized nanochip replaced one of your neurons. Imagine that this single microscopic silicon chip perfectly reproduced the same actions that the now gone neuron used to perform, receiving electrochemical signals over time, summing the inputs in a fashion a little bit like a leaky integrator, and then sending out signals to other neurons via electronic synaptic connections. If you had this one nanochip in your brain, replacing a single biological neuron out of billions, would you still be you? Do you think you would still have the conscious mind that you have now? The answer seems obvious: of course you would. A single neuron doesn't play a strong enough role among the billions in your brain for its electronic replacement, or even its complete absence for that matter, to make you somehow no longer human. But what if you had thousands of those silicon nanochips replacing real neurons? What if you had a stroke and some far-future doctors replaced your entire visual cortex with a multi-layered motherboard of a billion nanochips, all of which did exactly what those dead visual neurons used to do? Or what if it was your prefrontal cortex that was replaced in that fashion? Would you be a robot then? Would you no longer have your human mind, even though the same statistical pattern of electrical and electrochemical signals was traveling around inside your skull, and the same patterns of behavior were being produced by your body? By the same logic, if a robot was designed from the beginning to have billions of silicon nanochips with trillions of electronic connections, where the connectivity closely resembled that of human brains, and its humanoid body produced behaviors that were similar to those of humans, would society be right in classifying it as a mechanical slave machine that doesn't deserve civil rights? Okay, enough with the questions. If you need to mull over those ideas for a few minutes, go ahead and put this book down, but not for long; we have work to do. I'm going somewhere with this.

The gradual artificialization of humans is not some far-future fiction. It is already happening. Philosopher Andy Clark has written extensively about how we are all cyborgs to some degree, not just because we use *mechanical* technological developments, such as eyeglasses, smartphones, prosthetic ears, and prosthetic limbs, to augment our perception and our action but also because we use *informational* technological developments. For example, language itself functions as a kind of technological development that we use to deliver information across spatial distances and across temporal spans that our normal sense organs cannot reach. Thanks to language, we can learn about things that we never sensed, and we can deliver instructions to produce actions that our own limbs won't have to carry out. Humans have already been using that particular sensorimotor-augmentation technology for many centuries. Not only will a mechanical prosthetic (like an ear or an arm) do some of your sensing or acting for you, and thus become part of *who you are*, but an informational prosthetic (like a story you hear or an instruction you give) will also do some of your sensing and acting for you—and similarly become part of *who you are*.

The gradual development of artificial intelligence also is not some far-future fiction. It is already happening as well. The key insight that computer scientists have learned over the decades is that, in order for an artificial system to really be intelligent, it needs to be *pro*actively interactive, not merely *re*actively interactive. It can't just respond to prompts and requests. It needs to be able to initiate its own acquisition of information. Therefore, rather than a "brain in a vat," it needs to be a robot. The field of computer vision learned this the hard way in the 1970s and 1980s, while they were pointing cameras at still images and engaging in a futile struggle to develop computational algorithms that might segment and distinguish one object from another in these still pictures. As it turns out, humans and other animals aren't always very good at that either. Most biological users of a visual system, like you and your pet and animals in the wild, move their heads around a little bit to access multiple angles on a 3-D visual scene in order to identify one object as separate from another object. And once computer vision researchers such as Ruzena Bajcsy placed their cameras on top of mobile robots so they could proactively access visual input from multiple angles, all of a sudden that problem of image segmentation got a lot easier. This was a way to make the computer vision system *proactively interactive*.

Computer scientist James Allen figured this out with conversation bots as well. Rather than have the speech recognition system resign in abject failure when it failed to figure out one of the words in the sentence you just spoke to it, Allen programmed his TRAINS scheduling system to simply ask for clarification of the misunderstood word. It sounds simple and obvious, but until the late 1990s, most speech recognition systems simply threw up their hands when they failed to recognize a word. Instead, James Allen made his system proactively interactive, such that it would determine which word in the spoken sentence went unrecognized and prompt the human with intelligent queries as to what that missing word might have been. When you think about it, that's exactly what humans do when they fail to hear a word in a sentence that one of them speaks to the other. Humans don't just robotically respond with, "I'm sorry, can you please repeat that entire sentence?" No, we respond with something like, "I'm sorry, did you say '*cheese* burger' or '*three* burgers?'" Conversation then continues smoothly, despite background noise—because we humans are proactively interactive. Automated speech recognition systems are getting better and better every day in part because they are becoming proactively interactive.

Having some form of robot body helps the artificially intelligent system be proactively interactive, because it can autonomously (by its own guidance) explore its environment. It isn't forced to rely solely on programmers making time in their schedules to tell it what it needs to know. Instead, the robot can do much of its learning on its own by trial and error, a bit like a human child does, except that it's even more mobile and doesn't have to sleep. In much the same way that Andy Clark has been drawing inspiration from robotics research to improve our understanding of embodied cognitive science, a number of roboticists have been drawing inspiration right back from embodied cognitive science to inform their design of robots. For example, in the 1990s, roboticist Rodney Brooks demonstrated that much of the intelligence that emerges from a robotic agent is already inherent in the environment itself and is brought to bear by bodily actions of the robotic agent. His insect robots self-organized their group behavior in a way that looked remarkably well planned, even though there was no actual plan programmed into them. The emergent coordination among the robots took place because the environment itself already had some intelligence baked into it. This intelligence included where the lights in the room were pointed, where the obstacles on the walking surface were placed, the rigidity that those obstacles exhibited when a robot limb touched them, and the rigidity that one robot exhibited when another robot bumped into it. Together, these kinds of environmental factors naturally funnel the set of behaviors of those simple robots into producing similar and seemingly coordinated behaviors. Many of your own daily behaviors are actually a lot like those of Brooks's simple robots. A great deal of your everyday "autopilot" actions do not require much internal planning and goal-directedness. The goal-directed behaviors that do stand out in your daily activity involve sequencing those smaller "autopilot" actions in a goal-directed fashion and then allowing them to run largely by themselves.

Numerous other roboticists have continued to draw inspiration from embodied cognitive science. For example, Deb Roy programmed his robot to combine the meaning of a color word, such as "green," along with the motor programming of the cameras to point toward objects in the visual scene that reflect green light. So, the meaning of "green" in his robot doesn't just refer to some abstract set of chromatic features; it also refers to how the robot's body interacts with the world when that word is heard. And the word "heavy" is encoded along with the expected sensory feedback that the robot arm experiences when it grasps and lifts a heavy object. Deb's robot has embodiment programmed right into its language system, in a fashion much like what Freidemann Pulvermüller discovered in human brains. Roboticist Giovanni Pezzulo argues that, although robots can do some smart things simply by reacting to the stimuli in their environment, true intelligence comes from developing an anticipation of the possible states of the world that would result from certain bodily actions and then having an overarching goal that guides you to choose wisely among those options—being not merely reactive but also proactively interactive. Angelo Cangelosi argues that when the robot has this kind of proactive approach to its environment, it interacts with its environment in a fashion that extracts the intelligence inherent in that environment—especially when that environment contains other intelligent agents (human or robot). By designing robots that learn from their interaction with the environment in an embodied fashion, Pezzulo and Cangelosi are paving the way for a field of *developmental* robotics.

A key component of that developmental process is the sociality of the robot. A robot can learn much from interacting with the inanimate objects in its environment, but it can learn even more from interacting with the *people* in its environment. Social roboticist Cynthia Breazeal first discovered this when she began designing robot heads that had cute cartoon-like faces. Even when the robot's language system wasn't performing any comprehension at all, merely responding with turn-taking and producing occasional facial reactions and some "ooohs" and "aaahs," many humans were happy to tell their whole life story to this robot. In fact, later they would report that they felt like they had a successfully interactive conversation with the robot, even though it never actually spoke. It's almost like a Clever Hans effect, where the infamous horse Hans appeared to be able to perform arithmetic (such as "Tap out four plus seven with your hoof"), but it was actually the human trainer who would give him body-language signals for when to stop tapping. In this analogy, it is the robot that is the trainer, giving the right body-language signals at the right time, and the conversation itself is the horse—looking a fair bit smarter than it really is. The body language from Breazeal's robots gave people the social cues needed to feel comfortable in a conversation, and even though the people were doing all the actual talking, they felt as if it was a genuine back-and-forth conversation. Having these simple social-reactive routines is crucial for giving today's social robots the basic tools for encouraging humans to assist them in extracting information from their environment so they can learn to be intelligent. This kind of accidental Clever Hans effect in social interaction is not some illegitimate trick. It is something that we all naturally do to each other on a regular basis. We all give each other subtle, often unconscious, body-language signals that allow us to cocreate a conversation. A listener nods her head when she is following and agreeing with the speaker, wrinkles her brow when she doesn't understand something, and even completes the sentence of the speaker on occasion. And when you don't know what the other person is talking about but you're just waiting for them to finish their boring story so you can politely walk away and go talk to someone else at the party, what do you do? You nod your head, so they will get to the end of their story with minimal interruptions. You Clever Hans, you.

This intelligence, of an artificial kind, is made possible not because it is hardwired into the "mind" of the robot. This intelligence is achieved because (1) some of it already exists in the *environment* that the robot is interacting with and learning from, (2) some of it is *programmed* into the robot as a set of overarching goals (which can be modified through learning), and finally, (3) some of the intelligence already exists in the particular construction and the particular *materials* of the robot's mechanisms for collecting information from the environment (its sensors) and for moving around and altering that environment (its effectors). This third component is especially important for understanding the embodiment of robotics. It emphasizes how the shape, materials, and overall morphology of the robot's body actually do a kind of intrinsic computation for its interactions with the environment: morphological computation. For example, biorobotocist Barbara Webb built a robot cricket that mimics how a female cricket pursues the mating call of a male cricket. In this simple robot, the two tracheal tubes (for receiving sound) are able to determine the direction of the sound source, so they can drive the motor to move in that direction. However, the specific shape and diameter of those tracheal tubes provides this directional information only for the particular sound frequency that the male cricket actually produces. Because of the morphological properties of the tracheal tubes, sounds at other frequencies simply don't provide this directional information. Therefore, the robot cannot help but ignore sounds that are not at the same frequency as the mating call. Importantly, this selectivity is not the result of any clever software algorithm or neural filter but instead simply results from the physical shape of the robot's tracheal tubes. That's morphological computation.

In fact, your own ears perform some morphological computation for your ability to determine the direction that a sound is coming from. The particular shape of the cartilage that forms your outer ears is doing some of your hearing for you. The ears on the side of your own head capture the sound waves that pass by in a very specific way that is a little different from the way other people's ears capture those sound waves. Your brain has adapted to the unique pattern of sound filtering that your outer ears perform. Therefore, if your outer ears were replaced with those of a very different shape, you would no longer be able to tell where in space different sounds are coming from. It would take a few weeks for the neural networks in your brain to adapt to the new and different morphological computation that your new ears are performing as they catch sound waves on the side of your head.

Another example of morphological computation comes from decades of developing humanoid hands for robots. Years were spent programming the visual system of the robot to perfectly identify the size of an object, at varying distances, so that it could deliver precise instructions to its hand to grasp the object. If the measurement was one millimeter off, the object could be gripped too tightly and get crushed, or the object could be gripped too loosely and get dropped. Sensory feedback mechanisms were designed to quickly tell the robot's motor system when to let up on the pressure, and that improved performance. However, truly delicate objects, such as a light bulb, were still routinely broken by robot hands, because the sensory feedback loops weren't fast enough to modulate the force of the actuators on the hand for something that delicate. You can pick up a light bulb without crushing it. Why can't a robot? The genius insight, which might seem obvious in hindsight, wasn't to improve the software or to redesign the electronics but instead simply to implement some passive compliance on the robot's fingertips. Passive compliance is a fancy technical term for a layer of soft rubber. How thick a layer and how soft a material is actually very important, and that's because these layers of rubber on the fingertips are physically performing some crucial computations for the robot's interactions with the world. This particular kind of morphological computation has spawned an entire new field of "soft robotics," where passive compliance allows the motor commands to the robot's effectors to be much less precise because the robot's soft body surface can naturally accommodate the imprecision.

In fact, much like those robots, your own limbs rely heavily on morphological computation, passive compliance, and also tensegrity. Tensegrity refers to when the balance of *tension* forces that are distributed throughout a system are responsible for the *integrity* and stability of that system; hence, *tens-egrity*. A tensegrity object holds itself together because its internal pushing forces and pulling forces are balanced against each other. In your body, a substantial amount of morphological computation is actually carried out by the tensegrity network of fascia, or connective tissue, that threads its way between your muscles, tendons, bones, and ligaments. When you're eating meat, you call it gristle, but when it's inside your limbs, we call it fascia (pronounced "fasha"). This fascia takes physical forces that are introduced at one part of a limb and transmits them to another part of that limb almost instantly, and without needing to send an electrochemical signal. Imagine pushing on one side of an unusually dense gelatin mold. The force of that pressure travels to other parts of the mold very quickly. Just like that, fascia can transmit information about physical forces, produced by actuating a muscle or by receiving tactile input, across portions of tissue to activate neurons all the way on the other side of that finger, hand, or foot. This way, when a limb moves or external pressure is introduced to the skin surface, a rich and complex pattern of neural activation from all over that limb is delivered to the spinal cord, not just a few spikes from the neurons where the specific action is happening. Patterns of force information, transmitted throughout the network of fascia, balance against one another a little bit like a tensegrity object. Thus, even if you did think your body was just an encasement for your mind, you'd have to admit that this encasement is doing an awful lot of intelligent computation.

By making their robots reactive, proactively interactive, goal-directed, and morphologically computational, roboticists have *taught* cognitive scientists at least as much about embodied cognition as they *learned* from them. In their mission to become more human-like, robots are becoming social, soft, and developmental. They seek out interactions with people. Their bodies are becoming more like people's. And their minds are growing like people's. It's enough to make one wonder about the ethical considerations of such scientific advances. But hold on. As I said earlier, I'm going to have to ask you to bookmark that thought until the middle of chapter 8. No, don't thumb your way forward to chapter 8. Be patient. Let your unconscious mind mull it over for a while until we get there. For now, your job is simply to take the following lesson from robotics: the mental computations that make up a mind are happening not just in the brain (or the CPU) but also in the body.

To the Brain-and-Body

It seems clear that your body does a lot of your thinking for you. If you didn't have the body that you have, your mind would be significantly different. If your brain were placed into a different body, you would have a different mind. *Who you are* would be different. And if you didn't have a body at all, just a brain, your mind would be barely recognizable. The scientific evidence for the role that the body plays in the mind is overwhelming, from cognitive psychology, to linguistics, to neuroscience, to artificial intelligence, and everywhere in between. In cognitive psychology, there are dozens of examples of how various cognitive processes cannot help but activate their associated sensory and motor processes, and dozens of examples of how sensory and motor processes routinely influence the way cognitive processing works. In linguistics, there are myriad examples of how we use metaphors of the body (and how it uses space) to think and talk about all kinds of important concepts, everything from anger, to algebra, to state capture. While comprehending language, your body and brain cannot help but partially activate some of the very same actions that are associated with what you're hearing or reading. In neuroscience, there are so many examples of how the brain encodes perceptual input in conjunction with the motor output to which it relates. The biology of brain-and-body unmistakably shows that how you recognize events in the world is couched in terms of your own ability (or inability) to physically participate in events like that. Finally, in artificial intelligence, recent examples definitely attest to the importance of a robot body doing some of the critical information processing for an artificial agent, even before the "brain" of that robot gets a chance to work on the sensory input. The future success of artificial intelligence depends heavily on a close integration of the robot's software with its hardware, a tight merging of its ability to reason with its ability to act, and a unity between its "brain" and its body. Embodied cognitive science has taught roboticists much about how their artificial agents should be built, and recent advances in artificial intelligence and robotics have much to teach us about how our own brains and bodies work.

Chapter 3 showed you that the continuous, fluid flow of information back and forth between hundreds of different brain areas makes it impossible to treat any single brain region as the sole determinant of who you are. Whoever you really are, you are *not* just your prefrontal cortex. Using that same kind of logic, this chapter has shown you that the continuous, fluid flow of information back and forth between your brain and your body makes it impossible to separate your brain and treat it as the "ivory tower" for your mind. Your mind extends beyond your brain. It is coextensive with your body. So, if you had to draw a circle around the physical material that makes up *who you are*, a couple of chapters ago you might have drawn that circle around your frontal lobes. But a chapter later, the scientific evidence encouraged you to expand that circle to include the whole brain. In this chapter, the scientific evidence is encouraging you to expand that circle to include your body as well. *Who you are* is your brain-and-body, at the very least.

Directions for Use

Vilayanur Ramachandran, whom you will read about in chapter 5, has written extensively about how your body's interaction with the world determines how your brain cooperates with your body. In fact, in Ramachandran's medical practice, he has encountered patients with dysfunctional cooperation between the brain and body, and he was able to resolve some of these dysfunctions via clever tricks that the patient can use in his or her interaction with the world. A version of one of those clever tricks, which healthy brains-and-bodies can also enjoy, is the two-foot nose experiment, and this experiment is your "directions for use" in this chapter.

Start by closing your eyes and tapping your left hand with your right hand. You can tell that it's you doing the tapping, even though you don't see it, because the sensory input that your brain receives from your left hand is well timed with the motor command that it delivered to the right hand (and the sensory feedback received by that right hand as well). That is why it's pretty hard to tickle yourself with your own fingers, even in your most ticklish spot—wherever that is. What makes tickling ticklish is that your brain doesn't know exactly when and where the tactile pressure is going to happen. But when your brain does know this, the tickling just isn't ticklish.

Now, with that in mind, allow me to provide the instructions for your two-foot nose experiment. The first thing you will need is two other people to join you in this experiment. Among the three of you, find the pair who are the closest in height. One of them will be "the experiencer," and the other will be "the nose." The experiencer will be blindfolded and stand right behind the nose, with the two of them facing the same direction. These two people should be within a couple of inches of the same height. The third person stands to their right and is "the experimenter." The experimenter takes the blindfolded experiencer's right hand, closes it into a fist, and then pulls the index finger out to extend all by itself, as if they were pointing at something. Then, the experimenter holds the experiencer's right hand with their own right hand and gently strokes the nose of "the nose" while simultaneously using their own left hand to gently stroke the experiencer's own nose in an identical manner. The experimenter should stroke and tap in an unpredictable fashion, almost as if trying to tickle those two noses. When the experiencer's index finger is being guided to tap the bridge of the nose's nose, the experimenter is also tapping the experiencer's own nose in the same spot at the same time and in the same way. When the experiencer's index finger is being guided to stroke the side of the nose's nose, the experimenter is also stroking the experiencer's own nose in the same spot in the same way. This carries on for about a minute, so you can gradually trick the experiencer's brain into thinking that the tactile sensory input that she is feeling on the tip of her index finger (about two feet away from her face) is causally connected to the highly correlated tactile sensory input that she is feeling on her own actual nose. The result, which works well for about half the people who try it, is a compelling sensation that your nose is extended two feet away from your body. Basically, your brain knows that your body belongs to it primarily because of the systematic correlations between sensory inputs and also motor outputs. Therefore, when you stack the deck with the fake correlations in this experiment, you can trick the brain into thinking its nose is two feet long.

5 From Your Body to Your Environment

> Stop light plays its part,
> so I would say you've got a part.
> What's your part? Who you are,
> You are who, who you are.
> —Pearl Jam, "Who You Are"

Whether it happens on a gold-plated toilet seat as for some pretend billionaire, in a public bathroom as for you and me, or in the outdoors as for Henry Miller, there is something that we all do several times a day that unavoidably gets us in touch with our bodies: dispose of bodily waste. But not every environment is appropriate for this waste disposal ritual (Henry Miller notwithstanding). This ceremony involves finding a suitable place to put the waste. The relationship between that bathroom environment and your body provides you the opportunity to do something that you would never do in public: fully relax your bladder and bowels. The sight and/or feel of a toilet right below you is a powerful stimulus, is it not? Try not to be so shocked. It doesn't need to be a forbidden topic of discussion, in principle. It's just that since we primates evolved to find the smell of urine and feces displeasing, we developed social norms to treat pee and poo as though they were taboo. But, like the children's book of the same name says, everybody poops.

Decades ago, when I was junior faculty at Cornell University, I was at a dinner with some social psychologists, and I was describing to them what it meant for cognition to be embodied. One of my senior colleagues was grooving with this embodiment idea and said, "It reminds me of a haiku I saw penciled on the inside of the stall in our department's bathroom." The haiku went as follows:

> I sit on the pot
> and relax from deep within
> poop comes out of me

Imagine sitting on a toilet in the stall of a public bathroom and seeing that haiku scrawled on the door right in front of you. The morphological computation taking place between your butt cheeks and the toilet seat (along with the paper toilet seat cover probably) forms a surprisingly intimate interface between you and the world, one that allows you to do something that has become very private. My colleague explained to us how this haiku helped him understand the embodiment of cognition. Unable to hold back, I ran to him and hugged him, and told him how warm it made me feel inside that he liked my haiku.

Now that you've made it through chapter 4, you should be ready for chapter 5. You know the drill by now. If chapter 4 did to you what it was supposed to do, then you've embraced the idea that your entire body is a very important part of "who you are." But is that all there is to who you are? What about the tools you use, the food you eat, the things you do, and, oh, the places you'll go? Are those things just part of some external context that surrounds the real you, or is there such a continuous, fluid flow of information back and forth between your body and the things and events in your environment that you might as well include them as part of who you are—just like you did with your brain and body in chapter 4? Is that toilet you're sitting on part of who you are?

In this chapter, I'm going to show you a collection of scientific findings indicating that the information-bearing functions of your brain-and-body (that is, processes that carry and convert information) are so deeply intertwined with the information-bearing functions of the objects and events in your environment that you really ought to expand your definition of who you are to include your immediate environment as well. The information that makes you who you are includes not just the information carried inside your body but also the information carried by your surroundings. With every millisecond of external sensory input to your eyes, ears, skin, nose, and mouth, that environmental information is constantly being converted into brain-and-body information.

Sensory Transduction

For some traditional cognitive scientists, the "magic of the mental" starts with sensory transduction. Sensory transduction is sometimes roughly pinpointed as happening at the surface of the body and is treated as a very special process that might separate what is outside the mind from what is inside it. However, this can be a peculiar place to draw the line because it's not actually very special (fish, worms, and cockroaches do it, too), and it doesn't actually happen at the surface of the body. This peculiar distinction includes the peripheral nervous system along with the brain as constituting the mind, but not the rest of the body: not the optics of the eye, the shape of the ear, the skin, the muscles, or the fascia. All that morphological computation, from the previous chapter, carried out by the network of gristle in your hands, arms, legs, and feet is completely ignored. Sensory transduction is where mechanical (and other nonelectrochemical) forces impinge on neurons on the surface of the body, and that mechanical force is converted (or transduced) into an electrochemical signal. At a superficial glance, it can indeed appear to be a pretty special *qualitative conversion* from one medium of information to a completely different one, which might make it a good candidate for drawing a line between the physical and the mental. One example of sensory transduction is pressure on the skin that physically deforms a mechanoreceptor neuron under the dermis, and the deformation causes the neuron to produce action potentials that travel eventually to the spinal cord. That's one way to transduce a set of mechanical forces (deformation of the skin and its embedded neurons) into an electrochemical signal.

Another kind of sensory transduction is when sound waves travel into the ear canal and rattle your eardrum. The information pattern intrinsic to that sound, which allows it to be recognized by a listener, is already present in the mathematics of the frequencies of those waves of air pressure. That information pattern is preserved in the mechanical rattling of the eardrum, which in turn rattles a sequence of three tiny bones in your middle ear, the last of which taps on your cochlea. Inside the cochlea is a fluid that responds to the tapping by propagating waves across a series of thousands of hair cells. They're called "hair cells" because they have long, thin feelers that extend out from the cell body. The fluid waves in the cochlea continue to preserve most of the information pattern that was in the airwaves.

These fluid waves tickle the hair cells, slightly deforming their cell bodies and causing them to produce action potentials (electrochemical spikes) that travel along the auditory nerve on their way to the auditory cortex. That physical deformation of the hair cell body is what causes the patterns of mechanical vibration in the middle ear to be transduced into an electrochemical signal. Much like the mechanoreceptors in your skin, when something gently bends or squeezes these neurons, they send signals down their axons. But what's important is the information pattern. The information pattern of that sound was approximately the same when it was in the air as when it was in the cochlear fluid, as when it was in the sequence of action potentials traveling along the auditory nerve. The fact that the *medium* of transmission changed is far less important than the fact that the *pattern* of information was generally preserved at each transition.

Sensory transduction is impressive, but it's not some unique, unprecedented event of biophysics. Now that neuroscientists understand it, it's a relatively pedestrian phenomenon. Consider vision. When you experience visual perception, it is because electromagnetic radiation (visible light) lands on your retina and a process of photoisomerization slightly changes the molecular structure and shape of chemicals in the photoreceptors, which causes them to produce action potentials that travel down their axons, eventually sending signals along the optic nerve, into the brain. Those retinal photoreceptor neurons are transducing an *electromagnetic* signal into an *electrochemical* signal. That's not much of a qualitative conversion from one medium of transmission to another. They're pretty similar.

Now consider smell. The chemoreceptor neurons in the nose are transducing a *chemical* signal into an *electrochemical* signal. That's not much of a qualitative conversion from one medium of transmission to another either. How about electroreception? Certain fish, such as sharks and rays, have sensors that can detect electric fields (which travel reasonably well in saltwater). Their sensory neurons are transducing *electric* fields into *electrochemical* action potentials. That's not much of a qualitative conversion from one medium of transmission to another.

Imagine holding your smartphone in your hand, or if you have it nearby, don't just imagine it, do it. Concentrate on the pressure of its plastic case on your fingertips, thumb, and palm of your hand. If you buy into chapter 4's message and acknowledge that the physical deformation of the skin and fascia in your hand is part of the natural computation that makes up your mental experience, part of who you are, then you should seriously consider also buying into the idea that the phone case itself is part of who you are as well—because its physical forces are directly causally connected to the physical deformation experienced by your skin and fascia. The morphological computation taking place at the surface of your skin is not that different from the morphological computation taking place at the surface of your phone case.

Let's start at the point of sensory transduction and work our way backward. One step in the cause-and-effect chain for why you feel the smartphone in your hand is that a particular set of mechanoreceptor neurons in the skin of your hand are being deformed and are thereby sending electrochemical signals to your spinal cord, which then sends electrochemical signals to your brain. That's the sensory transduction talking. But that sensory transduction happening in your skin is just one step in the chain of cause and effect. Immediately preceding that step is the passive compliance, or squishiness, of your skin, the "soft robotics" of your human hand. The way your skin deforms from the external physical pressure of the smartphone's edges is directly responsible for which mechanoreceptor neurons get deformed and by how much. Therefore, the pattern of information inherent in the timing of those neurons sending their signals is a direct result of the pattern of information inherent in the way your skin is getting squished. If you had softer hands or harder, calloused hands, your phone would feel different in your hand because the passive compliance would produce a different pattern of squishing in the skin and then your mechanoreceptor neurons would produce a different pattern of signaling. That's your morphological computation talking. But let's continue to follow that chain of cause and effect backward. What caused that particular deformation pattern on your skin? The information pattern inherent in the shape and hardness of your smartphone case (and the amount of force that your hand muscles are exerting to maintain their grip) is directly responsible for the information pattern inherent in the physical deformation of the skin. If the morphological computation of your skin is part of who you are, then why isn't the morphological computation of your smartphone case part of who you are? Those two information patterns, one intrinsic to the plastic and the other intrinsic to your skin, are incredibly similar to one another, and one is the immediate cause of the other. That's your smartphone talking. It's not just a cute metaphor to say that your smartphone is "an extension of your body." It's kind of true.

Now, imagine eating an apple, or if you have one nearby, start eating it. Concentrate on how it feels in your mouth to have your teeth slice into it and crunch it into smaller bits. Mechanoreceptors in your mouth convert this force pressure into electrochemical signals. Taste buds on your tongue now convert some of the apple's molecular information pattern into electrochemical signals, and your nose can now smell that apple, too. That apple is on its way to becoming part of you, inarguably. When people say, "You are what you eat," they're not just rattling off a hackneyed phrase. It's very true. The proteins, vitamins, carbohydrates, and, yes, the nonfood chemicals that you consume don't just work their way through your body and then exit. Many of them stay there and become part of who you are, at least for a little while. By carefully choosing your foods, you are carefully choosing who you are. When you think about the journey that this apple takes, you might ask yourself, "Is there some point at which the apple definitely becomes part of who I am, whereas immediately beforehand it wasn't?"

Let's start at the point of digestion and work our way backward along that chain of cause and effect, kind of like what we did with the smartphone in your hand. Upon being digested, it seems safe to say that the apple has become part of you. Its sugar, fiber, and vitamins get extracted and become part of what your body does. Right before that, the crushed-up apple bits were already being somewhat digested by the saliva in your mouth, so they were kind of already part of you by then. And right before that, your mechanoreceptors and taste buds were doing their sensory transduction thing, so those information patterns were becoming part of you even then. Now think about this. The information pattern inherent in the shape and hardness of the apple itself was directly responsible for the morphological computation that your teeth, lips, and mouth carried out during that very first bite. Just like the smartphone in your hand, the apple at your lips has an information pattern intrinsic to it that is incredibly similar to the information pattern morphologically computed by your mouth. The apple's hardness and shape directly causes the deformation of your lips and mouth and produces the force feedback on your jaw as you munch down on it. At what point does the apple transition from being not part of you to suddenly being part of you? I'm not sure that any of those points in the chain of cause and effect can be adequately defended as the discrete transition point. Perhaps you should think of it as a gradual transition from the apple being slightly part of who you are, even when it's just in your hand, to then becoming more and more part of who you are as you gradually wrap yourself around it.

Now that you're thinking about some of the fine-grain details of the causal chain of events that leads to an external object or event becoming part of *who you are*, it should feel rather difficult to choose a specific link in that chain to designate as the succinct transition point where something changes from being external to you to being internal to you. In order to defend the notion that sensory transduction is the place where "the nonmental becomes mental," one would essentially have to claim that electrochemical signals are somehow uniquely special, and until an information pattern has been converted into that particular medium of transmission, it's not part of a mind. Chemical signals cannot be mental? Electromagnetic signals cannot be mental? Electrical signals cannot be mental? But electrochemical signals are somehow obviously mental? By that logic, Zenon Pylyshyn's thought experiment from chapter 4, with the nanochips replacing your neurons, would require a traditional cognitive scientist to conclude that an electronic nanochip neuron replacement process *would* indeed make your brain less of a mind.

Alternatively, one might argue that it's not that electrochemical signals in particular are somehow special but simply that the process of transducing a stimulus from one format of energy to another format of energy for the purposes of sensory perception is what makes it into "mental content." This argument has holes as well. For example, take computer vision. In recent decades, computer scientists have been developing smarter and smarter robots that take patterns of light into their camera sensors and convert that *electromagnetic* radiation into... well, more *electromagnetic* signals on a silicon chip. There's no qualitative transduction from one medium to another happening there (except perhaps from photons to electrons). When future humanoid robots are seeing and intelligently interacting with their visual environment, a traditional cognitive scientist would have to conclude that those robots have no mental experience of their visual world because no qualitative sensory transduction took place, whereas those same robots apparently do have a mental experience of their auditory environment, because their hearing system does involve sensory transduction.

In the end, it seems foolhardy to place a crisp boundary between the physical and the mental at the point of sensory transduction. It inappropriately separates the peripheral nervous system from the crucial information-bearing processes that are carried out by the skin, the muscles, and the fascia: the morphological computation that human bodies naturally perform. The biochemical details of sensory transduction, where sensory stimulation is converted into neural signals, have become well understood. Nothing magical is happening there. Sensory transduction is not a mystery, and it's simply not a viable candidate for where to draw the line between the physical and the mental.

Ecological Perception

Information patterns don't only come *into* your body. They also come *out* of it. The outgoing physical energy exerted by your body (and the accompanying information pattern) is constantly being converted into new incoming brain-and-body information patterns. Let's ignore external stimuli entirely for a second and just think about how your body moves. Close your eyes and bend your elbow several times in a way that doesn't bump into anything around you. You can't *see* your arm moving. You can't *hear* your arm moving. So how is it that you can perceive that your arm is moving? Part of it is mechanoreception in the crook of your elbow. The skin there is being squished a bit and then released each time you bend your elbow. But most of it is proprioception happening throughout the muscles and joints of your arm. Proprioceptors are neurons that send signals to the spinal cord, telling it when muscles are flexed and when joints are rotated. The reason you can tell that your elbow is bending, and even approximately where it is in space, despite your eyes being closed, is because proprioception is keeping track of the locations of your limbs. Basically, your motor cortex sends a signal to flex a muscle by a certain amount, and proprioceptors send a feedback signal confirming that the muscle was indeed flexed by approximately that amount.

In the 1940s, James J. Gibson realized that visual perception actually functions as a form of proprioceptive sense as well. Your visual system doesn't just see what objects and events are out in the world. Your visual system can instantly tell how your head (and the rest of your body) has moved in space. When your motor cortex sends a signal to your legs to walk forward, your visual system sends feedback signals to the rest of your brain confirming that you are indeed getting closer to objects in front of you. That's proprioception telling you where your body is in space.

Gibson's insight came to him while he was assisting the US Air Force in helping pilots land planes more safely during World War II. He realized that the motor output that a pilot's hand places on the control column (or the "stick") directly alters the path of the plane, and that directly alters the flow of visual input that the pilot sees through the cockpit windshield. The pattern of information flowing into that pilot's visual cortex is *yoked* to the pattern of information flowing out of that pilot's motor cortex to his hand—not unlike how two oxen are yoked together with a wooden crossbar (called a yoke) when they pull a plow together. Where one ox goes, so does the other. (In fact, some airplane manufacturers call the stick the yoke because it controls wing flaps on the left and right sides at the same time.)

Across all the cognitive and neural sciences, people have a tendency to focus almost exclusively on the way perceptual input causes motor output. You hear something, and you respond. You see something, and you react to it. The interesting thing about this plane-landing insight that Gibson had was that it turns this around completely. When a pilot is landing a plane, his motor output is not merely a reaction to his visual input. His motor output is also a direct *cause* of his visual input. His hand on that stick is in charge of the flow of visual information that goes into his visor, just as when you walk forward, your legs are in charge of the flow of visual information that goes into your eyes.

J. J. Gibson realized that a great deal of information useful for knowing where your body is in space, what direction your body is going in, and how your body is adapting to its environment is already present in that optic flow of visual information landing on your retinas. Rather than conceiving of perception as a process of generating internal representations of the external world, Gibson conceived of it as a process of directly mapping sensory input onto motor output, which then directly alters the next sensory input in a manner determined by the laws of physics and the equations of optics. In Gibson's ecological perception, the objects in the environment do not need to be *represented* internally as static images and symbols in the brain in order to be perceived; they can be perceived directly via sensorimotor interaction with those objects themselves. Indeed, perception itself was seen by Gibson not as something that happens *inside* the brain but as something that happens in the dynamic interaction *between* organism and environment.

Gibson insisted that vision evolved not as a sensory process but as a *sensorimotor* process. In fact, it is worth noting that none of the sessile (immobile) animals of the sea, such as sponges, barnacles, and coral, have eyes that focus light. Since they don't move around, they don't need to sense *patterns* of light. Animals that move around usually have eyes with lenses, and animals that don't move around don't have eyes with lenses. And get this: an animal that moves around in its early life cycle has an eye with a lens, but then when it becomes sessile in its later life cycle, it no longer has that eye at all. The sea squirt, which has a primitive eye with a lens when in its larval stage, transforms into its sessile existence by attaching itself to a rock and then absorbing its own eye and brain for their nutrients. Unable to move around for the rest of its life, it no longer needs that eye or that brain. The primitive nerve network that remains is sufficient for the sea squirt to filter food out of the water it pumps while attached to that rock. Not even evolution is interested in vision for perception's sake. Evolution rewards vision for action's sake.

In fact, when you look at how your own eyes work, it becomes clear that they are not at all designed to build an internal model or representation of the world around you. But that's what traditional cognitive psychologists assumed in the 1980s. By assuming that the brain developed a detailed internal 3-D model of the visible world, they set themselves the task of determining exactly how the brain patches together the little snapshots that the eyes pick up as they flit about the scene. When your head is relatively stable, your eyes typically move around in quick jerks, called "saccades." *Saccade* is French for jerk; the verb, not the noun. As your eyes saccade from one location to another, they can move faster than 360 degrees per second. (That is, they move so fast that, if the eye muscles allowed it, your eyes could do a full 360-degree rotation in their sockets in less than one second. Luckily, they only do that in horror movies.) After moving at that incredible speed for a few dozen milliseconds, they stop on a dime. These stops are called "fixations." During a fixation, the brain can collect stable visual input for a few hundred milliseconds before the next saccade typically launches the eyes to point at another location in space. However, this stable visual input has high acuity (like an HD TV) only in the central one degree surrounding where the eye is pointed. Therefore, in the 1980s, cognitive psychologists set about the task of figuring out how these tiny high-definition images from each fixation can be pieced together like a puzzle to generate the internal 3-D model of the scene that they assumed was there, somewhere in the brain.

In order to piece those snapshots together like a puzzle and create an internal representation of the entire visual scene, you would need to store the location of each image, so the brain would have to record the angle and distance of each eye movement (and subtract for any head movements) that led to a fixation that collected an image. Then it could know where to place each piece of the puzzle. If the patterns of information in the eye-movement command in the frontal cortex, and in the proprioception fed back to the brain by the eye muscles themselves, were sufficiently precise, then it could be done.

One of the early experiments to explore how this might be done involved presenting a 5×5 array of dots on a TV screen in the nearby periphery of the person's field of view. Initially, 12 of those 25 dot locations were lit up on the screen. Then, when an eye movement was made to the center of the array, the display changed to have 12 different dots lit up. If the person's brain could piece together the first 12 dots in the periphery from a second ago with the new 12 dots after the eye movement, then they should be able to see which twenty-fifth dot never got lit up. Sure enough, the results showed that people could do it! They were good at it. They could fuse together the two visual snapshots obtained from different eye fixations. For a very short while, cognitive psychologists saw this as powerful evidence for the idea that the brain aggregates these puzzle pieces of images into a full internal model of the scene. And that's why we feel like we are aware of the full visual scene in front of us, because we have a detailed model of it represented in our brains.

But that enthusiasm lasted less than a year. It was quickly discovered that the TV screen in the experiment had been emitting some faint phosphorescent persistence (a slight glow) of the initial 12 dots when the display changed to the new set of 12 dots. Thus, the visual persistence that allowed people to fuse the first set of 12 dots with the second set of 12 dots wasn't in the brain of the observer but instead on the screen itself! Let that be a lesson: scientists can make mistakes, too. Luckily, they can correct them as well. When cognitive psychologist David Irwin and his colleagues did the experiment with LEDs instead of a TV screen, people were no longer able to fuse the two images accurately. Several other perceptual psychologists similarly found that the effect did not replicate when screen persistence was prevented.

One of those perceptual psychologists was Bruce Bridgeman. He wasn't just correcting the methodology and proving that some cool effect was a mistake. He had a theoretical perspective that suggested effects like this shouldn't happen. His theoretical perspective was ecological perception. Can you guess who handed this perspective down to him? That's right, James J. Gibson. In the 1960s, Bridgeman was an undergraduate at Cornell, working with Gibson. He learned that when you hypothesize that the brain is computing some internal representation, you better have pretty good evidence because, as we've seen, much of that computation can actually be carried out by the body itself and how it interacts with the environment that it's in. Bridgeman did a senior thesis with Gibson (which was later published in 1987, after Gibson's death). He then went on to earn his PhD at Stanford with Karl Pribram (the guy in chapter 3 who defined the Four Fs of evolved behaviors). From the 1970s until well into the twenty-first century, Bruce Bridgeman's research program explored the role that eye movements play in organizing our perception of the world around us as a stable environment—rather than as a random collection of snapshots between each eye movement. Unlike many cognitive psychologists, he didn't *assume* that this was achieved by generating an internal 3-D model of the visual scene.

Along with perceptual scientist Larry Stark, Bridgeman explored a careful study of (1) exactly how much information about where and how far the eyes were supposed to move is carried by the eye-movement command in the frontal cortex and (2) exactly how much information is carried by the proprioception of the eye muscles themselves about where and how far the eyes actually moved. We can demonstrate the basics of this experiment right here and now. Find a spot on this page on which to fixate, and then close one eye. While maintaining fixation, not moving your eyes, use your index finger to very gently and slowly press on the eyelid of your open eye. It should look like the world moves a little bit, even though your eye stays fixated on the spot you chose. Why does this happen? It is because your eye muscles are naturally fighting back the displacement that your finger is producing. Therefore, they send proprioceptive feedback to your brain, telling you that your eye muscles flexed. Even though the image on your retina hasn't changed much at all, since your eye muscles are compensating for your finger's force, your brain interprets that proprioceptive information as evidence that the world has moved. With this kind of experiment, what Bridgeman and Stark showed is that the amount of information in this proprioceptive signal and the amount of information in the eye-movement command in the frontal cortex simply do not add up to enough to give sufficiently precise information about where and how far movements of the eyes have gone. Even if the brain wanted to piece together these snapshots into one big jigsaw puzzle image, an internal model of the scene, it wouldn't know with sufficient precision where all the pieces belong.

Bridgeman had proven that the information pattern in the eye movements is not sufficient to accurately reproduce an internal model of the visual world inside the brain. If there can't be an accurate internal model of the scene, why do we *feel* as if we have one? In 1994, Bridgeman and his colleagues made an unusual proposal. They suggested that perception begins anew with each eye fixation. There is no stitching together of multiple snapshots in visual memory because, after all, we don't really need it. If a reason arises to access information from a specific location in your field of vision, rather than accessing some stored memory of that information in your brain, you can just point your eyes in that direction and access the information from the world. The world itself can function as your visual memory.

While we may feel as if we have in our minds a full-mosaic understanding of our visual surroundings, it is mostly an illusion. This should begin to sound familiar, from the change blindness experiments in chapter 1. For instance, if someone asks you what color shoes your friend is wearing today, you might feel like you knew the answer all along as you respond to the person, "black." But during the half second between their question and your answer, you probably made a saccade toward your friend's feet and perceptually accessed the color information so quickly and absentmindedly that you didn't even realize you didn't know the answer before your eyes retrieved it for you. This kind of theory about the apparent visual stability of our perception would lead one to predict that you could make a surreptitious change to someone's visual environment and they wouldn't even notice it. For example, imagine that a really good magician changed your friend's shoes from black to blue without your noticing, and then someone asked you what color your friend's shoes were. According to Bridgeman's theory, you would spend half a second looking at their shoes and then answer "blue," once again feeling as if you knew the answer all along. It's as if you would be *blind* to many of the *changes* in your visual environment. That's right; Bridgeman's theory presaged, predated, and predicted the entire change blindness literature (discussed in chapter 1). Like J. J. Gibson before him, Bruce Bridgeman taught the cognitive and neural sciences that we may not need to assume that the brain does a whole lot of internal representation to accomplish visual experience. The visual scene can represent itself just fine, thank you very much.

The Action-Perception Cycle

At the beginning of this chapter, we focused on information patterns that go *into* the observer. Then we focused on how ecological perception teaches us to attend to information patterns that go *out* of an observer as he or she moves around, which immediately changes the way information goes in. In this section, the focus is on how information also goes out of the body to *change things* in the environment. It's not only that the information *entering* the observer's body changes when he or she moves around the environment, but information *in the environment itself* changes when those actions alter the location, orientation, and shape of external objects. The brain-and-body's information is constantly being converted into environmental information, which is then soon converted into new brain-and-body information.

Even just picking up an object and turning it around in your hand performs information-bearing operations on that object that are similar to the kinds of cognitive processes you could perform on an internal neural representation of that object. For example, close your eyes and imagine the book (or other reading device) that is in your hands, and then mentally rotate that imaginary book in your mind so that it is upside down. Go ahead, do it. Now, with your eyes open, use your hands to physically rotate this book (or reading device) upside down (unless you're reading this on a desktop computer; please don't turn your desktop computer upside down). These two cognitive operations, turning the book upside down in your mind or upside down in your hands, performed very similar transformations of the information in front of you. As another example, think about when you try to solve a wire puzzle, one of those toys where you're supposed to remove the wire ring from the tangle or release the metal ball from its cage. In principle, you could generate a full 3-D model of the wire puzzle in your brain and then simply close your eyes, apply the mathematical theory of knots, carry out some "mental rotations" of the image, and try out some potential solutions to the puzzle. But that's not at all what you do, is it? You don't perform *mental* rotations of the wire puzzle in your brain; you perform *physical* rotations of the puzzle in your hands to try out potential solutions. In these situations, you are literally thinking with your hands.

There are a variety of actions that you can perform on an object, depending on the properties of that object and the properties of your body. Those places where the potential functions of that object match the abilities of your body are what J. J. Gibson called "affordances." An affordance is a relation between the organism in question (e.g., a pilot, a businessman, or a larval sea squirt) and potential actions that the environment affords the organism in question. An affordance exists neither inside the organism nor inside the object with which the organism is interacting. An affordance arises between the two. It is a relation. The information-bearing properties that an affordance provides, the cognition that it carries out, happen at the junction of organism and environment. That said, when a person perceives an object, his or her brain is likely to exhibit some evidence of the affordances made available by that object. Therefore, a *record* of the affordance may show up inside the organism's brain. For instance, when you look at a chair, part of how you recognize it as a chair isn't just that the shape and contours are familiar but also that you know what your body can do with it. You can sit on it, lean on it, or stand on it to change a light bulb. Cognitive psychologists Mike Tucker and Rob Ellis showed that people are faster to make a simple judgment about the orientation of an object on a computer screen if the object's handle is on the same side as the hand they are using to make their response. Essentially, the affordance of grasping the handle of that object with the nearby hand (even though it's really just a two-dimensional picture of the object) speeds up the response to the object. Part of recognizing the object and determining its orientation involves acknowledgment of its most prominent affordances (in this case, grasping). Corroborating this observation, cognitive neuroscientist Eiling Yee recorded people's eye movements while they followed spoken instructions to look at an object. She found that they sometimes accidentally looked at other objects that have similar functions (even though those objects don't look similar or have similar-sounding names). Evidently, even just hearing the name of an object activates some of the neural population codes associated with the actions that you can carry out with that object (i.e., its affordances). As a result, other objects with those same affordances sometimes draw your attention briefly. These findings show that as an organism and its environment mutually negotiate their affordances, the brain-and-body of that organism is likely to reveal signatures of this negotiation process. The brain does not *contain* affordances, but it plays an important role in how they arise.

When an organism takes advantage of an affordance and carries out an action on the environment, the environment is often changed in a way that immediately alters the sensory input to that organism. It becomes a loop, where each action alters the next perception. In 1976, the father of cognitive psychology, Ulric Neisser, called it the action-perception cycle. Neisser was in the same psychology department as the persuasive J. J. Gibson at Cornell University, so it may come as no surprise that Neisser wrote the book *Cognition in Reality* almost as a recanting of the scientific mission he had laid out a decade earlier in his book *Cognitive Psychology*. Rather than treating the study of cognition as having the goal of figuring out how internal representations are generated in the mind, Neisser was now suggesting that internal representations are less important than he once argued. He suggested that the field of cognitive psychology should instead focus on how cognitive operations happen in the loop created by actions that alter perceptions, which then produce new actions, which then alter perceptions, and so on. This action-perception cycle no longer treats a *stimulus* as though its only purpose is to produce a *response*, and it no longer treats a response as though its only purpose is to react to the most recent stimulus. Instead, the action-perception cycle acknowledges that it isn't so easy to unravel the chain of cause and effect to see what started it. The stimulus is not the sole cause of the response, and in fact the response is often causing the stimulus. Cognition, and the mind, is not merely what's happening after the stimulus and before the response. The mind emerges among the entire incessant loop composed of brain, body, and environment. Sometimes your most insightful cognitive operations happen *outside* your brain and body as you modify the physical objects you are using to think.

For example, cognitive psychologist Sam Glucksberg gave a difficult puzzle to people and watched their hand movements carefully while they tried to solve it. The puzzle—first introduced in 1945 by Karl Duncker—involves a candle, a matchbook, a box of thumbtacks, and the instruction to mount the candle on the wall and light it. The puzzle is quite difficult, and usually only about 50 percent of people discover the solution. Spoiler Alert: The solution requires realizing that the box holding the tacks is functionally flexible enough to do more than just hold tacks; it can be the mounting shelf that you tack to the wall. Most of the time, people simply reach into the cardboard box, pull out some thumbtacks, and start trying in vain to build something with the tacks alone. Glucksberg suspected that if you could trick people into *not* pigeonholing the box as nothing more than a container for the thumbtacks, then they would find the solution more readily, so he blindfolded them. As the experiment's participants blindly felt their way around the objects, their hands often accidentally bumped into the side of the cardboard box. Eventually, these accidental touches of the cardboard box helped people discover the solution, and 80 percent of them found the answer! They also did so twice as quickly as the sighted participants. One of the participants even said plainly, "I touched the box and got the idea to use it." This is an example of a cognitive insight, about how to solve a difficult problem, triggered not by thinking deeply about the problem but by physically interacting with the environment. The morphological computation of the hands touching the morphological computation of the cardboard box brings to bear the passive compliance that the cardboard affords for sticking thumbtacks through it and the rigidity that it affords for supporting a candle on the wall. Essentially, cognition can happen anywhere that information is being converted and transmitted, whether that's between brain regions in your skull or between the surface of your skin and the object in your hand.

Using a similarly difficult puzzle—also designed by Karl Duncker—Betsy Grant and I recorded people's eye movements while they looked at a diagram of the puzzle. We found that not only is *touching an object* a way to gain insight from your interaction with the world, but simply *looking at an object* can do that, too. In Duncker's tumor-and-lasers problem, most people trying to solve it eventually give up or run out of time. Imagine that you are a doctor dealing with a patient who has an inoperable stomach tumor. You have lasers that destroy organic tissue at sufficient intensity. How can you cure the patient with these lasers and, at the same time, completely avoid harming the healthy tissue that surrounds the tumor? While trying to solve the problem, people suggest all manner of imperfect solutions, such as carving a small incision in the stomach lining to aim the laser at the tumor, but that would harm the healthy tissue. After wrestling with these imperfect ideas, people usually go silent for a few minutes and appear to be stuck at an impasse. Then, out of the blue, about a third of people will suddenly burst out with the correct solution. In the technical literature, it's called the aha moment. That's really the technical term for it. In fact, people really do blurt out "Aha!" when they have this insight. Spoiler Alert: The solution to this famous problem involves converging multiple weak lasers, each of which travels harmlessly through the healthy tissue and then meet at the tumor, where their heat combines and is sufficient to destroy the tumor. (Actual contemporary radiation therapy for cancer is slightly similar to this scenario, but it uses gamma rays instead of lasers and routinely damages healthy tissue no matter how hard you try not to.)

In our first experiment, we showed people a simple diagram with a disk in the middle (the tumor), surrounded by an oval (the stomach lining), and recorded their eye movements while they guessed at possible solutions. When our experiment participants were 30 seconds away from finding the solution, our eye-tracking data showed that they were focusing especially on the stomach-lining oval. By contrast, people who were 30 seconds away from giving up or running out of time did not pay any special attention to the stomach lining. Much like Glucksberg's blindfolded participants with the candle problem, we suspected that people weren't just finding the solution in their brains and that's why they looked at that particular part of the diagram. Maybe it was the other way around. Perhaps looking at that part of the diagram helped spur the cognitive insight to have the lasers *converge* on the tumor. So, on Dick Neisser's suggestion, we designed a second experiment, where the diagram on the computer screen had a simple little animation added to it: the stomach-lining oval blinked one pixel thicker three times per second. (Daniel Richardson programmed the animation for us.) This unavoidably drew people's eye movements to the stomach lining, and instead of one-third of the people solving Duncker's tumor-and-lasers problem, *two-thirds of them* discovered this elusive solution! A simple perceptual trick—not that different from Glucksberg's blindfold—drew people's attention to a component of the diagram that helped them gain cognitive insight into a rather abstract concept of convergence to solve this problem.

Importantly, you don't always have to wait for your environment to give you these perceptual tricks. In the right situation, you can make them happen for yourself. Cognitive scientists David Kirsh and Paul Maglio explored how people spontaneously take advantage of these perceptual tricks themselves "on the fly" while they play a video game. When you're playing the game *Tetris*, you're a little bit like a grocery bagger at the checkout stand. You get different shapes of blocks, one after another, and you have to rotate them by pressing a button to fit best into the blocks that are at the bottom of the screen. As you progress through the game, it gradually goes faster and faster until it becomes difficult to determine, in the time available, what is the best rotation for the current object to fit into the shapes at the bottom of the screen. By watching *Tetris* players carefully, Kirsh and Maglio observed that most expert players were modifying their environment on a millisecond timescale to carry out the important transformations of information on the computer screen instead of inside their brains. They suggested that rather than looking at the next block and *mentally* rotating to see which orientation would fit best at the bottom of the screen, these *Tetris* experts were *physically* rotating that next block by pressing a button several times really quickly. This way, no time was spent making a mental representation of the block in their brains and then going through the neural computations to mentally rotate it 90 degrees and then 180 degrees, and so on. Instead, each 90-degree rotation took place on the computer screen in a couple hundred milliseconds with the press of a button. These expert *Tetris* players were making their environment perform some of their cognition for them, and that's an important part of why they did better than the other players. It's an important part of what made them experts.

When your brain is performing some of your cognitive processes for you and your environment is also, it can become difficult to distinguish where the mental *you* ends and the rest of the world begins. That's exactly the point of the action-perception cycle. Rather than thinking of the brain-and-body as one system that interacts with another system (e.g., the environment), cognitive scientists are beginning to take to heart the scientific mission of Gibson, Neisser, and Bridgeman and analyze the organism and environment as one system. It's not an easy analysis, but it can be done.

One interesting little experimental laboratory scenario in which an organism becomes statistically "coupled" with an external object, such that they function together as one system, is in pole balancing. Physicists refer to this as the "inverted pendulum problem." Curiously, the longer the pole, the easier it is to balance. For example, if you try to balance a pencil on the tip of your index finger, you won't be able to do it for more than a few seconds. But if you try to balance a two-foot long, half-inch thick wooden dowel on the tip of your index finger, you'll be able to keep it balanced for at least a dozen seconds—and much more with some practice. This task involves a very tight action-perception cycle, where the next few dozen milliseconds of vision and touch feedback are used to immediately guide the adjustment of your arm and your stance to keep that pole balanced. Neuroscientist Ramesh Balasubramaniam has studied this process carefully with Tyler Cluff, recording nearly every millisecond of change in the position of the top of the pole, the bottom of the pole (where the finger is), and the postural center of pressure where the two feet are supporting the body. Over the course of several days of practice, their experiment's participants improved from balancing the pole for about 25 seconds to balancing it for almost two minutes straight. To see how this improvement in performance showed up in the statistical patterns of the body's movements coordinating with the pole's movements, Balasubramaniam and Cluff analyzed the data streams from the center of pressure (showing how the body posture sways over time) and from the bottom of the pole (showing how the base of the pole sways over time).

When pole balancers are just learning, they tend to be a bit rhythmic and regular in their finger movements and in their postural sway, in part because they are oscillating between a little too much in one direction and then a little too much in the other direction, and then back again. But as they get better at the task, these finger movements and postural movements each become less rhythmic and less regular, because the time spans of momentary stability are longer and more frequent. Most notable in Balasubramaniam and Cluff's data was the fact that, as people improved at the task, the coordination *between* their finger movements (i.e., the bottom of the pole) and their postural movements increased dramatically. When they analyzed not just the postural data stream and not just the pole's data stream, but the combination of those two data streams together, they found that practiced pole balancers were statistically best described as *a single system* comprising pole and balancer. When your behavior gets statistically coupled with an object in the environment (especially a tool that's in your hand), you and the tool become one. That external object becomes part of *who you are*.

It is important to realize that the action-perception cycle that binds organism and environment together involves a continuous, fluid flow of information from the actions produced by the organism, to the resulting changes in objects in the environment, to the resulting changes in sensory input to the organism, to the resulting next actions from the organism, which changes the environment further, ad infinitum. That organism is you, and that environment is also you. It's a constant loop that continues throughout your every waking second. Your eyes are moving at least three times per second, collecting new visual input as they go; your neck turns your head every few seconds, affording a new range of visual inputs; your hands are moving tools and objects around; your speech system is sending signals to other people; and the neural population codes in your brain never stop changing their activation patterns. Never. I know it's exhausting just thinking about it. Take a deep breath and let it out slowly. You need it. You probably haven't really done that since chapter 3.

When Objects Become Part of You

When a tool in your hand "becomes part of you," it's not just a metaphor. And it's not just a statistical description of the motions of your body and the motions of the tool. It's real. Your brain makes it real. For example, neurons that respond specifically to objects that are within reach of your hand will also respond to objects that are close to a *tool* that's in your hand. Neuroscientist Atsushi Iriki recorded the electrical activity of individual neurons in the parietal cortex of a monkey, a brain area right between the somatosensory cortex (for touch sensation) and visual cortex. Some of these parietal neurons have visual receptive fields that cause the neurons to become activated when visual stimuli are presented right next to the monkey's hand—no matter where that hand is in the monkey's field of vision. Basically, these neurons are responding to visual stimuli that are recognized as being within easy reach of that hand. If an object is close enough to that hand to afford a quick grasp, then these visual neurons respond. Upon finding these neurons in the monkey's brain, Iriki mapped the shape of that receptive field *before* training the monkey to use a tool and then again *after* training with a tool. The tool was a small rake that allowed the monkey to pull a distant piece of food closer so he could then grasp it with a hand. Before tool training, the receptive fields for these neurons spread out from the monkey's hand a couple of inches in every direction. After tool training, and with the rake in hand, the same neurons now had receptive fields that stretched out to surround the rake as well and spread out from it a couple of inches in every direction. Essentially, the monkey's brain was treating that rake as part of the monkey's body: an extended hand, as it were.

This general effect has been observed with human behavior as well. When people are holding a reaching tool and getting ready to use it, their distance estimates of objects are compressed. The object appears a little closer to them because of the simple fact that they're holding a reaching tool that makes the object seem less "out of reach." Cognitive psychologists Jessica Witt and Denny Proffitt briefly flashed a spot of light on a table and then asked people to touch or point to where the spot had been. Interspersed among those touch/point trials were some trials where the person was simply asked to estimate how far away the spot was (in inches). There was a fair bit of variation in their guesses, but on average they were pretty close to accurate, except when they were holding the reaching tool (a 15-inch orchestra conductor's baton). When they were holding the reaching tool and prepared to use it, a spot that had been 39 inches away was perceived as having been only 35 inches away. That's a 10 percent reduction in perceived distance just because they were holding a baton in their hand.

Witt and Proffitt theorized that this effect was a result of the brain generating a mental simulation of the reaching movement and thus tricking itself into thinking the object was closer and therefore more accessible when a reaching tool was held. To test this idea, they did a follow-up experiment where they showed the baton to participants and told them to merely *imagine* using it to reach for the spot of light. When these participants reported the perceived distance of the spot of light, once again there was a substantial compression of that estimated distance. They didn't even have a tool in their hands. They merely imagined reaching toward the spot with the baton, and that motor simulation caused them to underestimate the distance of the spot. Even if the brain just *pretends* that the tool is part of its body, then the tool is part of its body.

There are also other ways to make external objects become part of your body. Recall the "directions for use" in chapter 4, where you get someone's nose to sort of feel as if it might be two feet long. Twenty years ago, cognitive scientists Matthew Botvinick and Jonathan Cohen demonstrated that you can trick the brain into thinking a rubber hand is part of its body. They had people place their right hand palm-down underneath a table and then placed a rubber hand on the tabletop above the real hand. Then they stroked and tapped the knuckles and fingers of the rubber hand in synchrony with the stroking and tapping of the real hand, invisible under the table. When the experiment's participants *saw* the touching of the rubber hand and *felt* the same places and timing of touching on their real hand, they began to feel a bit as if the rubber hand was part of their body. In fact, when people are asked to indicate the location of their unseen real hand under these circumstances, their estimates are often shifted almost halfway to the location of the rubber hand. Psychologist Frank Durgin showed that it doesn't even have to be actual touch on the hands. When a mirror is aligned just right to make the rubber hand look like it's exactly where the real (hidden) hand is located, the light from a laser pointer traveling along the rubber hand is enough to give people an illusory sensation of warmth, and even touch, on their real hand.

This unusual observation actually has medical applications. Around that same time, neurologist Vilayanur Ramachandran was exploring this kind of phenomenon with amputees who experience phantom limb pain. Phantom limb pain is when an amputee feels excruciating pain in the limb that has been amputated. It might seem impossible, but it makes sense when you think about how the brain codes for that limb and how the brain reorganizes itself upon losing that limb. For instance, if an arm is amputated just below the elbow, groups of neurons that used to code for the hand obviously no longer receive sensory input from the mechanoreceptors in that hand. Over time, some of those neurons gradually develop connections to nearby neurons that have been coding for the elbow, which is still receiving sensory input. Sometimes those connections can cause the brain to think that the hand has somehow moved up right next to the elbow. Your brain knows full well that if your right hand were curled up so much as to be close to your right elbow, it would be incredibly painful. (Don't try this at home.) So, the brain naturally generates a pain response. If the missing limb were still there, some movement of it would quickly allow the brain to figure out that the hand is not at all curled up like that. However, with the limb missing, there's no way for the brain to use proprioception to figure this out. However, with visual input, it can. Ramachandran had the genius idea to place a mirror next to the amputee's intact limb. When the patient sits in the right position and the mirror is set at the proper angle, the reflection of the intact limb looks to the patient just like a copy of the missing limb, and in a location where that missing limb would naturally be. Movements of the intact limb are visually processed by the patient's brain as copycat movements of the missing limb as well. Thus, if a patient is feeling pain in their phantom right arm, watching a mirrored reflection of their left hand clench and unclench a fist can train their brain to realize that the (missing) right arm is not at all contorted in a manner that should cause pain. For cramping and other muscular pain in the phantom limb, Ramachandran's procedure is remarkably effective.

Whether they are tools, toys, or mirror reflections, external objects temporarily become part of who we are all the time. When I put my eyeglasses on, I am a being with 20/20 vision, not because my body can do that—it can't—but because my body-with-augmented-vision-hardware can. So that's who I am when I wear my glasses: a hardware-enhanced human with 20/20 vision. If you have thousands of hours of practice with a musical instrument, when you play music with that object, it feels like an extension of your body—because it is. When you hold your smartphone in your hand, it's not just the morphological computation happening at the surface of your skin that becomes part of who you are. As long as you have Wi-Fi or a phone signal, the information available all over the internet (both true and false information, real news and fabricated lies) is literally at your fingertips. Even when you're not directly accessing it, the immediate availability of that vast maelstrom of information makes it part of who you are, lies and all. Be careful with that.

When You Become Part of Your Environment

It's not just objects in your hand or on your face that become part of who you are. You make external objects in your environment perform perceptual and cognitive operations for you all the time in your everyday life. For instance, some people put their keys on a hook near the front door, so that whenever they are about to leave the house, the keys are there staring them in the face so they don't forget to take them. Thus, on your way out of the house, rather than having to rely solely on your brain to remember to get your keys, the environment itself can "remember for you" and provide a perceptual reminder at just the right time.

Can you remember when you moved into your current kitchen? You probably made specific choices as to where to put various appliances and which drawers to use for which items. In fact, a professional kitchen is carefully designed to have everything in just the right place to optimize the sequence of events that go into preparing various dishes. The exact spatial arrangement of those objects carries information, and this information guides and streamlines the way those objects are used to make meals. People use these perceptual tricks not only to arrange their environment in a permanent sense, but they also use them "on the fly." Next time you go to the grocery store, watch the bagger at the checkout stand. It's almost as if he's playing a new version of *Tetris*. While he is making sure to put heavy things in the bottom of a bag, you will often see him set aside the lighter objects in a buffer zone on the counter. That spot on the counter serves as an external memory for items that still need to find the right-shaped home in a bag. Later, when the right spot opens up in the top of a bag, he grabs that light object again and places it in the bag. David Kirsh refers to the use of these information-bearing spatial arrangements of objects as "jigging your environment." In woodworking and manufacturing, a "jig" is a term used for any custom-made device that is designed to guide the work that is performed on an object in a standardized fashion. The jig can be used again and again to make that identical bit of work go quicker and easier every time it is performed. When you've informationally jigged your environment just right, your environment does a lot of your thinking for you, so your brain doesn't have to. That fact is perhaps the key observation to take home from this entire book. The fact that those living and nonliving systems outside your body frequently "do some of your thinking for you" is not a cute metaphor. The science showcased throughout this book strongly suggests that it is literally true.

The human brain has a remarkable amount of its computational resources devoted to coding for relative spatial locations of objects, and it frequently lets the objects or information in those external locations themselves serve as the operands in various cognitive operations that take place outside the brain. Throughout the cortex, there are dozens of retinotopic maps (for vision), spatiotopic maps (for reaching), and tonotopic maps (for hearing). Your brain is like a real estate agent placing its emphasis on the three most important aspects of any property: location, location, and location. If your brain keeps track of *where* an object is, then it only needs to attach a little bit of meaning to that location, so it knows when and why to look there. The "where" of the object is stored in the brain, but most of the "what" of the object is stored in the environment. We attach bits of meanings to objects and locations all the time. This is part of how we jig our environment, how we put our minds out into the environment, how we become part of our environment.

It is true that we learned from the work of Bruce Bridgeman that the brain does not combine these relative locations, gathered from different eye fixations, into one internal neural model of the visual world. Nonetheless, the brain is still pretty good at keeping track of the approximate locations of a few objects of interest at any one time. For example, Zenon Pylyshyn tested people's ability to keep track of a target subset of moving balls on a computer screen and found that the most people can usually follow accurately is about five. If asked to follow the movements of six balls, out of ten, most people would lose track of that sixth ball, and sometimes the fifth ball as well. Imagine, for a moment, being in an experiment like this. Ten white balls show up on the computer screen, arranged in a circle. Five of them flash a couple of times to indicate that those are the ones you are supposed to keep track of. Then suddenly they all begin drifting around the screen in semirandom meanderings. If you were a chameleon lizard, you could follow one ball with one eye and another ball with the other eye. But you're not. And if you follow one ball with both eyes, it actually makes it harder to keep track of the others. So, what you wind up doing feels a lot like what Zen masters refer to when they say "reach out with your senses." You don't point your eyes at any one ball. You stare straight forward and follow the target balls with your attention. Your eye muscles take a break, and your vision does its job without those muscles for a change. All five targeted balls feel somehow visually highlighted for you as their paths twist and turn among the other balls. The fact that you can do this shows how attuned your brain is to attaching relevance to objects as they move in space.

This attachment of relevance to external objects is a major part of how we use the information in our environment to augment our thinking. Pylyshyn calls this attachment a "visual index," like a virtual finger that continually points at the object so you know where it is and can query its information whenever you need it. Linguists refer to this kind of process as a "deictic reference," such as when you say "this" or "that" while pointing at an object. The object's name does not get mentioned in this kind of reference, but if a listener can see you pointing, they know what you are referring to nonetheless. Computer programmers call this a "pointer," a number that refers to an address in a memory database. The content in that memory address is not coded as part of the pointer's identity; only the address is. Just as with change blindness, if someone changes the content in that memory address, the pointer will still happily refer to that address, unaffected by the change in content. When Dana Ballard and Mary Hayhoe saw this pointer-like behavior in people's eye-movement patterns, they called it a "deictic pointer." They gave people a Lego task where they were shown a pattern of colored blocks and were instructed to make a copy of it using blocks they found in a different part of the board. If people routinely built internal 3-D models of the visual world in their brains, as traditional cognitive psychologists once assumed, then you could easily carry out this task by staring at the original pattern of blocks until you memorized it and then simply pick up new blocks and place them in the right arrangement without ever having to refer back to the original pattern. But that's not what people did in this task. By recording people's eye movements, Ballard and Hayhoe were able to discover that people frequently look back at the original pattern of blocks while placing each new block in the copied pattern. People were naturally letting the world be its own representative rather than building a detailed representation of it in their brains. Most of the time, their eyes would land on a block in the original pattern, then move to a same-colored block in the resource area, and then, while the hand grabbed that new block, their eyes would sneak another peek back at the original pattern and then move to the copied pattern to guide block placement. Ballard and Hayhoe surmised that eye position served as a kind of "deictic pointer" that allowed the brain to use the original block pattern as a content-addressable memory. All the brain had to do was store the "address" of each block and point the eyes at the right one when it needed to access the content of that address, obtaining information about the block's color or its relative location. It's like that famous quotation attributed to Albert Einstein: "Never memorize something that you can look up."

So much of the information you use on a daily basis—the everyday stuff you always know—is not actually stored in your head. You jig it into your environment. Therefore, one could argue, your immediate environment is part of your mind. Your brain-and-body has a natural tendency to treat objects and locations in space as the holders of information. Those objects and locations are the addresses. They hold the content. And your brain-and-body uses deictic pointers to access that content to perform cognitive operations. Even when those locations no longer hold that content, we still tend to point to them anyway. Daniel Richardson and I observed exactly that when we recorded people's eye movements while they tried to recall the spoken content of a talking face that told a short fact in one of four corners of a computer screen. People saw a talking face tell a short fact in each of the four corners of the screen, and then those four squares on the screen went blank. When a disembodied voice asked about one of those facts, people frequently moved their eyes to the now blank square on the screen that, several seconds ago, had contained the right talking face for that fact. This tendency to move the eyes to the correct location of a now gone bit of information while trying to recall it has been replicated in numerous versions of Richardson's original experiments. Even though people can easily see in their peripheral vision that the information they are trying to recall is no longer present in that location, they still move their eyes to that location anyway.

Have you ever walked to another room in your home with the intent of retrieving an object but when you got there you had forgotten what it was you wanted to get? Yeah, me too. When you find yourself standing dumbly in that room not knowing why you came there, what do you do? That's right, you walk back to the room you started from to see if that can help reactivate your memory. As you look around that first room, you might notice an object that was somehow associated with what triggered your plan to go to the other room. Thus, despite your brain-and-body not remembering the plan, your brain-body-environment has successfully remembered it. Your brain-and-body relies heavily on accessing the content of objects and locations in the environment in order to do its everyday thinking, so much so that sometimes when that brain-and-body has moved to another room that doesn't have those objects and locations to point to, it's almost as if it has lost part of its mind! Perhaps it has. Perhaps the continuous, fluid flow of information between brain, body, and environment—producing a heavy interdependence among one another—makes it impossible to draw a crisp boundary between any one of them. Perhaps you should think of your brain, your body, and your immediate environment as the combination of physical material that constitutes your mind. Is that who you are?

Open Minds and Closed Systems

Psychologist Timo Järvilehto has proposed treating "the organism-environment system" as the appropriate target of cognitive analysis instead of treating the organism itself (such as a human being) as the topic of interest for scientists of the mind. Indeed, if what makes up a mind is the combination of a brain, a body, and an environment, then it would make sense to study that whole triad together rather than extracting any particular part of it and trying to analyze it in isolation. It is well established that the brain-and-body is a dynamical system, a system that is best characterized by how it changes over time, not by any particular steady state that it might briefly exhibit. Therefore, the brain-and-body can, in principle, be studied with dynamical systems theory. However, most dynamical systems mathematical tools have been developed for "closed systems," systems in which all the factors and parameters have been accounted for. Nothing from the outside influences a closed system. Its behavior can be studied, measured, and simulated, in a vacuum. This complicates matters when you acknowledge the obvious fact that the brain-and-body is not actually a closed system. As evidenced throughout this chapter, the environment has a massive influence on the brain-and-body, and vice versa.

Computational philosopher Jeff Yoshimi and mathematical biologist Scott Hotton are quick to point out that, even though the human brain is obviously an "open system," the vast majority of scientific advances using dynamical systems theory have been made with "closed systems." That is, when scientists build a computational simulation of a complex dynamical system, they usually make it closed. They do not allow uncontrolled and unmeasured input signals to enter the system from outside it. It's like a fish tank, with the temperature, humidity, and food delivery all perfectly controlled. The scientists determine the equations that the system will use and give it some starting values, and the equations use those starting values to make new values. Then those new values go into the same equations again to make yet more new values for the next time step. The system feeds back on itself this way for hundreds or even thousands of time steps to produce complex behaviors, and the scientists statistically analyze those behaviors. A great deal of advancement in our understanding of nonlinear dynamical systems has resulted from the study of such closed systems.

For example, as nonlinear dynamical systems theory evolved into complexity theory over the course of the previous century, scientists figured out that unpredictability (to the extent of appearing random) can emerge from nonlinear feedback loops that have absolutely no randomness in them. Take, for example, the logistic map equation. It is a very simple equation that calculates a value for y on the left side of the equals sign and then, for the next time step, you plug that new value into the y's on the right side of the equals sign. Rinse and repeat. As you record each newly calculated value of y, you find unusual patterns over time. In the equation $y=2y(1-y)$, you can start with a y value anywhere between 0 and 1 and the y value will always gradually work its way to 0.5 and stay there. Once $y=0.5$, calculating $y=2y(1-y)$ just makes it equal 0.5 again. In the equation $y=3.2y(1-y)$, you can start with a y value anywhere between 0 and 1 and the y values will gradually work their way to alternating between 0.7995 and 0.5130, in perpetuity. Finally, in the version of the logistic map equation $y = 4y(1-y)$, you will get values that are so unpredictable that it was used as a random number generator in the 1950s. When *apparent randomness* emerges from a system that you know has no actual randomness in it, it's called "chaos." The study of closed systems, such as the logistic map equation, has greatly advanced our understanding of chaos and complexity theory. But what about *open* systems, such as your brain-and-body?

By comparing the dynamics of open and closed systems, Yoshimi and Hotton have developed an abstract mathematical framework that may provide the stage on which our mathematically rigorous understanding of closed dynamical systems, such as the logistic map or a fish tank, can transition toward a mathematically rigorous understanding of open dynamical systems, such as your brain or your body. When you acknowledge that your brain is an open system, it means you're acknowledging that there are properties outside it that play a crucial role in its function. It means that even if you knew every neuron's activity state and even every neurotransmitter molecule's location, you still wouldn't be able to predict your brain's function. If your brain is an open system, as shown in chapter 4, then you have to include your body as some of the relevant parameters defining its behavior. And if that brain-and-body is still an open system, as shown by the present chapter, then you have to do what Timo Järvilehto suggests and include the environment's parameters as part of the system. Once you've included equations and parameters for the brain, the body and the environment, it is possible that you've included enough properties in this model of the mind to treat it as a closed system and finally run the simulation. Essentially, if you make the fish tank big enough, you just might have a realistic description of what the mind is.

Cognitive scientists Harald Atmanspacher and Peter beim Graben call this "contextual emergence." They have studied systems like these in intense mathematical detail and designed computational simulations of how a mental state can emerge from a neural network. They find in these mathematical treatments that being in a particular mental state requires more than just for the neurons to be producing a particular pattern of activity. The context of that simulated brain, such as the body and its environment, also need to be accounted for in order to develop a statistically reliable characterization of the mental states that the mind takes on. Hence,

From Your Body to Your Environment

based on these computational simulations, the brain by itself is not sufficient to generate a mental state. A mental state, such as a belief or a desire, only emerges between a brain and its context; that is, its body and the environment. It does not emerge in the brain alone.

So the brain is an open system. That's essentially what we learned in chapter 4. Here in chapter 5, we have learned that your brain-and-body is also an open system. In order to provide a fair description of how that system works, we have to track down some of the important outside factors that are becoming part of that open system. By doing so, you incorporate the environment into the system of interest, into that mind, into *who you are*. What you are studying has become an "organism-environment system."

The Organism-Environment System

Based on the many different experimental and computational findings described in this chapter, it seems clear that your environment does a lot of your thinking for you. Be thankful. This includes everything from the food you eat, to the eye movements you make, to the tools you hold, to the smart devices you use, to the rearrangements of your environment that you carry out, to the impressive degree to which we rely on the environment to remember information for us. Your environment is intimately responsible for making you be you. When you move from one environment to another, your mind is a little different. *Who you are* is a little different. The scientific evidence for the role that the environment plays in the mind is overwhelming. Instead of building an internal mental representation of the environment and performing neural cognitive operations on that, we often perform our cognitive operations out in the world, using our eyes, our hands, and objects in the environment. Much of our cognition takes place outside our bodies. We think with our hands. We think with our tools and smart devices, and those tools and smart devices become part of us.

Some readers may experience a little difficulty accepting these ideas: the idea that when you straighten your work desk, it improves the way that desk does some thinking for you; the idea that when you manipulate an object out in the world, that physical manipulation performs a cognitive operation that is part of your mind; or the idea that when you hold a well-practiced tool in your hand, it literally becomes part of you. These ideas go against centuries of Western thought about what the mind can be made of.

But if you find yourself resisting these ideas, ask yourself, "Why am I resisting?" Are there real logical and scientific grounds for rejecting these ideas that outweigh the dozens of scientific findings that you just read about in this chapter, or are you rejecting these ideas just out of an intuitive discomfort with them? If your resistance to these ideas doesn't have sufficient logical and scientific evidence backing it up, then you may want to try—even just as a temporary experiment—letting go of that resistance and embrace the scientific findings in this chapter and their conclusion that important aspects of your environment are physically and literally part of *who you are*. Try it on like a hat, just as a brief test. Set an alarm for 24 hours from now, and until that alarm goes off, accept these ideas as fact. The external environment really is part of your mental content. Let yourself believe it. Look in the mirror and see how you look wearing this new hat. Don't worry, tomorrow your alarm will wake you from this temporary spell, and then you can make a fair comparison of how things feel with and without this new definition of self.

Chapter 4 showed you that the continuous, fluid flow of information back and forth between your brain and your body makes it impossible to treat your brain as the sole determinant of who you are. Whoever you really are, you are *not* just your brain. Using that same kind of logic, this chapter has shown you that the continuous, fluid flow of information back and forth between your brain-and-body and the environment makes it impossible to treat your body as the container of your mind. Your mind extends beyond your body. It is coextensive with your environment. So, if a few chapters ago you had to draw a circle around the physical material that makes up who you are, you might have drawn that circle around your frontal lobes, but a chapter later, the scientific evidence encouraged you to expand that circle to include your entire brain. In chapter 4, the scientific evidence encouraged you to further expand that circle to include your body as well. Now, this chapter has provided a wealth of scientific evidence encouraging you to expand that circle yet again, to include your immediate environment. *Who you are* isn't just your brain-and-body. It's your brain-body-environment.

Truly embracing this outlook on life can be transformative. If your environment is part of who you are, this means that when your body dies, some of you is still alive in the environment. The physical jigging of the environment that you did in your home, the mental jigging you did in your work, and the social jigging you did with your family and friends are all information-bearing jigs that you placed into the environment that will still be there after your body is gone. And as we've seen, those jigs are part of who you are.

In fact, research shows that those jigs that you've placed into your environment bear enough information for a neutral observer to be able to perceive some of your personality in them. For instance, psychologist Sam Gosling had a group of untrained observers look at the offices of people they didn't even know and try to estimate various personality traits based solely on how the office space was laid out. To get a "ground truth" of those personality traits, Gosling and his team collected responses on those personality traits from the office occupants themselves and also from their close friends and associates. When Gosling compared those "ground truth" personality traits to the personality estimates based just on viewing the offices, he found a remarkable correspondence. Not all personality traits were transparent in a person's office layout, but the neutral observers of the offices were reliably better than chance at estimating the occupant's level of extroversion, conscientiousness, and openness to experience. Those office occupants have somehow managed to exude those personality traits out into their office space so visibly that even when they are absent the office all by itself reflects those personality traits.

As a by-product of their everyday activities, living organisms leave behind what Gosling calls "behavioral residue." For example, various species of termites build giant mounds of dirt that can span 30 meters wide, with complex tunnel structures inside, and if the termite colony dies out, that unoccupied dirt mound has enough intelligence built into it that another species of termite just might take it over and use it for themselves. When many deer populate a forest, they will often shape well-traveled trails that snake their way through the woods, revealing their migration patterns. Carnivores collect carcasses of their prey in their dens, telling the tale of their hunting and feeding habits. And humans, evidently, make surprisingly transparent imprints of *who they are* onto their working and living environments. Think about the imprints that you are making on your various environments right now: work, school, home, and elsewhere. Are you leaving behind good behavioral residue or bad? Do your best to make sure that most of the behavioral residue you leave behind is good residue, because that's *who you are* when your brain-and-body isn't there anymore—and apparently everyone can see it.

Directions for Use

This exercise is intended to prepare you psychologically for the transition from this chapter to the next. Chapter 6 is still about including the environment as part of your mind. However, rather than focusing on the *inanimate objects* in your environment, as chapter 5 did, it will focus on the *people* in your environment: your friends, your family, your co-workers. If they are part of your environment, and you are busy including your environment as part of who you are, are *they* part of who you are? And are *you* part of who they are?

I want you to explore this question here in a wordless fashion by performing a little experiment that will take this question and seek an answer that doesn't use language. Find someone close to you. This can be a family member, a good friend, a lover, or a spouse who's all three of those. Stare into that person's eyes while they stare into yours. Get up close, almost nose to nose, for five minutes straight—no talking, touching, or giggling. Give yourself over to them wordlessly, touchlessly, and let them give themselves over to you the same way. In his book *Eyes Wide Open*, Will Johnson calls this "eye gazing." While you are staring into this other person's eyes, silently ask yourself—ask the universe—whether you are temporarily and partially becoming *one system* with them. But don't listen for a "yes" or a "no." The answer won't come that way. Just listen, without any preconceived notions of how you might categorize the answer. When that nonlinguistic answer comes to you, you will know it. And then your five-minute timer will go *ding*.

6 From Your Environment to Other Humans

Who do you want to be today?
Who do you want to be?
—Oingo Boingo, "Who Do You Want to Be"

In the 1970s, an American candy company began an advertising campaign for peanut butter cups. In the television commercial, one person is walking down a busy city sidewalk while eating a chocolate bar, and another person is walking toward them while eating peanut butter from a jar. When they bump into each other, their foods collide. "You got your peanut butter on my chocolate!," one of them decries, while the other one complains, "You got your chocolate in my peanut butter!" —as if somehow one of those things was right and the other was wrong. As it turns out, they both find the flavor combination pretty darn tasty. Something very similar to that candy collision happens when your own action-perception cycle collides with someone else's. When some object in the environment becomes part of your cognition and is also part of someone else's, the two cognitions get combined. This is *who we are*. For instance, both of you might be looking at a map, pointing at it, and talking about it together. Or you might be watching a television show together and sharing observations about it. Or maybe you can recall the first name of a Hollywood celebrity and your spouse remembers the last name for you. Does that object (or event) out there in the environment belong to *your* cognition or to the *other person's*? As if somehow one of those things was right and the other was wrong. No one decries, "You got *your* cognition in *my* cognition!" That's because it is a natural phenomenon that we all engage in all the time—and it's usually pretty darn tasty.

Now that you've made it through chapter 5, you should be ready for chapter 6. You could almost write this opening paragraph yourself by now, couldn't you? If chapter 5 did to you what it was supposed to do, then you've embraced the idea that your organism-environment system is "who you are." The food you eat, the objects you look at, the tools you use, and the modifications you make to your environment are all part of your cognition, part of your self. But is that all there is to who you are? What about those "objects" in your environment that have brains of their own: the *people* in your environment? Is that where you draw the line? Or is there so much physical overlap and such a continuous, fluid flow of information back and forth, between *your* organism-environment system and *their* organism-environment system that you might as well include them—at least a little bit—in your definition of *who you are*? And can they include you—at least a little bit—in their definition of who they are? Thanks to chapter 5, you're already including inanimate objects in the environment as part of your selfhood. Do you really have a logical excuse to categorically exclude those objects that just happen to have hearts and brains inside them?

In this chapter, I'm going to show you a long list of scientific findings indicating that the information-bearing functions of your organism-environment system are so deeply intertwined with the information-bearing functions of other people in your environment that you really ought to expand your definition of who you are to include those people as well. The *fluid back-and-forth sharing of information* between you and other people is often so dense that it can be difficult to safely treat you and them as separate systems. The information that makes you *who you are* includes not just the information carried inside your brain, body, and nearby objects but also the information carried by the people around you.

You Are What You Eat

People always say, "You are what you eat." Usually they mean something like if you eat bad food, you'll have bad health, or if you eat junk food, you'll be a junk person. But there's another way to take that aphorism. Not only is it true that the food you eat will cause you to have some of that food's characteristics, but it's also true that the food you eat is something of an indicator of the cultural group to which you belong. The food you eat doesn't just *cause* some of who you are; it also *signals to others* something about who you are. People from every culture are often wary of trying "foreign food" because it looks and smells unfamiliar. Preferences in the tastes of food are actually some of the most deep-seated prejudices one can encounter. Some people won't even try a small taste of a different culture's signature dish if it looks unusual to them—apparently forgetting the fact that millions of humans just like them enjoy that dish on a daily basis. Things that you're unfamiliar with often appear a bit dangerous and threatening, and sometimes these things are "an acquired taste." There are several interesting foods that I didn't like when I first tried them, but I was brave enough to try them a second time, in some cases years later, and was frequently surprised to find that I liked the food on the second try.

In the Middle East, there are millions of people who have never tasted peanut butter. Many of them would find a peanut butter cup disgusting on their first try. In the United States, there are millions of people who have never tasted vegemite or marmite. Many of them would find their first vegemite sandwich disgusting. Moreover, there are a wide variety of cultural and religious practices that forbid people in certain groups from eating particular foods at all. Vegans don't eat meat or dairy foods, Buddhists don't eat meat, Muslims and Jews don't eat pork, and Catholics don't eat warm-blooded meat on Fridays. What people eat and what they don't eat often reveals to others a little bit about what kind of person they are.

For example, Americans who eat almost exclusively "diner food" probably belong roughly to an approximate category of people, just as people in China who eat exclusively Chinese food are similarly pigeonholing themselves into a narrowed experience of the world—and I don't mean just "the world of cuisine." Trying foods from foreign cultures can be scary, but also very rewarding. It can expand your horizons. Just as a racially prejudiced person can begin to soften their position when they find themselves accidentally enjoying the company of a few friendly people from the group they've been prejudiced about, a person with cultural prejudices about foreign food can likewise begin to soften their stance about foreign cultures when they find themselves accidentally enjoying the taste of some unusual cuisine. It's not so scary anymore once you discover that it is actually pretty darn tasty.

When I was ten years old, I remember seeing the Greek food carts at the Santa Cruz Beach Boardwalk, with their huge rotating cylinder of gyro meat, from which the cooks would carve off strips and place them in pita bread with tomatoes, onions, and tzatziki sauce. That exotic tree trunk of beef and lamb sausage, mixed with savory spices, smelled intoxicating to me. It looked like "dinosaur meat" to my young brain and reminded me of the awesome Cave Train Adventure ride at the Boardwalk, where you travel back in time and see animatronic dinosaurs. But when I asked my parents if I could try a gyro, they told me I wouldn't like it. I went for years seeing Mediterranean food as a strange foreign cuisine that I shouldn't expect to like—until I was 15. One summer day, I took my lunch break from the ice cream shop where I worked and walked to a hidden corner of the mall, where I knew a Mediterranean restaurant was tucked away. I felt as if I was doing something illicit. It felt like a clandestine mission to try this foreign food without permission. I didn't have a lot of money, so if my parents were right and I didn't like it, I would have wasted a precious $5. It was scary.

But once I bit into that gyro, I found myself enjoying it. It was pretty darn tasty. I was suddenly in love with gyros. I went back to that dingy little gyro shop for lunch at least once a week all summer long. More than three decades have passed since that summer, and massive industrial globalization is now allowing many small towns and many different countries to experience a much wider variety of cuisine options than ever before. With this enjoyment of "foreign foods," many of the citizens of these countries are opening not just their stomachs but also their hearts to those "foreign cultures." Most metropolitan cities do an impressive job of naturally attracting diverse multicultural cuisines to their restaurant industries. For example, some of the best Indian curry that I've ever tasted was in restaurants in London, England. Some of the best Thai food that I've ever tasted was in restaurants in Munich, Germany. Some of the best Ethiopian food I've ever tasted was in restaurants in Washington, D.C. (Full disclosure: This may in part result from the fact that I've never been to India, Thailand, or Ethiopia.)

You know who *did* try the cuisine in India, Thailand, and Ethiopia—and many other countries as well? TV host and cookbook author Anthony Bourdain. What Bourdain was doing with his books and TV show was remarkable. Everybody loves food, so Bourdain found a way to trick viewers who might not normally be very interested in cultures on other continents into finding themselves accidentally enjoying the introduction he provides via an exploration of their food. Sometimes shocking, sometimes delectable, and sometimes a curious mixture of both, the cuisines that Bourdain sampled for us provide an intriguing window into the rest of that entire culture. My favorite quotation from him is: "Walk in someone else's shoes, or at least eat their food." When we see these people enjoying food that is *unfamiliar* to us but doing so in a manner that is very similar to the way we enjoy the foods that we already love, we can both embrace the commonalities among our cultures and rejoice in the interesting differences. Watching Bourdain enjoy strange food in other countries and have enlightening conversations with people brings it home to you—whoever you are—that it's a big world out there, and it's not all that scary. In fact, much of it is pretty darn tasty.

Who Your Family and Friends Are

If it is true that you are what you eat, then expanding your cuisine horizons to include a bigger variety of dishes will make you a bigger person—perhaps both figuratively and literally. Of course, it is not just the different foods that you are getting familiar with when you try out other cultures but also the different people. Many of our most influential experiences—meals, entertainment, community events—are shared with, and partly shaped by, other people. Whether you're having a conversation with someone, attending a lecture, coordinating on a project at work, watching TV, or just reading a book like this one, it usually involves information that is being shared among two or more people. Clearly, it isn't just food that gets inside you to make you into who you are; it's also the actions and words of other people that get inside you. There's an old Spanish proverb that goes, "Dime con quien andas, y te dire quien eres": Tell me with whom you walk, and I'll tell you who you are.

A growing body of cognitive science research is showing that when two people cooperate on a *shared* task, their individual actions get coordinated in a way that is remarkably similar to how one person's limbs get coordinated when he or she performs a *solitary* task. For example, take the solitary task of waggling your two index fingers up and down in opposite sequence. When your right index finger is lifted up, your left index finger is pointing down, and then they trade places, again and again. Try it out. It's easy. Instead of perfect synchrony, where the fingers would be doing the same thing at the same time, this is a form of syncopation, where the fingers are doing opposite things at the same time. Dynamical systems researcher Scott Kelso discovered that people can do this relatively quickly and stay in antiphase (each finger moving opposite to the other) for a while. However, when they speed up to a really fast finger-waggling pace, they tend to accidentally fall into an in-phase pattern. The fingers unintentionally slip into a pattern where they are both moving up at the same time and then both moving down at the same time. Kelso showed that this accidental tendency toward coordinated finger movements, even when you're trying to produce anticoordinated finger movements, fits into a mathematical model that describes how two different subsystems can often behave like one system. In this case, one of those subsystems is your left hemisphere's motor cortex trying to do one thing (raise your right finger) and the other subsystem is your right hemisphere's motor cortex trying to do the opposite (lower your left finger) at the same time. Then they reverse that pattern, again and again. Those two brain regions have several neural connections between them, and those connections make it hard to maintain this anticoordinated movement pattern when it ramps up to high speed. The *fluid back-and-forth sharing of information* between these two brain regions causes them to slip into synchrony. Instead of functioning independently of each other, these two different subnetworks start behaving like one network, doing the same thing at the same time, such as lifting both index fingers at the same time and then lowering them at the same time. They get in sync, even when you're trying to have them stay out of sync.

When two different systems, such as your left motor cortex and your right motor cortex, get tightly coordinated like that, it makes sense to treat them as two *subsystems* that can be analyzed together as *one system*. Rather than one neuromuscular system for your left hand and a different one for your right hand, it temporarily becomes one bimanual neuromuscular system that is making both hands coordinate with each other. Kelso and his colleagues developed a simple scale-free mathematical model (the Haken-Kelso-Bunz model) that describes how *any two subsystems* that are interacting can maintain antiphase syncopation under certain circumstances but will tend to slip into in-phase synchrony under slightly different circumstances. When one subsystem's activity is being continuously transmitted to the other subsystem, and vice versa, they have a tendency to become so interdependent that they are better described as one system, not two. This continuous transmission between two subsystems could be conducted via the corpus callosum, which connects those left and right motor cortices to each other; by linguistic transfer between two people talking to each other; or by visuomotor signals between two people dancing together. Almost any medium of transmission will do.

For example, psychologist Richard Schmidt extended Kelso's experiment to take place with one person's left leg swinging from the edge of a table that they're sitting on and another person's right leg swinging from the same table. Just like Kelso's two index fingers, the two separate people are supposed to swing their legs in antiphase with one another. One of them swings her leg forward (away from the table) while the other swings his leg backward (under the table). Then they reverse, again and again. And just like Kelso's two index fingers, as these two people speed up their antiphase leg movements, they tend to accidentally slip into an in-phase pattern. They wind up swinging both legs forward at the same time and then both legs backward at the same time. Clearly, this is not the result of a corpus callosum connecting one person's motor cortex and the other person's motor cortex. Evidently, neural connections are not the only thing that can provide the fluid back-and-forth sharing of information that synchronizes two different subsystems. Your own action-perception cycle can share information back and forth with someone else's action-perception cycle, in a manner that makes the two of you synchronize and behave a little bit like *one system*.

In fact, *anything* that moves to-and-fro between two extremes can show a tendency to fall into synchrony with a similar nearby thing that is also moving between those two extremes—as long as there is a properly conductive medium of transmission between them. Over 300 years ago, Dutch mathematician Christiaan Huygens invented the pendulum clock and quickly discovered that when two of them are resting on the same wooden beam, their pendulums have a tendency to gradually fall into synchrony with each other. In his book *Sync*, mathematician Steven Strogatz details this tendency toward rhythmic synchrony in everything from electrons, to brain cells, to rivers, to our sun.

Clearly, people do it, too. Not only will their legs fall into in-phase synchrony with each other when they're trying to swing them in antiphase, but their entire bodies and brains also will tend to exhibit similar patterns of activity when they're just having a simple everyday conversation with each other. Psycholinguists Martin Pickering and Simon Garrod have suggested that essentially the same kind of spreading of information that goes on *inside* one speaker's brain (e.g., to make them think of "red," then "rose," and then "petals") is also going on *between* two speakers in a conversation. This is part of why we can often finish the other person's sentence for them when they pause, or we know exactly what they *meant* to say even though they misspoke or trailed off.

A dramatic example of this spreading of information across two brains comes from an ingenious experiment conducted by Anna Kuhlen. She recorded brain activity (with EEG) of a woman telling a story and made a video of the woman at the same time. She also did this with a man telling a different story. Then she superimposed the two videos on each other, making a kind of ghostly image of both talking heads and an overlay of the audio tracks from both stories. When you watch and listen to this video, it can be slightly distracting, but if you decide to pay attention to one of the speakers and ignore the other, you can do it. That's exactly what Kuhlen asked her next set of experiment participants to do, *while she recorded their brain activity*. Thus, she had several minutes of brain activity from the female speaker, from the male speaker, from listeners who were told to focus on the female speaker, and from listeners who were told to focus on the male speaker. As predicted by a spreading-of-information type of account, listeners who focused on the female speaker produced brain activity that was statistically correlated with the brain activity of the female speaker (and less correlated with the brain activity of the male speaker), and vice versa for listeners who focused on the male speaker. The strongest correlations had a time lag of a dozen seconds. Think of it this way: the speaker's brain would produce activity that corresponds to thinking about a particular element of the story, then they would turn that thought into a spoken sentence stretching over the course of a dozen seconds, and then the listener's brain would turn what it heard into neural activity that was remarkably similar to the brain activity that the speaker produced 12 seconds ago. Essentially, listening to someone tell you a story makes your brain do some of the same kinds of things that the speaker's brain was doing. Since Kuhlen's findings, this "two-brain approach" in neuroimaging has been growing rapidly.

But you don't have to be recording brain activity to see these correlations between speaker and listener emerge. If two brains are highly correlated, then it's a safe bet that the movements carried out by the bodies attached to those brains are also going to be correlated. My students Daniel Richardson and Rick Dale recorded a speaker's eye movements while she looked at a grid of six pictures of cast members from the TV show *The Simpsons* and told an unscripted story about her favorite episode. Just the audio recording of that story was later played over headphones for listeners who were also looking at that grid of six pictures and having their own eye movements recorded. They couldn't see the speaker, but they could hear her voice. Over the course of the few minutes of story, there was a remarkable correlation between the eye movements of the speaker and the eye movements of the listeners, with a lag of about 1.5 seconds. Think of it this way: The speaker would look at a cast member's face and a half-second later say that character's name. Then, the listener's brain would take a half second to process hearing the name and then another half second to direct the eyes to that cast member's face. There's your 1.5-second lag. Interestingly, listeners whose eye movements had a higher correlation with those of the speaker also answered comprehension questions faster at the end of the experiment. Listeners whose eye movements were more synchronized with the speaker's eye movements appeared to have absorbed the information from the story in a more efficient manner.

All this means that, when processing language, the listener (or reader?) will tend to exhibit some brain activity and eye movements that are similar to the brain activity and eye movements produced by the speaker (or writer?). That's right; while you are reading this paragraph, your brain activity and eye movements are probably somewhat correlated with the brain activity and eye movements that I produced while proofreading this paragraph. No matter who you are, no matter when you read this, a time slice of you and a time slice of me are briefly "on the same page," both literally and figuratively, both physically and metaphysically. (And if Oprah Winfrey were to read this paragraph, then your brain would get briefly correlated with *her* brain!)

This physical and mental coordination is even more impressive when the two people are simultaneously present with each other and contributing equally to the conversation. With the help of Natasha Kirkham, Richardson and Dale expanded their eye-tracking experiment to work with two people in an unscripted dialogue. They put two eye-tracking machines in two separate rooms and had two people view the same painting while talking with each other over headsets. Imagine being in this experiment. You're alone in a room looking at a picture, but you get to talk to someone else on your headset who is in another room and looking at the same picture. It's a bit like calling up a friend on the phone and watching a live sports game on TV together, like my buddy Steve and I do. While their eye movements were being recorded, these experiment participants talked about the painting that they were both looking at. When you analyze the transcript of the conversation, you see numerous interruptions and completions of each other's sentences, because an important part of how we understand what a person is saying is by anticipating what's coming next out of their mouth. Importantly, since the two people in the conversation were anticipating each other's words and thoughts so closely, that 1.5-second lag in the correlation between eye movements disappeared. People were looking at the *same parts* of the *same image* at the *same time*. Their eye movements had slipped into genuine synchrony, not unlike Kelso's waggling fingers and Schmidt's swinging legs.

When two brains get correlated by a shared conversation, it's not just the brains and eyes that produce coordinated behavior. It's the hands, the heads, and the spines, too. The two bodies get coordinated in ways in which they weren't coordinated before the conversation began. For example, human movement researcher Kevin Shockley had two people talk to each other about a shared puzzle while they stood on force plates that record their postural sway. Even when you think you're standing still, your body's center of mass still sways back and forth by a few millimeters. And when Shockley's participants were talking with each other about this puzzle, their postural sways became correlated with one another. When a pair of participants talked with *other* people about the puzzle, their postural sways were no longer correlated with each other. Some of that coordination of postural sway may have resulted from hand gestures they made together or head nods they made to each other in the back and forth of the conversation.

When cognitive scientist Max Louwerse focused on the gestures and nods that people produce during a conversation about a shared puzzle, he found that specific kinds of gestures showed correlations at different lag times. For instance, coordinated head nods happen about one second apart from one another. Essentially, one person proposes something and then nods her head as a form of requesting confirmation from the other person, and then the other person nods back to give that confirmation. Social mimicry, however, happens on a different timescale. People often mimic one another without realizing it, like crossing your legs in a meeting after someone else crossed their legs, interlacing your fingers and resting them on the back of your head after someone else did, or rubbing your chin after someone else did. Louwerse found in his experiment that when one speaker touches their face, the other speaker is also statistically likely to touch their face—not one second later like the head nods but more like 20–30 seconds later. This accidental motor coordination tends to induce more empathy and rapport between the people involved. The shared conversation not only improves the processing of shared motor *output* but can also improve the processing of shared perceptual *input*. How do two people "share perceptual input," you ask? Good question. Cognitive scientist Riccardo Fusaroli examined the conversation transcripts of two people performing a shared perception task with about a hundred experimental trials. On each trial, each person carried out a difficult visual discrimination task and then discussed their *individual* responses to reach a consensus for a *joint* response to the visual stimulus. Occasionally, their individual responses differed, and the joint response would require that one of them change their answer. Imagine doing this experiment with a close friend. You see a brief flash of several visual stimuli and have to decide which stimulus was the odd one out. You make your guess by pressing a button, and then you find out that your friend made a different response. Now the two of you have to decide together which of you was right. Fusaroli found that some of the conversation transcripts didn't show much correlation in their use of expressions of certainty (for example, one person might use phrases like "confident," "slightly unsure," or "completely uncertain," while the other used phrases like "I dunno" and "maybe"). In those pairs of people, the joint responses tended not to provide any improvement compared to the responses of the best individual responder in the pair. For example, sometimes a very accurate responder can give in too often to a less accurate responder, and their joint responses can be *worse* than the very accurate responder's individual performance. However, Fusaroli found that when a pair of participants showed strong coordination between them in their expressions of certainty, using the same kinds of phrases, their joint responses outperformed that of their best individual responder. Essentially, the pair is developing their own miniature language that allows them to calibrate each other's level of confidence in their perception of a particular visual stimulus. This linguistic coordination made their two brains act like one system, with better performance than either brain alone. I guess two heads really are better than one—particularly when they are "on the same page" about how to talk about what they are seeing.

A bit like the mathematics that we saw with the synchrony studies earlier in this chapter, these correlations between two people mean that if you knew what kinds of gestures, brainwaves, eye movements, and word choices one speaker in a conversation was making, then you could make moderately good predictions about what kinds of gestures, brainwaves, eye movements, and word choices the other speaker would be making. This is because that collection of brainwaves and behaviors (emanating from two people) is becoming *one system*. A good conversation is not really a collection of individuals taking turns sending complete messages to each other. They're not sequentially adding messages to the dialogue. That's a bad conversation. A good conversation involves people *simultaneously cocreating a shared monologue*. When you get together with your friends and family, a good conversation is essentially a way for a group of brains and bodies to get synchronized with each other.

Who Your Co-workers Are

Of course, there's more to coordinating with people than simply doing the same things around the same time. Sometimes coordination involves doing the *opposite* thing at the same time. Think about lifting a large table to move it from one room to another. Usually one person stands at one end of the table and the other person stands at the other end, and they're facing each other. They synchronize the lifting of the table, but then one person walks *backward* while the other person walks *forward*. Their legs are doing the opposite thing at the same time in order to coordinate on this shared task. If both tried to walk forward, no progress would be made. When coordinating with someone at work, you're sometimes *complementing* rather than *mimicking* the other person's actions.

Importantly, in order to accurately *complement* another person's actions, you first need to *anticipate* that other person's actions. You have to be mentally simulating their upcoming actions and how those actions will go along with your upcoming actions. Social neuroscientists Natalie Sebanz and Guenther Knoblich were able to show this complementarity that emerges between two minds while two people carried out a shared task on a computer. Their reaction times and their brain activity revealed that even though person 1 was only in charge of response A (with its own button press), they were nonetheless mentally taking into account response B, which was entirely person 2's responsibility. That's how teamwork works. You can't just do your part of the job and nothing else. In order to time it right, you have to anticipate how your part fits in with what other people are doing. Recall how chapter 3 showed us that different parts of the brain can't function like independent modules and get the job done right. Instead, they have to engage in a back-and-forth sharing of information so that the visual system can help the language system solve its puzzles and vice versa. Thus, those different subsystems begin to function as one system. By the same token, two or more people working together on a shared task have to engage in a back-and-forth sharing of information that forces each of them to understand a little bit about what the others are doing. They become, to some degree, one system.

For example, in a real-world scenario, cognitive anthropologist Edwin Hutchins studied how the crew of a ship worked together, and he determined that each crewmember served a unique cognitive function for the whole. It was almost as if each crewmember was a brain region, and this conglomerate brain was what planned, navigated, and steered the actions of the ship. By complementing each other's actions and serving as cognitive mechanisms for each other, the crew of a ship is able to make smart decisions and act on them in a timely manner, almost like a single mind. When a group interacts toward a shared goal, that *group* can be studied with many of the same cognitive analysis tools with which an *individual* human mind is studied.

Cognitive scientist Paul Maglio has applied these principles to analyzing the workplace and improving its efficiency. By identifying the different roles played by people in a business office, diagramming the duties and relationships of those roles, and closely following the time course of their interactions, Maglio and his colleagues at IBM helped define the field of service science. As more and more of what you pay for in the marketplace these days is a *service*, rather than a *good* that you can hold in your hand, the science of optimizing the development and delivery of that service is a crucial field of discovery and design. Service science research does for the services that a business provides what traditional operations management research does for the goods that a business provides. However, it does it in a different way. Service science looks at a business as a complex system made of people and technological devices that interact with one another to provide value to customers. Using tools inspired by complex systems theory, service science is able to develop analyses and simulations of that business to discover where missteps arise and how to fix them.

For instance, Rob Barrett, Paul Maglio, and others examined in detail the steps that led to an IT systems administrator, we'll call him George, failing to solve a particular network configuration problem. Over the phone and via text messaging, George described the problem to his supervisors and co-workers, and they tried to help. But George had misinterpreted a particular network port as intended to send data in one direction when it was in fact intended to send data in the opposite direction. The firewall prevented that direction of data flow, but none of George's helpers could see the computer screen he was looking at, so they didn't realize the misunderstanding under which he was operating. The problem resisted being solved for quite some time, despite having numerous minds working on it, because every time George explained the problem he gave a slightly inaccurate description. Finally, one of his co-workers brought up the same screen display that George had been looking at (and misinterpreting) and realized the error. Sometimes, with certain arrangements of people and technologies, multiple heads aren't better than one. Identifying those imperfect arrangements and reconfiguring them into better arrangements is what service science aims to do.

By building complex simulations of the real thing, these kinds of business management approaches can test out different arrangements of those people without risking any damage to the service, the product, or the business office itself. A particularly illustrative example of a simulation like this comes from a scene in the movie *The Founder*, which details the early development of the McDonald's fast food restaurant. One of the original McDonald brothers (played by Nick Offerman) conducts multiple simulations of different kitchen arrangements on a tennis court. He draws different food stations on the court in chalk, with paid employees miming their different jobs. This allows him to identify when things are inefficient and redraw the kitchen in chalk in a new configuration. It takes many attempts at different chalk arrangements in this simulation of a fast food kitchen, but eventually he finds a configuration that works smoothly and quickly. Thus, for better or worse, fast food was born.

In studying who your co-workers are and how you collaborate with them, cognitive anthropology and service science draw on findings from cognitive science and complex systems theory to treat the entire workplace as one big cognitive system. You and your co-workers, and the technologies around you, are the puzzle pieces that make that puzzle fit together, that make that hive mind work successfully. By cooperating on the projects of that office or factory, you and the other workers create a service or product that has value, and that value is what keeps the company in business. Recent insights from service science point to a process of *adaptive comanagement* that allows business innovation to emerge from the convergence of knowledge from diverse backgrounds. Rather than relying solely on a vertical hierarchy, where a worker is managed by a boss who is managed by another boss, adaptive comanagement encourages the integration of multiple minds for decision-making and maintains an openness to self-assessment procedures. The application of these insights from cognitive science in the real world is paving the way for genuine improvements in business management practices.

Who Your Social Group Is

Perhaps insights from cognitive science can help pave the way for genuine improvements in society as well. Accurately understanding your relationship with the rest of society is crucial for knowing who you are. Clearly, it's not just your long-gone ancestors who make up part of who you are. It's the living people around you, too. *Who you are* overlaps with your friends, your siblings, your co-workers, and even people you barely know!

All humans have a tendency to join groups of people with whom they feel they have something in common. Right now, you are a member of several groups. One group is your family, with whom you have some genetics, an upbringing, and a love bond in common. Another group is your friends, with whom you have some common interests, which might include hobbies, a sport, a religious affiliation, or neighborhood relations. Yet another group to which you probably belong is that of your co-workers or fellow students. When you are embracing your belongingness to a particular group, it is called your "in-group." If it turns out that two groups to which you belong are in conflict with one another, you may find yourself in the position of having to choose which one is your "in-group" and which one is your "out-group" for that moment. It can be a very difficult choice, and once your in-group memberships are determined, the consequences can be rather powerful.

People often derive a great deal of emotional satisfaction from their membership in a group. We often feel as if the group to which we belong "has our back." You don't betray your fellow members in a group. We get a sense of belonging to something larger than ourselves in a group, and a group can often become more than the sum of its parts. The group can perform better on a task than the sum of its members' individual performances on that same task. Social psychologist Roy Baumeister finds that most people have a "need to belong" to some kind of group. Moreover, he suggests that a key component that allows a group to outperform its individuals is that each member develops a differentiated role in the group. When too many people in the group are too enmeshed with the group identity and not individuated, the group can find itself underperforming or malfunctioning. Groupthink can take over and lead the group to endorse a plan that the majority of its own individuals do not actually prefer. By contrast, when each individual simultaneously embraces their membership in the group and also finds a recognized and valued role to perform in the group's mission, then the evidence suggests that the group will indeed produce its optimal performance.

Group membership has its benefits. In fact, psychologist Arie Kruglanski and his colleagues have used questionnaires to conduct several experiments that show membership in a group can reduce your anxiety about your own inevitable death. Belonging to something that is larger and longer lasting than your own body-and-brain seems to confer a sense of significance that nearly everyone craves, and it slightly eases the fear surrounding the conclusion of one's own individual physical existence.

However, there is a dark side to group membership as well. Kruglanski's research also shows that this reduction in death anxiety makes it easier for a person to commit to being a martyr for the group's cause. If the group's overarching cause is the eradication of the out-group, then you suddenly have a version of terrorism on your hands that has become particularly insidious throughout the world. For all its benefits, forming an in-group carries with it some serious drawbacks as well. Whenever we define our in-group, it almost unavoidably requires us also to define an out-group—at least implicitly—and we members of groups have a very bad habit of concocting exaggerated criticisms of members of the out-group. Members of in-groups often generate prejudiced negative stereotypes of the personality characteristics of members of the out-group, and sometimes we're not even aware that we have these negative stereotypes. Social psychologists Mahzarin Banaji and Anthony Greenwald have studied this kind of implicit prejudice for decades. With evidence from questionnaires, reaction-time experiments, neuroimaging studies, and more, they point to the cognitive "blindspot" that we have for our own social biases. Even when we think we are aware of social injustices and make efforts to rectify them, we often slip into accidentally biased perspectives on in-group/out-group issues. The groups could be gender groups, race groups, ethnic groups, religious groups, socioeconomic status groups, age groups, language groups, or sexual-orientation groups. There is always going to be a group you don't find yourself easily belonging to, and when you think about that group, talk about that group, or make decisions that affect that group, you are likely to have some implicit biases and insensitivities that you don't even know about. Banaji and her colleagues have even discovered different brain activity patterns when you are evaluating someone in your in-group compared to when you are evaluating someone you perceive as part of your out-group. Think about this for a minute: when you choose not to imagine yourself in someone else's shoes, because you think of them as belonging to an out-group, your brain functions differently when you are judging them. If that sounds unfair to you, then I have a solution for you: don't think of other humans as part of your out-group. Put them all in your in-group, and put yourself in their shoes. Perhaps eat some of their food too.

Also from this general research framework came the implicit association test, which reveals these negative stereotypes even in people who explicitly report having no prejudices about the out-group. In one version of this implicit association test, Russell Fazio had participants view an image of a face briefly flashed on the computer screen for a third of a second and then followed it with a positive written word (such as "wonderful") or a negative written word (such as "disgusting"). The task was simply to press one button if the adjective was positive and another button if it was negative. Among his Caucasian participants, Fazio found that when the briefly presented face was that of an African American, positive words induced longer reaction times and negative words induced faster reaction times. (Among his African American participants, this pattern was reversed. However, in some studies, disenfranchised groups exhibit the same bias against their own group that the majority group exhibits.) The way these kinds of findings are often interpreted is that even though these participants reported having no bias against other races, their reaction times reveal a mental readiness (at the millisecond timescale) to associate negative words with an out-group face. Even when flashed for just a third of a second, the out-group face seems to prime (or spread some activation to) the idea of negativity, such that negative words are recognized a little faster. Social psychologists Melissa Ferguson and Michael Wojnowicz extended these findings to a computer-mouse task, where Caucasian participants were simply asked to click a LIKE button or a DISLIKE button when they saw the words "cancer," "ice cream," "white people," or "black people." For the word "cancer," people made straight computer-mouse movements to the DISLIKE button. For the word "ice cream," people made straight computer-mouse movements to the LIKE button. And even though practically everyone clicked the LIKE button for the phrase "black people," the average computer-mouse trajectory curved a little bit toward the DISLIKE button—more so than with the phrase "white people." This result suggests that even when people are explicitly reporting their friendly attitude toward a certain out-group, the way that they do it has—hidden in the dynamics of the action itself—subtle evidence of implicit prejudice.

Why do these implicit prejudices exist in the first place? Oftentimes, the root excuse for a prejudice is the idea that the out-group has somehow injured or taken advantage of the in-group. At the end of the fictional film *Imperium*, in which the FBI infiltrates a white supremacist group, Toni Collette's character has a memorable quotation: "When it comes down to it, there really is only one essential ingredient to fascism: victimhood." We often think of fascists as the victim*izers*, but whenever fascism has taken hold of a group, it started with a sense of victim*hood* among the soon-to-be-fascists. When people feel they have been unfairly victimized, they are often ready to lash out irrationally. In his book *How Fascism Works*, philosopher Jason Stanley itemizes the sociopolitical recipe that goes into the burgeoning of a fascist movement, a key ingredient of which is an "us and them" mentality. Some politicians can finagle their rise to power by stoking the embers of fear and resentment in the majority culture of the minority cultures in that nation and turning them into a veritable bonfire. For instance, in the United States, certain groups of racist white people feel so victimized by the government's advances in racial equality that they think they are warranted in exacting violence on innocent bystanders who belong to their out-group—and some American politicians have harnessed that hatred for their benefit. In parts of Europe, certain groups of extremist Muslims feel so victimized by their racially motivated exclusion from parts of society that they think they are warranted in exacting violence on innocent bystanders who belong to their out-group. Palestinians and Israelis each feel victimized by the other and use this as justification for violence and oppression. Sunnis and Shiites throughout the Middle East are each so focused on how a recent attack victimized their in-group that they seem to forget that the attack was a response to one of their own attacks on the out-group (which itself was a response to a previous attack, ad infinitum). When a group obsesses over its victimhood (real or imagined), it becomes willing to relinquish many of its democratic freedoms in order to exact revenge on the out-group that it assumes is responsible for it. The sacrifice of those freedoms is where fascism comes in. It is exactly what happened in a democratic Germany in the 1930s. A majority of the German people were convinced by Hitler's fake news rhetoric that the country's low employment rate was the fault of the Jewish community, so they voted him into power because he told them that he alone could fix it. A number of pundits at the time genuinely thought that, once in power, Hitler would pivot and soften his rhetoric and rule more sensibly. They were wrong.

Who Your Society Is

Perhaps the lesson here is to not let yourself feel victimized, because that is exactly what can lead you to victimize others. Whether your sense of victimization is real or imagined, pursuing a "tit-for-tat" strategy in these kinds of feuds can gradually lead to mutually assured destruction. In mathematical studies of large-scale social interactions, one thought experiment that has figured strongly is the prisoner's dilemma and its variants. The original prisoner's dilemma, developed by game-theory experts in the 1950s, involves two criminals who have been arrested and are being interrogated in separate rooms. If they cooperate with each other and both keep their mouths shut, then both get minimal culpability and mild punishment for the crime. If each betrays the other (or defect on each other), then they both get moderate culpability and moderate punishment. However, if one of them implicates the other while the other keeps his mouth shut, then the snitch gets no punishment at all and the silent cooperator gets the maximum punishment all by himself. This one-way betrayal is clearly the optimal solution for a purely selfishly motivated agent in this scenario. Therefore, it is quite remarkable that it is not the typical strategy. The prisoner's dilemma can be generalized to everyday interactions of all kinds, where most of us tend to have a social compact among ourselves that relies on some form of cooperation and relatively minimal betrayal. Studies show that people often prefer to cooperate with each other, even when it means that they won't personally achieve as much as if they had betrayed the other. Thank goodness, right?

In iterated versions of a prisoner's dilemma, where the game is played again and again with the same two players, computational simulations have found that a "tit-for-tat" strategy is among the most optimal strategies. Whatever was done to you last time, do that back to the other player this time. (Or, in more detailed cases, evaluate the statistics of your past experiences and estimate the likelihood that your next interaction will try to take advantage of you, and then behave accordingly.) However, once a few betrayals have happened, the tit-for-tat strategy can fall into an unending sequence of betrayals (or defections), with no chance at cooperation whatsoever—just like a feud where each group thinks they are the victim and it gradually leads to mutually assured destruction. When you expand this kind of game to involve not just two players but hundreds or thousands of players jockeying with one another for success, the process can get very complex. In this case, a generalized tit-for-tat type of strategy can still be useful, but it can lead player B to take what player A did to him last time and do that to the undeserving player C. Philosopher Peter Vanderschraaf has conducted computational simulations of agents interacting in a two-dimensional arena where each is trying to achieve success in a game similar to the prisoner's dilemma. Rather than being about criminals trying to minimize their punishment, these versions of the prisoner's dilemma focus on individuals who are trying to maximize the payoff from their interactions with one another. The math is generally the same: mutual cooperation brings moderate benefit to both parties, mutual betrayal brings minimal benefit to both parties, and a one-way betrayal brings maximal benefit to the betrayer and a loss for the cooperator. In Vanderschraaf's simulations, groups of cooperators naturally coexist and can actually influence occasional betrayers (he calls them "moderates") to pursue cooperation instead. If there are no *constant* betrayers in the game (he calls those "dominators"), then constant cooperation spreads like a happy contagion to all corners of the arena. However, if there are even just a few *constant* betrayers in the arena, their negative influence prevents those occasional betrayers from becoming constant cooperators and eventually converts even the constant cooperators to become occasional betrayers. The simulations suggest that without additional preventative measures for betrayal and domination behavior—such as legal restrictions and social conventions—a natural tendency toward cooperation is not at all what emerges in a society. Quite the opposite. Gradually, one by one, cooperators get betrayed by betrayers, and then they become betrayers themselves.

This tendency toward betrayal may result in part from a fundamental asymmetry that arises when you compare the learning experiences available to cooperators and betrayers. A constant cooperator can be converted into an occasional betrayer by being betrayed enough times. Essentially, the cooperator finds himself frequently witnessing his interaction partner do better than he does. You can't really blame him for trying a different approach. But the inverse is not quite true: a constant betrayer cannot easily be converted into an occasional cooperator, because if he has never cooperated, then he has no idea that mutual cooperation has a better payoff than mutual betrayal. A constant betrayer never witnesses his opponent do better than he does. Therefore, he may never feel the need to try a different strategy. A constant betrayer may never witness wealth getting *created* at all; he may only see it getting carved up in a zero-sum game. By contrast, an *occasional* betrayer in these cooperation games can in fact learn that mutual betrayal is sufficiently unpleasant that it should be avoided when possible. One improvement for the tit-for-tat strategy that game-theory experts, such as Robert Axelrod, have promoted is to add a small chance for forgiveness after a betrayal. After you've been betrayed, consider the slight possibility of forgiving the other person for it and cooperating next time anyway. In iterated simulations of the prisoner's dilemma, that can sometimes bring a long-term feud to an end and return the system to cooperation once again. A tit-for-tat strategy with a small chance for forgiveness has been proven to be more optimal than a tit-for-tat strategy by itself.

In an expanded version of a prisoner's dilemma game, complexity scientist Paul Smaldino had large groups of simulated agents cooperate with or betray one another in the acquisition of scarce resources that are used for survival and reproduction. Much like the prisoner's dilemma, in his simulation, mutual cooperation generated more food resources than did mutual betrayal, but a one-way betrayal resulted in lots of food resources for the betrayer and a loss for the unfortunate cooperator. Similar to Vanderschraaf's simulations, in Smaldino's simulations, the cooperators tended to form spatially conglomerated groups in the game arena, and the betrayers tended to prey on the cooperators on the edges of those groups. In Smaldino's simulations, agents who failed to obtain enough resources eventually died. His simulations revealed that, early on, those cooperators who belonged to small groups were completely exploited by betrayers and starved to death in droves. The betrayers were taking advantage of them like nobody's business and then multiplying like rabbits. However, the larger huddled masses of cooperators were able to survive, as they created a wealth of resources within their group. And when the betrayers ran out of small groups of cooperators to abuse, they themselves began to starve. When left to their own devices, the betrayers could only steal from each other with minimal payoff, and they gradually died out entirely. Once the open space was no longer riddled with betrayers, the groups of cooperators were finally able to grow in number.

There are perhaps some life lessons that can be extracted from this brief glimpse into game theory and simulations of group interaction. We are a society of mostly cooperators, sprinkled with a handful of betrayers and victimizers. The research suggests that we have a few different ways to deal with these victimizers in our midst. We can become victimizers ourselves and thereby stop being taken advantage of, but that will typically lead to mutually assured destruction. Instead, we can turn the other cheek on occasion, forgiving our transgressors, and wait patiently for those *occasional* betrayers to learn that cooperation has a better payoff in the long run. Perhaps then we can also couple that occasional forgiveness strategy with an avoidance of any interaction at all with the *constant* betrayers. Just let those privileged piggish predators feed on their own kind until they go extinct.

Who Your Nation Is

The research on social interactions among family, friends, and co-workers suggests that your brain-and-body often gets coordinated with other brains and bodies in a way that might suggest they are functioning like *one complex system*. Unfortunately, the research on social group formation suggests that we have a nasty tendency to build walls that cordon off our own complex system, the in-group, and separate it from other groups of people, the out-group. If you expand your sense of self to include only those whom you deem to be your in-group of cooperators and assume all members of the out-group are constant betrayers, you are guaranteed to be wrong. The out-group could not possibly consist only of constant betrayers, because a group composed entirely of constant betrayers eventually consumes itself. That apparent out-group wouldn't still be there if they were all constant betrayers.

For several decades in the twentieth century, the complex system that is our planet's geopolitical atmosphere was dominated by two nations: the Soviet Union (with its support for communism) and the United States (with its support for capitalism). If one of them was your in-group, then the other had to be your out-group. But after the Soviet Union collapsed in 1991 and Russia's stock market crashed in 1998—with numerous Russian oligarchs absconding with billions in wealth (seeking money-laundering opportunities everywhere)—the world was no longer bipolar in its geopolitics. Over these past few decades, it has become widely multipolar. Instead of two major world players setting the agenda for everyone, there is now a group that includes China, the European Union, the United States, Japan, the United Kingdom, Brazil, Canada, India, Russia, and others.

This shift from bipolar to multipolar geopolitics has been confusing and painful for many state leaders. In 1982, right before he died, Herman Kahn (the inspiration for the movie *Dr. Strangelove*) predicted that our then bipolar world would become multipolar and therefore much less stable. What nations did he predict would be the main players on the multipolar stage? The United States, China, Japan, Germany, France, Brazil, and the Soviet Union. Not a bad prediction from the man who almost single-handedly mentally prepared two generations for nuclear apocalypse. Since the collapse of the Soviet Union, Russia has fallen out of the top ten in gross domestic product, but it remains an influential player on the world stage nonetheless. A number of experts in international politics have written in depth about this complex geopolitical transition from bipolar to multipolar dynamics. There are good things and bad things about it. Terrence Paupp called it the rise of the global community, Fareed Zakaria referred to it as the post-American world, and Bernhard-Henri Lévy laments America's abdication. Gideon Rachman and Edward Luce call it an easternization that is happening across the globe and a retreat of Western liberalism, respectively. Political scientist Daniel Woodley points to the larger role that multinational corporations, instead of governments, will have in developing international policy. Noam Chomsky still suggests that the United States rules the world. A decade ago, Zbigniew Brzezinski and Brent Scowcroft were already warning against the worrisome growing reaction among some in the United States to pull away from that global community and instead embrace protectionism and nationalism, to huddle with their own narrow in-group. Recently, a growth of that same attitude has been tracked in the European Union as well, linked to immigration statistics.

Extreme nationalism leads people to think that, somehow, their particular nation (and perhaps culture) is more important than all others. To help avoid a mindset like that, which too often leads to conflict and war, the "America First" crowd and the "Brexit" crowd might benefit from some adjustment of their in-group/out-group boundaries. The research described in this chapter pretty compellingly shows that the people you spend time with greatly influence who you are. In fact, in terms of the continuous back-and-forth information flow between you and them, they *are* part of who you are. As a result of this cognitive interdependence, it stands to reason that if you had been born into a different culture or nation, you would inevitably be a different person than you are today.

For example, if you had been born in a different nation than your actual home nation, you might perceive your actual home nation as the out-group. Remember, whatever group you perceive as the out-group is someone else's in-group, and that group wouldn't still be around if it wasn't composed mostly of cooperators of one sort or another. If your in-group is mostly cooperators and that out-group is mostly cooperators, and your respective missions are not the destruction of each other, then why the heck aren't these two groups joining forces together to help ward off the stray marauding betrayers out there? It's not that difficult to use a person's actions (instead of their skin color or their nationality) to determine whether they belong in our in-group of humanity. As Martin Luther King Jr. proclaimed decades ago, it is the content of their character that should determine a person's membership in humanity, not the color of their skin.

Many of us too often allow the out-group to get defined for us by inaccurate markers, such as skin color, belief systems, or false rumors. Have you ever noticed how politicians, the people who determine public policy, sometimes have a hard time understanding why legal protections are needed for disenfranchised groups (e.g., African Americans, single mothers, the disabled, or members of the LGBTQ community)? The reason they have a hard time understanding the need for such protections is because they don't have firsthand experience with someone in their perceived in-group being subjected to that disenfranchisement. Most politicians have an ingroup that is so narrowly defined that it does not actually include all the people for whom they are implementing public policy, all the people they are being paid to represent. The proof for this observation comes—time and again—when a politician suddenly finds himself with a family member who *does* belong to a disenfranchised group. When one of their children marries an ethnic minority or comes out as gay, suddenly the politician becomes aware that public policy protections for that social group are sorely lacking. But if politicians were to start out with the mindset that their entire constituency is their in-group, then perhaps we wouldn't have to wait until they just happen to acquire a member of a disenfranchised group as part of their family. Perhaps politicians would be better at pursuing the best interest of everyone they are charged to represent.

When it comes to civil rights protections, the formation of in-groups and out-groups is a complicated issue. One of the most common methods by which a category of people finds some niche in which they feel as if they are being treated fairly (whether that category is race, gender, sexual orientation, nationality, or belief system) is for them to collect in assemblies of only their own group. It makes sense for an oppressed or disenfranchised people to seek refuge among their brethren, but they should be careful not to contribute to their out-group status by retreating too much from mainstream society. A strategy of saying to mainstream society, "We're here. We're different. Get used to it," can be very effective. What you don't want to do is play into the hands of the "separate but equal" policy that ran the American South in the 1950s. That definitely did not work. Organizational psychologist Robin Ely might refer to that antiquated policy as a "discrimination-and-fairness" perspective for group formation and teamwork. Her research clearly shows that it doesn't make for productive, cooperative behavior. Rather, her research on group problem solving suggests that an "integration-and-learning" perspective is the most successful mindset for a culturally diverse team to adopt in its approach to cooperative teamwork. This emphasis on both learning and integration has much in common with the "adaptive comanagement" framework for service science. Ely's studies clearly show that culturally diverse groups of people that are well integrated and ready to learn from each other make for more successful problem-solving teams than do homogenous groups of people all of whom come from similar backgrounds. This is true not only in business but also in sports. For example, the French soccer team has been in the finals match of three of the last six World Cups of soccer, and their success is largely attributed to their racially and culturally integrated team. For the past 40 years, the French government has made a concerted effort to promote healthy immigration, and one of the showcased results has been an extremely competitive World Cup soccer team. It's hard to argue with those results. The life lesson here is that, instead of collecting in assemblies of your own similar-looking and similar-thinking group, you should take advantage of the problem-solving benefits and performance boosts that cooperative intercultural teams can provide.

As a self-declared nation of immigrants, the United States has been a centuries-long, occasionally flawed experiment in just that. In 1776, the Declaration of Independence separated the American colonies from England. It declared that those 13 colonies (which would eventually grow into the United States) were independent of their former sponsoring country. Maybe what the United States needs now, and what perhaps every country could use, is a Declaration of *Inter*dependence: a statement that firmly avows a nation's commitment to cooperation and fair treatment among all genders, all ethnic groups, all political positions, and all belief systems within their nation. Over the past several decades in the United States, there has been a gradual expansion of civil rights to African Americans, women, the disabled, and the LGBTQ community. This expansion has slowly made progress toward including all these various groups as part of the in-group of recognized and protected members of the nation. It's not perfect yet, but it's made remarkable advancements. Many other countries have been making similar self-improvements as well in their movement toward a Declaration of Interdependence. Maybe each nation needs one. As a matter of fact, there already exists one from which they could copy and paste. The Charter of the United Nations, signed in 1945 in San Francisco, reads a lot like a worldwide Declaration of Interdependence for interactions among nations. It dedicates the UN to fostering *international* cooperation in the pursuit of equal rights and self-determination for all peoples from all nations. Now, all we need is for each nation to individually dedicate itself to fostering *internal* cooperation in the pursuit of equal rights and self-determination for all its peoples. How hard can that be?

Who Your Species Is

In chapter 3, it was the back-and-forth sharing of information between your different brain areas that showed the interdependence of those brain areas and thus forced you to treat your entire brain as a core engine of *who you are*. In chapter 4, that same kind of back-and-forth sharing of information, a powerful interdependence between your brain and your body, encouraged you to accept the idea that your brain-and-body is one complex system that generates *who you are*. Then, chapter 5 showed a similar informational interdependence between your body and the environment, enough that it makes sense to treat the brain-body-environment as a single system that makes up *who you are*. But it's not just inanimate objects that are in your environment, is it? There are other people in the environment as well. The same kind of interdependence, the same back-and-forth sharing of information between people, analyzed at multiple levels in this chapter, should encourage you to see groups of people, perhaps all peoples of the world, as forming one complex system that makes up *who you are*.

This one complex system of humanity has been around for many thousands of years, evolving slowly in terms of biology and quickly in terms of culture. Evolutionary biologist Joseph Henrich points to evidence that human culture and human biology are coevolving, such that biological evolution can sometimes look as though it is responsible for a cultural innovation, and cultural evolution can sometimes look as though it is responsible for a biological innovation. For example, evidence suggests that hundreds of thousands of years ago, our hominid ancestors mastered fire and began to cook meat. It is likely that this cultural change in diet (i.e., cooked meat instead of raw meat) is responsible for significant evolutionary changes in the digestive tract of *Homo sapiens*, you and me. Our bodies are now relatively dependent on cooked food. (For instance, some studies show that a purely raw food diet can increase the risk of hormonal imbalances and heart disease.) Henrich shows compelling demonstrations that biological and cultural evolution are inextricably woven together, working in concert to change our brains, bodies, and environments. One of the most important of those cognitive innovations, maintained equally by our innate cognitive predispositions and our social morays, is—you guessed it—cooperation. We humans are not the only animals who can be described as cooperative, but we do appear to have mastered cooperation, most of the time, in a fashion that substantially outstrips the cooperative behaviors of ants, fish, bats, wolves, and nonhuman primates.

Irrespective of which particular culture or which particular nation you belong to, we all belong to the species *Homo sapiens*, and we all lean toward cooperation. We are all human. Of course, each of us should be allowed to embrace our specific culture and be patriotic toward our specific nation. However, the scientific research suggests that, in our patriotism, we should perhaps promote our nation's *cooperation* with other nations rather than its *betrayal* of other nations because, in the end, we all belong to one much larger nation as well: the nation of humankind. Rather than defining a subset of the human population as your in-group, and thereby treating all other human beings as members of an out-group, you could treat humanity itself as your in-group. Imagine Native Americans, African Americans, Caucasian Americans, Mexicans, South Americans, Africans, New Zealanders, Australians, Canadians, Icelanders, Europeans, Russians, Middle Easterners, Asians of all kinds, every human on the planet as part of your in-group. Be thankful. Just as you are expanding your in-group to include someone you might not have included earlier, there's someone else reading this book who is now finally including *you* in their in-group. I know I do. We all actually have a great deal in common: we all rely on similar proteins and vitamins, and clean air and water, to survive, and according to Roy Baumeister, we all have a "need to belong." Maybe *each* member of humanity can feel that belongingness with *all* of humanity.

Indulge me now, and let's perform another breathing experiment. Slowly take a deep breath now, and as you feel the air going into your lungs, imagine that all your relationships are like slack ropes connecting those people to you. With each breath, as your lungs expand, those ropes lift off the ground and become taut, strengthening the network. As you exhale, those connecting ropes stay taut. With each new breath, new ropes rise off the ground and become strong connections. In some cases, these are connections to people you don't even know, but, no matter how distant, you have an influence on them, and they have an influence on you. Take another breath, and feel more ropes lift off the ground and become strong. You are a node in this multibillion-person network, connected by all those ropes. Your positive contributions to this complex system help keep the entire thing healthy. They help keep *you* healthy because this network is who you are, who *we* are. Don't let go of that. Keep breathing, my friend.

Directions for Use

Even some candy companies, of all people, have realized that sharing is a good idea. Remember when a double-size candy bar was called "king size?" Any single human who ate that entire king size candy bar was not acting like a king; they were acting like a dummy. The amount of refined sugar in a king size candy bar is practically poisonous for your body to take in at one sitting. Rather than calling a double-size candy bar "king size," some candy companies have recently taken to calling those extra-large candy packages "share size." That makes much more sense. Share your extra-large candy bar with your cooperative friends. Don't eat the whole thing yourself like a dummy apex predator bent on self-destruction!

A key part of cooperation is sharing, and if you have enough money to buy this book, then you have enough money to share a little bit of it. Millions of people around the world simply do not have enough money and resources to feed their family or get health care. They would not be able to buy this book, because they need that money to buy food and medicine. This means that if you are lucky enough to be able to buy this book, or even just lucky enough to have a well-off friend who loaned it to you, then you are in better shape than millions of other people who have almost no money and all their friends are in the same situation. I think you can show your cooperativeness and share a little bit of your bountifulness, don't you?

I'm sure many of you may already give generously to your preferred charities, but you can always give a little more. Whatever your financial status, your assignment for this chapter is to find a charity to which you have not donated and send them some money or volunteer some of your time. I trust you to choose the right amount. You can do this. After all, since you are part of the human race, and the human race is part of who you are, you are actually giving to yourself.

7 From Other Humans to All Life

> Far away, you were made in a sea.
> Just like me.
> —Red Hot Chili Peppers, "Parallel Universe"

Perhaps treating all of humanity as *your family* is relatively easy—at least compared to treating all life on Earth as your family. If you have successfully expanded the definition of your in-group, and even of your selfhood, to include all other humans, then this is progress worthy of celebration. But what about other life forms?

Many people have a dog or cat (or some other pet) that clearly knows it is a member of the family. Maybe that can be the "foot in the door" that allows you to open yourself to including nonhuman life in the definition of *who you are*. These domesticated animals cuddle with us, play with us, and sometimes even eat with us. These animals become important parts of our lives. They become part of who we are. For example, my parents-in-law have a small chi-poo (Chihuahua-poodle mix) named Rosie. She helps her owners wake up in the morning, stands waiting for them to come home at the end of the workday, and reminds them when it's time to go to bed. It took a little while for Rosie to get used to my joining the family; she was very protective of her "sister" (my wife). But now that she has accepted me, I belong to her—and I mean *belong to her*. When we sit on the couch, she likes to climb up on my chest and have me scratch behind her ears while she gazes contentedly into my eyes. If we get up to say goodbye, she barks in protest. And if I kiss my wife in front of Rosie, oooh, she gets jealous and growls at us. She seems to think I'm her man. A familial bond that crosses species boundaries is a prime example of the many ways in which the stuff that makes you *you* can get woven together with the stuff that makes up other living things as well.

These extended familial bonds are where chapter 7 will test you. Since you made it through chapter 6, you should be ready to give chapter 7 a fair shake. If chapter 6 helped you expand your definition of who you are to include *all of humanity*, at least tentatively or halfheartedly enough to propel you to the next chapter, then you just might be able to come along with what chapter 7 is designed to do to you. But you should brace yourself.

You now know that you are not just a frontal lobe of selfhood surrounded by some impressive circuitry for perception and action, housed inside a biological robot body. That circuitry and that body are part and parcel of your selfhood. Chapters 2–4 clearly showed that you would be a different you if you had different circuitry and a different body. This is because the continuous back-and-forth flow of information between all those parts makes it impossible to draw a logical partition between body parts that are you and body parts that aren't you. By the same token, chapters 5 and 6 clearly showed that your body has a similar continuous back-and-forth flow of information with the objects and people in your environment. Therefore, once again, you cannot really draw a logical partition among those bodies and objects and say that one part is you and the other parts are not you. In this chapter, we expand that scope even further to include *nonhuman life forms* that are in your environment. Some of them are moderately cooperative with us, some of them have hearts and brains like us, some of them share a substantial amount of DNA with us, and some of them are constantly exchanging life-giving molecules back and forth with us. How could they not be part of *who you are*?

The Human Microbiome

You have a bad habit of casually thinking that your body has some kind of boundary between it and the rest of the environment. So do I. But we actually know better. In previous chapters, we already saw a number of ways that this boundary is permeated by external forces. Information gets into your eyes and ears to change *who you are*. Physical pressure deforms your skin, and your brain reconfigures itself, when you grasp a tool—and that changes *who you are*. Other people interact with you, and "get in your head," in a way that changes *who you are*. And, of course, food crosses that boundary regularly, resulting in an external object becoming part of *who you are*. But there are other things crossing that boundary that we haven't even talked about yet. What about the microscopic bugs that literally penetrate your skin because they're so small that they can crawl into your pores? Sometimes we call them germs.

Before Louis Pasteur's experiments in the nineteenth century, the idea that microscopic germs could cause infections and disease was not well accepted. Rather, Aristotle's theory that bugs were spontaneously generated by nonliving material was highly influential even into the 1800s. For instance, Aristotle was convinced that scallops were spontaneously formed by sand alone. I'm not kidding. Aristotle never conducted any controlled experiments to produce this theory. I guess his theory of spontaneous generation was just a spontaneous intuition; perhaps not the best way to go about knowledge acquisition. Pasteur's controlled experiments showed clearly that, once sterilized, a broth of nutrients would only grow new bacteria if it had access to the outside air (which provides the germs that would then multiply in that nutrient-rich broth). Throughout the 1800s, it took almost a century for all the defenders of spontaneous generation to finally give in to the mounting experimental evidence. I guess some paradigm shifts take longer than others. But the proof is in the pudding, or in the broth as it were. One of Pasteur's sealed-glass flasks of sterilized broth is on display at the London Science Museum. It's been there for well over a century, and not a single thing has spontaneously grown in that clear liquid all this time.

Your body is a little bit like Louis Pasteur's glass flasks, except that your microbiome was never fully sterilized (no matter how thoroughly you may think you bathe), and you are one of the experiments that was *not* sealed off from the environment. Microscopic bacteria, viruses, and fungi have been a part of your body since you were born. Sometimes they have caused you to get ill, but more frequently—unbeknownst to you—those microbes have protected you from getting ill.

The theory of germs was right to discard the theory of spontaneous generation, but it was wrong to allow people to slip into the comfortable assumption that all germs are bad and that a healthy body is somehow germ-free. Your body has more bacterial, viral, and fungal cells in it and on it than there are human cells. Usually, this jungle of human and nonhuman cells, called "your body," self-organizes wonderfully, but not always. When you get an infection, it's not necessarily because some foreign bacteria that wasn't there before has suddenly invaded your body. Often, that particular bacterium has always been present in your body. It is just a recent imbalance of its population relative to the population of other microbes, fungi, and viruses that results in the infection. In appropriate amounts, those same infecting bacteria—in concert with the other microbes, fungi, and viruses—are actually good for your body. They help you digest food, they influence the glucose levels in your gut, they help keep your immune system active, and they keep each other in check as well. You can't live without them. Without all your microbes, you wouldn't be you.

Immunologist Rodney Dietert went through his own personal paradigm shift as his research gradually revealed to him that how he thought disease worked is not quite how disease actually works. He points to a wide variety of scientific interventions (e.g., antibiotics in humans, antibiotics in farm animals, pesticides for crops, genetically modified organics, pollution from factories) that have conspired to so dramatically change our planetwide ecology that the human microbiome is now exposed to a vastly different environment compared to only a few decades ago. Is it any wonder that many of us have a human microbiome that is out of balance? Noncommunicable diseases, such as cancer, diabetes, heart disease, inflammatory bowel disease, and serious food allergies, are twice as common today as communicable diseases. Our vaccines and antibiotics have almost eliminated several communicable diseases that spread from person to person, such as cholera, polio, and tuberculosis, and our pesticides and genetically modified foods have improved crop production in a way that shows promise for reducing hunger worldwide. But these scientific interventions have come with a price. Of course, the solution is not to eliminate the vaccines, antibiotics, and other interventions, because that would bring back all those communicable diseases. However, a careful examination of the intricacies of how these interventions interact with one another and how they interact with the human microbiome is in order. We have to think in terms of the *system* of natural and technological influences; we have to use "systems thinking."

For instance, it might not be the pesticide that a farm uses that is specifically interfering with your gut's ability to process that vegetable that gives you gas or hives. Rather, it might be the genetic modification that the seed for that vegetable underwent—in order for it to survive the pesticide that the farm uses—that is causing your intestinal distress. You can wash the pesticides off that vegetable as vigorously as you want, but you won't change its DNA back to the old version of the vegetable that your body was accustomed to. In a case like that, the pesticide is a "distal cause" of the health ailment—influencing it indirectly—and the genetic modification is the "proximal cause" of the health ailment, directly resulting in your allergy to the way that new version of the vegetable expresses its proteins and vitamins. My wife suffers from food allergies, and she says that her doctor suggested she eat organic vegetables. He told her, "Don't just look for a sign that says 'organic.' Look for veggies that look a little bit ragged, and maybe the leaves have a few insect-eaten holes in them." These are veggies that haven't been inundated with pesticides and also probably haven't been genetically modified in a laboratory (unless nearby GMO crops have accidentally cross-pollinated with it). A good rule of thumb to go by is that the veggies that were clearly safe for the *bugs* to chew on a little bit are the veggies that are safe for *you* to eat, which makes sense because, after all, you are mostly made of bugs.

What it all comes down to is a balancing act. If you get really sick and your doctor suggests that you take some antibiotics, perhaps ask if there are any alternative measures that could be pursued first, or seek a second opinion. If there are no alternatives, and the second opinion agrees with the first, then take those damn antibiotics so you can avoid needlessly dying. Doctors trained in Western medicine are right more often than not. Just be ready to respond to the havoc in your human microbiome that those antibiotics are going to wreak. Add some good probiotic foods to your body to balance things out again, such as yogurt, kombucha, dark chocolate, and pickled and fermented veggies. Your bugs will love you for it. As Rodney Dietert says, "We have met the microbes, and they are us." Take care of them, and they will take care of you.

In fact, if you and your microbes are taking care of each other just right, then maybe you and your microbes should be seen as forming one entity, one agent, one system. When one species (e.g., a human and its cells) forms an interdependent symbiotic relationship with another set of species (e.g., a wide variety of microbes in the human's gut and elsewhere), the proper term for that tightly knit conglomeration of species is a "holobiont." Philosopher of science Alfred Tauber champions this idea as he analyzes the human immune system. Rather than focusing on the various microbes that your immune system fights, Tauber focuses on the myriad microbes that your immune system embraces and nurtures. Those microbes become part of the holobiont that you and they form together. Perhaps you and your microbes together form one creature: a holobiont.

The Mental Life of Nonhuman Animals

Nonhuman animals are no different. They are as riddled with microbes as we humans are. And just as with us, those microbes are usually balanced in a way that is actually helping the animal rather than hurting it. The reason this section header refers to "nonhuman animals" and not just "animals" is because it is important to remind ourselves that humans are animals, too. Don't ever forget that. Humans evolved from a dizzying array of similar animal species over the course of many millions of years. We are animals, you and I, and this section of the chapter focuses on how animals that are not humans have pretty interesting mental lives, even though they don't use a human language to prattle on about it.

Think of your favorite animal pet. Even if he or she is long gone now, you were once very close to that pet. How many times did you see that nonhuman animal obviously interpret a human situation intelligently? Maybe it could tell when you were sad or sick, and it cuddled you. Maybe your pet could tell when someone visiting the house had ill intent, and it never accepted that dubious person. Particularly dogs and cats, which have been selectively bred for thousands of years to be human-centric, routinely pay pretty close attention to what's happening among the humans in their home. In fact, domestic dogs routinely learn to recognize dozens of spoken words, in whatever language they were trained in. There is probably a dog somewhere that understands more German or more Japanese than you do!

I once had a cat that preferred to drink running water from a faucet. I would occasionally turn the bathroom sink on at a trickle for her. She would sit on the edge of the sink, tilt her head under the faucet, and lick at the water coming out. One day, this cat was on the bed next to me and saw my glass of water on the nightstand. Evidently, she was thirsty, because she walked over to my water and sniffed it. I took it away and said, "You have your own bowl of water, in the kitchen. This is mine." She looked at me, looked at the glass in my hand, and you could see the wheels turning in her head. Then, as cats do, she suddenly jumped off the bed, hurried to the bathroom, and jumped up onto the sink. Like a well-trained human, I followed her and turned the faucet on for her. Back in the bedroom, when she had looked at me and looked at my water glass, that tiny brain in her skull was clearly thinking and planning about water and its potential future opportunities at various locations throughout the house—objects that were not present and visible to her at that moment. Then she chose her preferred plan and went there.

Contrary to old-fashioned behaviorist psychology, many animals are not solely driven by immediate stimuli that induce immediate responses to those stimuli. Many nonhuman animals are able to string together a number of ideas in their head to produce behaviors that are complex, organized, and planned. This actually isn't surprising when you consider the fact that monkeys, cats, dogs, and mice have brains that are remarkably similar to human brains. The neural connectivity between cortical areas in a cat or a mouse is not that different from the neural connectivity between corresponding brain areas in a human brain. Neuroethologist Ádám Miklósi has even found that when dogs recognize other dogs' vocalizations and human voices, a specific region of their brain becomes active—similar to that seen in human brains. When those vocalizations take on different emotional properties, the neural activation pattern changes—similar to that seen in human brains. Neuroscientist Gregory Berns has identified a human face recognition area in dog brains as well, much like that identified in humans.

Similar brain connectivity and neural activity between human and nonhuman animals, again, should not be surprising when you consider that we humans share a great deal of DNA with many nonhuman animals. We share about 99 percent of our DNA with chimpanzees and bonobos. About 90 percent of a cat's DNA, about 85 percent of a mouse's, about 83 percent of a dog's, and about 60 percent of a chicken's or a fruit fly's can be matched up with human DNA. (These estimates vary slightly depending on whether one measures intron phase compatibility or synteny to measure the similarity between genomes.)

Because of this remarkable neural and genetic similarity between humans and certain other animals, it stands to reason that many cognitive abilities that we often think of as uniquely human are in fact enjoyed to varying degrees by a variety of nonhuman animals. For example, nonhuman primates use tools, chimps share their resources, ants grow their own crops, birds communicate with grammatical structure in their songs, octopuses solve puzzles, rhesus monkeys can learn to do simple addition, and the babies of all mammals and birds engage in play with one another. (However, evidently, the babies of most reptiles do not frequently engage in mutual play.)

It is worth noting that nonhuman animals also engage in some of the same kinds of group-based cognition that humans engage in, as detailed in chapter 6. Philosopher Georg Theiner surveys a few types of distributed/extended cognition exhibited by nonhuman animals that can be compared to the types of distributed/extended cognition that we humans often exhibit. One of the simplest forms is flocking behavior, such as that performed by birds or by a school of fish. The action-perception cycle of each individual animal is so intertwined with those of the other animals around them that the group "moves as one"—not unlike people exiting a movie theater or perhaps line dancing. However, such examples of *perceptually* distributed cognition are eclipsed by what Theiner calls the *socially* distributed cognition that social animals engage in when they use communication to engage in complexly coordinated plans and joint actions. Take, for example, when multiple honeybee scouts come back to the nest with information (conveyed in a "dance") about several alternative potential new nest locations. This dance-based information is aggregated among the different scouts as the hive gradually decides on one location for its new nest. Thus, in much the same way that humans can show a "wisdom of the crowd" when a social group aggregates its problem-solving abilities, honeybees can show a "wisdom of the hive" when they aggregate their nest-hunting tactics. And when a colony of insects gathers, integrates, and processes information as a single cognitive system, it can bear a remarkable resemblance to how a single human brain gathers, integrates, and processes information.

Based on these observations, it seems clear that we should include some nonhuman animals in our vaunted circle of "intelligentsia." A century ago, psychologists developed the notion of the intelligence quotient (IQ) as a single measure that would somehow encapsulate all aspects of what makes a person smart. Obviously, nonhuman animals would score terribly low on a paper-and-pencil IQ test. Most of them aren't even smart enough to hold the pencil properly. But seriously, in recent decades, psychologists have dramatically revised that notion of IQ to accommodate the various kinds of intelligence that different humans exhibit. Rather than treating intelligence as a one-dimensional measure in which you are either high or low, psychologist Howard Gardner has suggested that there are nine different ways to be intelligent. Some people are good at math, spatial reasoning, or the natural sciences. Other people are good at interpersonal interactions, philosophy, or self-knowledge. Still others are good at linguistic communication, music, or motor coordination. Nobody is excellent at all nine of these. Now that "intelligence" is understood to be a multidimensional construct, primatologist Frans de Waal is well justified in suggesting that animals that can't be measured by an old-fashioned IQ test may nonetheless exhibit some other forms of intelligence. He encourages us to think of these many different types of intelligence as branches of a tree, a little bit like an evolutionary tree. He even traces some of the building blocks of wisdom, generosity, and human morality in the observed behaviors of nonhuman primates. In the same way that it is inaccurate to treat some people as though they are summarily "smarter" than other people, it is likewise inaccurate to treat nonhuman animals as though they are summarily "dumber" than the human animal.

When you watch a video clip of an octopus solving a puzzle, the only reason it is surprising is because you didn't give that animal the benefit of the doubt, that maybe it can think. Or consider when you watch some elephants cooperate to rescue a baby elephant or when you watch the famous chimpanzee Ayumu outperform humans in a number memory task. Nonhuman animals are far more intelligent, in a variety of ways, than we give them credit for. It is not merely rude of us to categorically treat ourselves as intellectually superior to them; it is also scientifically inaccurate. Rather than bending the scientific facts to push forth a narrative that places humans in a category all by themselves—allowing us to unfairly neglect the mental lives of other animal species—perhaps we should embrace the actual scientific observations showing how we have a great deal in common with the rest of the animal kingdom.

You Are Coextensive with Nonhuman Animals

Now that you are perhaps ready to consider nonhuman animals as intelligent, let's consider ways in which their intelligence and our intelligence coexist with each other and perhaps even *coextend* with each other. Sounds crazy, right? The physical matter that makes up your brain and body extends over space only so much. That physical matter covers only so much ground, takes up only so much real estate, and the physical matter that makes up your pet animal takes up a different, nonoverlapping region of space. So, how in the world could you and a nonhuman animal (no matter how smart it is) be coextensive with each other? How could the spatial extent of who *you* are overlap with the spatial extent of who *it* is?

I will tell you how. In much the same way that you and your conversation partner get physically correlated with each other (as detailed in chapter 6), you and a nonhuman animal can become physically correlated with each other as well. You can function as one system, even if just for a little while. Think about when you sit on the couch with your dog or cat cuddling next to you. You probably repetitively stroke their fur with your hand several times. What do they typically do to you in return? They repetitively stroke your hand with a tongue or rhythmically knead your skin with two paws. Your action-perception cycle is getting entangled with their action-perception cycle. I suspect the stroked fur feels better to them than the licked skin does to us, but they're trying their best with what they have—even if they do occasionally bite your hand because they don't know any better. Then again, fellow humans do that too.

We also get correlated with our animal friends in ways that aren't under our voluntary control. Have you ever noticed your pet yawning after you just yawned? Or maybe you've noticed yourself yawning after your pet just yawned. Yawn contagion is a real thing, and the interspecies version of it is especially interesting. Of course, we've all noticed ourselves yawning after seeing someone else yawn, as if the other person's yawn is somehow contagious. This yawn contagion happens among nonhuman primates as well. In controlled laboratory settings, yawn contagion works among adult humans about half the time, and it works at least a third of the time among adult chimpanzees and stumptail monkeys. You play your pet chimpanzee a video clip of a chimpanzee yawning, and, sure enough, he shows an increased likelihood of yawning (compared to a video clip of a chimpanzee not yawning). Furthermore, chimpanzees are more likely to yawn if they see a chimpanzee from their own tribe yawn than if they see an unfamiliar chimpanzee yawn. And, apparently, tortoises do not exhibit contagious yawning at all, no matter how hard you try. But what about running this kind of experiment with a video clip of one species yawning and presenting it to a different species of animal? Is there cross-species contagious yawning?

When social neuroscientist Atsushi Senju and his research team performed multiple noisy yawns for 29 domestic dogs, 21 of those dogs yawned in response. In the control condition, where the same researchers performed silent nonyawn openings of the mouth, none of the dogs yawned. Domestic dogs, which of course have been selectively bred to affiliate themselves with humans, clearly see humans as enough a "part of the tribe" to find themselves susceptible to cross-species yawn contagion. Thus, on the very short timescale of you and your dog yawning at each other, and on the very long timescale of human culture coevolving with canine biology, there is a multiscale dynamic entrainment going on between human and dog (mutually influencing one another) that makes them behave a little bit like *one system*.

In fact, it turns out that you can even just play an audio recording of a human yawning and it will make a dog yawn, particularly if the recording is from the dog's owner. That's right. Your dog probably knows the sound of your yawn well enough that he or she would yawn in response to that sound all by itself and probably not to the sound of *my* yawn. Yawn, Yawn, Yaaawwn. Has reading about yawning made you yawn yet? It did me. Perhaps you and I are more correlated now.

"Tribe affiliation" that spans across species is not actually that uncommon. Mother tigers have been observed suckling baby leopards or even piglets. An orangutan might babysit some tiger cubs. A leopard might tend to a baby baboon. Cats, dogs, and ducklings that grow up together can wind up behaving like siblings. In fact, companion dogs are routinely used to help raise young cheetahs in zoos. This is especially common when you look at animal species that have been bred for thousands of years to serve human needs. Not only do dogs and cats fit this description, but so do horses. Anthropologist David W. Anthony draws on archaeological evidence to suggest that the development of horseback riding (with a mouth bit for steering) goes back as far as 4000 BCE, predating the invention of the wheel. That's more than six millennia of selective breeding to evolve horses into human-friendly animals. The relationship that a horse and rider develop together is remarkable. Any horseback rider will tell you that horses are extremely perceptive. Many of them will tell you that, like children, these riding horses crave structure—even if they won't admit it. When a horseback rider shows their horse that they are consistent, reliable, and steady in their training, the horse will accept their dominance. And all of this is achieved with only the smallest tidbits of linguistic communication. Most of it is body language, communicated in both directions. Comparative psychologist Karen McComb and her research team have been studying communication and sociality among primates, elephants, and horses for years.

In one study, she recorded the heart rate of trained riding horses while they were presented with photos of angry human faces and happy human faces. These trained horses experienced an increase in heart rate when they saw photos of angry human faces compared to when they viewed the happy ones. Clearly, they can recognize the emotional consequences of some human facial expressions. The coordination of body language between a horse and its rider is essential to maintain the motor coordination that is required for successful riding through rough terrain with obstacles.

We get coordinated with our animals in repeated activity cycles at the short timescale of seconds and minutes during petting, at the longer timescale of hours of coordinated play or horseback riding, at still longer timescales such as daily feedings and bedtime, and at the much longer timescales of social coevolution. Many of our own behaviors often get correlated with those of the animals around us. In fact, to become a member of a tribe of chimpanzees in Tanzania, Jane Goodall intentionally mimicked some of their behaviors in order to gain acceptance. In the Congo, Dian Fossey was eventually admitted into a community of gorillas in their natural environment, precisely because she grunted like them and snacked on the nearby vegetation like them. And numerous researchers have brought chimpanzees, bonobos, and gorillas into domestic environments to teach them sign language. While debate over whether these nonhuman primates can really string together multiple signs into a proper grammatical sentence still continues, it is uncontestable that they use these signed words to refer to events, ask for things, answer questions, and tell stories. (And, anyway, I can think of a few adult humans in politics who also can't seem to string together multiple words into a proper grammatical sentence.) The famous gorilla Koko lived to the age of 46, in the San Francisco Zoo with her mate, and she used over a thousand words in sign language and understood over two thousand spoken English words.

While observing those research programs, philosophical ethologist Dominique Lestel long ago pointed out that, from the perspective of cognitive anthropology, there are two ways to look at these nonhuman primates entering a domestic environment and learning sign language. One way to look at it is that the researchers have domesticated the nonhuman primates by teaching them sign language and some other human skills. However, another way to look at it is that the nonhuman primate has taken full advantage of the willingness of the researcher to provide them with domestic niceties, such as education in a range of skills, a variety of tasty foods, and a roof that doesn't leak—all for the low, low price of just playing along with their curious language games. Has the chimpanzee perhaps domesticated the human just as much as the other way around—a little bit like my cat training me to turn on the faucet for her? Rather than selecting one of these perspectives and assuming the other was flatly incorrect, Lestel encourages anthropologists and ethologists to embrace them both at the same time. Rather than building a research program that just focuses on how contact with animals affects humans, or a research program that just focuses on how contact with humans affects animals, why not build a research program that explores the shared lives that emerge among humans and other animals?

We rely on the animals around us as much as they rely on us. Dogs were domesticated over thousands of years, in part to stand guard against predators and enemies. Even a small dog can usually sound an alarm like nobody's business. By getting a watchdog to live at your house, you've "jigged" your environment to improve the way it protects you, and the dog has "jigged" *her* environment to get paid with food and cuddles for a pretty cushy job. Similarly, cats were domesticated over thousands of years, in part to fend off vermin. They try to earn their keep, even if sometimes their offerings of proof for their grisly work are not exactly welcome, flayed out on the kitchen floor. And even when you consider animals that aren't your pets, you can still find quite a few of them whose existence coextends with yours. Many of the medicines and cosmetics that you might use were tested on animals to make sure they won't harm you. Wool and leather have been very important forms of human clothing for centuries. And think of the meat you eat. Some of who that animal was is now part of who you are. Be thankful. Or, if you are a vegetarian, then the milk, eggs, and cheese that you consume are products of those animals that become part of your body. It is worth noting here that eco-friendly animal farming is proving to be more achievable than you might think. But even if you are a vegan, we can still point to the fertilizer that is used to grow the plants you eat. Where do you think that fertilizer comes from? Vitamins and minerals that are excreted by animals are used to feed the plants as they grow. When that plant has absorbed enough of those vitamins and minerals that the animals were so kind to produce for us, they become part of your body as you eat that vegetable. I hope it's an organic vegetable.

The Mental Life of Plants

These plants that you eat have lives, too, you know. There's a song by the rock band Tool, titled "Disgustipated," in which lead singer Maynard James Keenan preaches, "These are the cries of the carrots. ... Tomorrow is Harvest Day, and to them it is the Holocaust!" He then playfully suggests that we should save the carrots and "let the rabbits wear glasses." The song eventually descends into a transcendental mantra of "life feeds on life feeds on life feeds on life." In the same way that some of us may hesitate about eating an animal because it seems to have an awareness, we could almost as easily hesitate about eating plants for the same reason.

Plants react to their environment in complex ways that could genuinely be interpreted as evidence for a form of awareness. Charles Darwin's favorite plant, the Venus flytrap—which traps and consumes insects by suddenly closing its toothy mouthlike appendage and secreting digestive enzymes—is perhaps an extreme example of a perceptive plant. However, even everyday plants, which only react on the timescale of hours and days during growth, appear to have some kind of understanding of their environment and respond accordingly. They are perceptive. For example, a young sunflower plant will bend westward to follow the arc of the sun over the course of a dozen hours and then bend right back overnight to greet the sun in the east the next morning, almost as if it *remembers* where the sun will come up. In the nineteenth century, one of the fathers of experimental psychology, physicist Gustav Fechner, wrote an entire book about "the soul-life of plants." The original German title was *Das Seelenleben der Pflanzen*. Fechner thought there was ample evidence even then that plants have at least a simplistic version of consciousness, even though they don't have a brain—and he was not alone. Numerous botany experiments from that time were aimed at determining how intelligent and adaptive plants are in their response to changes in their environment, responding a little bit like animals. For instance, plants clearly grow in the direction of the light they are receiving. Therefore, it was obvious from those experiments that plants have a way of responding to the light source (phototropism) as if they "know" the direction in which they should grow. In 1908, Francis Darwin (son of Charles) wrote in the journal *Science*, "In plants there exists a faint copy of what we know as consciousness in ourselves." This younger Darwin conducted experiments with his father to examine how plants respond to mild injuries to the tip of a root. When you cut or pinch the tip of a root extending to one side of the root branch, the top of the plant will grow in the other direction for several days. Basically, the plant senses an unpleasant impediment in the soil and then leans itself toward growing away from that impediment. All of this is accomplished without the use of a brain.

Plants can also sense the direction of gravity and respond accordingly. When the Darwins grew plants in pots that were balanced sideways, the plant would gradually curve during its growth so that its leaves and stems were pointed upward instead of sideways. And when they dug up that plant, they saw that its roots had curved to grow somewhat downward as well, instead of sideways. It makes sense that plants would have evolved to do this because they need to "know" the direction in which to grow their roots for absorbing water and minerals and the direction in which to grow their leaves for absorbing sunlight. What's more, when a tree grows on the side of a steep hill, it needs to curve itself to grow straight up with respect to gravity if it wants to avoid falling down from its own weight. Trees that don't do that do not get to live very long, and therefore natural selection culled them a very long time ago. Plant researchers now understand that this ability for the plant to "sense" gravity (gravitropism) comes from the plant's cells responding to their individual compression in one direction or another. Gravity's miniscule squishing of a plant cell is detected by the cell, and it then sends chemical signals for growth in that direction, which is detectable in a matter of hours.

In his book *What a Plant Knows*, plant biologist Daniel Chamovitz wonderfully details how plants have their own ways of feeling, smelling, and seeing—but hearing, not so much. In fact, Chamovitz carefully documents the flaws in those old experiments claiming to show evidence that plants prefer classical music over rock and roll. Attempts to replicate those experiments have roundly discarded the hypothesis that plants prefer any genre of music over any other. However, the Venus flytrap can definitely "feel" in its own way. Its trigger hairs are calibrated perfectly to close that trap only on insects that aren't too big to fit and aren't so small that they could escape. Like Goldilocks's porridge, the size of the prey has to be *just right*. And a growing parasitic dodder vine can "smell" the difference between a nearby tomato plant and a nearby wheat plant. It detects the subtle difference in chemical gradients given off by both plants. Interestingly, it prefers the tomato plant and will decidedly grow in the direction of it when given the choice. When undergoing severe water deprivation, some plants will shift their energies to growing more roots instead of growing shoots and leaves, and almost all plants have a way of "seeing." When a plant senses the direction from which it is receiving light, it will respond selectively to certain wavelengths of light, almost as if it were "seeing" color. Certain wavelengths of light that the sun emits have an effect on the hormone auxin in a plant. Many other wavelengths of light do not have that effect on auxin. Therefore, a plant "sees" certain wavelengths of light and ignores other wavelengths, a little bit like how our own visual system works. When auxin reacts to the sunlight, it causes plant cells on the shady side of the plant's main stem to grow longer, while cells on the sunlit side of that same stem don't grow longer. The unavoidable geometric result of this asymmetric elongation of the stem is that it cannot help but curve in the direction of the light as it grows.

Not only can plants *sense* things like light, water, chemical gradients, and pressure, they also *remember* what has happened to them. I kid you not. It is generally accepted now that many plants engage in rudimentary forms of learning and memory. Stimuli that impact the plant at one time can then have effects on the plant's growth that only show up much later. For example, botanist Michel Thellier took the two embryonic leaves of a seedling Spanish needle plant and poked a few tiny holes in the left-hand leaf but kept the right-hand leaf undamaged. Several minutes later, he removed those two leaves from the main stem. He then clipped off the central bud, between where the baby leaves had been, and watched as the two lateral buds grew in its place over the course of a week. The lateral bud on the left side, where the baby leaf had been poked before being removed, was growing more slowly than the lateral bud on the right side. The Spanish needle plant seemed to be remembering that growth toward the left side encountered greater hazards, so it was investing less energy on that lateral bud.

A similar form of memory can be seen in growth of the main stem of a plant as well. When both of a seedling plant's first leaves are slightly damaged while it is in a nutrient solution, it does not change its growth pattern. However, if the plant is later transferred to a pure water (less nutritive) environment, it suddenly slows down the growth of its main stem. By contrast, a seedling that did not have its first leaves damaged and is then transferred from a nutritive medium to pure water does not change its growth pattern. It is as if the plant whose embryonic leaves were slightly damaged has remembered that it is in a hazardous environment, and the transfer to a less nutritive medium caused it to conserve its energy a bit. The plant hormone auxin is believed to play a role in these kinds of memory processes, but so are some of the same chemicals that brains use in order to carry out memory and learning: calcium, potassium, nitric oxide, and *glutamate*. Remember how we started chapter 1 with a glutamate neurotransmitter molecule helping your brain understand the sentence you were reading? Plants use glutamate, too. But when it's a plant doing it, some scientists prefer to call it a "stress imprint" rather than "learning" or "memory." Either way, it seems clear that the plant is displaying the ability to behave differently as a result of its past experiences, and to me that sounds an awful lot like learning and memory.

Not only do these nonanimal life forms "perceive" stimuli and "remember" past events, they also are pretty good puzzle solvers. Consider the slime mold. A slime mold is not actually a plant or a fungus, but it's close. Slime molds belong to their own kingdom of protists. In a cool, moist environment, a slime mold such as *Physarum polycephalum* can grow, or "stream," at the rate of about one millimeter per second, surrounding food sources (such as an oat flake) and secreting digestive enzymes to consume them. Given a sufficient food incentive, a slime mold can find its way through a maze. It can even find the *optimal route* through a maze that has a food reward at its exit by "smelling" the subtle intensity differences in the chemical gradients that different pathways provide. And the slime mold "remembers" where it has been. Everywhere the slime mold goes, it leaves behind a slimy layer of polymers, and when a growing tendril encounters this layer of goo, it turns and grows in a different direction, thus ensuring that it doesn't forage for food in a previously explored area. That goo is part of its behavioral residue, not unlike yours and mine. Professor of Unconventional Computing Andy Adamatzky has developed computer simulations of this process and likened the slime mold's navigation of a maze to parallel computing. The parallel computing powers of a slime mold are so powerful that when applied mathematician Atsushi Tero placed food sources in a petri dish in a spatial pattern that corresponded to the relative locations of train stations around Tokyo, he found that his slime mold branched out tendrils to interconnect all the food sources in such a way that it closely resembled the layout of the actual Tokyo rail system. In a matter of hours, the slime mold solved the optimal connectivity among these neighborhoods, a feat that the city of Tokyo took decades to figure out. Of course, slime molds aren't the only nonanimal life forms that can find their way through a maze of sorts. Every small plant in a forest or jungle has to navigate a constantly changing maze of light and shade shaped by the canopy of trees above it—and they usually solve this puzzle quite well, thank you very much.

To be sure, there are some quacks out there who have suggested that plants have thinking minds like ours, that they scream when injured, that they prefer classical music, and that they can read our thoughts, but I have been careful not to include them in the discussion here. There is no need to put on your tinfoil hat. What I have reported here are real and repeatable experimental results (as I have done in all these chapters). You can try these experiments yourself at home. Just be careful not to spill potting soil all over your kitchen when you're turning that poor tulip plant sideways to demonstrate its gravitropism.

Based on these scientific reports, it is unmistakable that plants react to their environment in intelligent ways that are sometimes reminiscent of how animals react to their environment. In fact, some botanists have even seen fit to (only slightly metaphorically) name their new field "plant neurobiology." They're not claiming that plants have neurons, of course, but the interactive manner in which various parts of a plant send hormonal signals back and forth to one another (much like synapses) to differentially influence growth patterns is quite reminiscent of a neural network. Plant neurobiologists make a convincing case that the information networks that are formed by a plant's cellular interactions and environmentally reactive chemicals are not that different from the information network that makes up an animal's nervous system.

You Are Coextensive with Plants

Now that you can get a sense that plants do indeed have lives—they are living, after all—and even respond intelligently to complex changes in their environments, does that make you feel a little closer to them? A little more *akin* to them? After all, they "perceive" and "remember" a little bit like you do.

We humans have coevolved with plants in much the same way that we coevolved with other animals. No species is an island. Like it or not, the physical material that makes up *who you are* is substantially correlated with the physical material that makes up plant life on this planet. The molecular processes that make plants so dependent on water to maintain their cell structures are essentially the same molecular processes that make us animals so dependent on water to maintain our cell structures. When plants breathe in carbon dioxide and breathe out oxygen, we breathe in that oxygen and give them back carbon dioxide in return. When you eat vegetables, those plants become part of your body. Their vitamins and minerals are fueling your ability to read this page.

The planet's surface essentially breathes with its trees (and ocean algae and peatlands), on a seasonal timescale. During the winter, carbon dioxide collects in our planet's atmosphere in large amounts in part because the leaves on the trees, which absorb the carbon dioxide, have mostly fallen. When those leaves grow again in spring, microscopic holes in the leaves suck in some of that carbon dioxide, hang onto the carbon part, and exhale some of the oxygen. During the summer, this outflow of oxygen from the planet's forests inundates our planet's atmosphere. Then, as autumn arrives, the leaves begin to fall, and carbon dioxide begins to amass in the air again. It's a lot like the surface of the planet is breathing, except that it inhales CO_2 and exhales oxygen, while we animals do the opposite.

In your lungs, the bronchi and their alveoli look a lot like trees, branches, and microscopic leaves, and they even function a bit like them, too. They take in oxygen from the air you breathe in, and they transport it into your bloodstream. Then you exhale carbon dioxide. Our respiratory waste products, CO_2 and water vapor, are exactly what the trees and other plants use as their respiratory nutrition. By the same token, we animals feed on the respiratory waste product of those plants: the oxygen that they exhale. This is not a lucky accident, of course. It is the product of hundreds of millions of years of animals and plants coevolving on Earth into this delicate balance. The animals and the plants that didn't participate in this mutual handshake have generally died out. These things happen. If we humans don't continue to participate well in this handshake, if we generate more CO_2 than the plants can consume for us, we may die out as well. So take a deep breath. I know, I ask you to do that a lot, don't I? Go ahead, it won't hurt. Breathe in. Now hold onto that nasty CO_2 that your lungs just generated. Don't let it out. Do your part to reduce the CO_2 in the atmosphere by not exhaling. Just kidding. Let that breath out slowly. There you go. In that breath that you just took in, those aren't new oxygen atoms that you just put into your bloodstream. They're used. Or, should we say, previously owned. Some of those oxygen atoms that you just converted into part of your bloodstream were once inside a plant's leaves (and before that they were inside some other animal's lungs). It seems inescapable that those plants are part of who you are, and you are part of who they are. You wouldn't be here if not for them. Be thankful.

Based on the scientific studies described in this chapter, it seems clear that an animal has an awareness that is at least similar to yours, and in fact so does a plant. When you assume that your best friend has an "awareness" or a "consciousness," you do this based not on any direct access to his or her mental experience. You do it based on your observations of their behavior. When they produce a behavior that looks similar to your own past behaviors, and you recall that you had a particular mental experience when you produced that behavior, then you feel justified in assuming that your friend is having a similar mental experience, or awareness, when they produce that familiar behavior. There is no legitimate reason why we should not also apply that same logic to other animals and even plants. When you see a cat sniff at a cactus and suddenly pull away from it because one of the needles poked its nose, you are entitled to conclude that the poor animal experienced an unpleasant sensation of pain. That is a form of awareness. And when a plant gradually pulls away from one side of the soil because one of its roots on that side got pinched by a nerdy scientist, you can be just as entitled to conclude that the plant experienced an unpleasant sensation of something like pain. It is a form of awareness.

If you go outside and hug a tree, you will give it some of your body's warmth, and the tree will be aware, in its own little way, of the warmth you are giving it. It will be aware of your presence, and it will give you, the tree hugger, some oxygen as a thank you. This little thermodynamic exchange that you are engaged in with the tree makes you and the tree a little more coextensive than usual. (Don't let your neighbors see you doing this.) Part of the energy and matter that make up your body has gone into the tree, and part of the energy and matter that make up the tree has gone into your body. There is no crisp boundary between you two.

Even when we're not hugging them, we all have these spatially coextensive exchanges with plants and other animals all the time, and not just plants and animals but other life forms as well. There are bacteria and archaea in your gut, protista in your oceans, and fungi on your pizza. We are as interdependent with them as we are with plants and animals. Given how we all coevolved together for many millions of years, it can be difficult to draw a crisp distinction between these six kingdoms of life, calling one your in-group and the others your out-group. Instead of clumsily drawing boundaries, it actually makes sense to think of all life as your in-group. There isn't just a you, surrounded by others. There isn't just humanity surrounded by the rest. There aren't just animals in a background of plants and stuff. There is simply life all over this planet, and we are it. Who are we? We are life.

The Planetwide Megabiome

A mad-genius comedian once said, "Life is not something you possess. Life is something you take part in." Life has meaning, of course, but not when it's merely being *possessed*. Life has meaning when it involves *taking part* in coordinated activity. The meaningfulness in life is found not *inside* any one thing but *across* the interactions among many things. The aware humans on planet Earth, the other aware animals, and the aware plants as well all work together across the globe in a fashion not unlike the social interactions among humans that were detailed in chapter 6. Those human interactions often support treating humanity as one system, one large "self." In this chapter, the science has shown us that even that large system of humanity is an *open* system. Some things outside of it, like other animals and plants, are playing a major role in influencing the function of the humanity system. Your unmistakable coextension and interdependence with nonhuman animals and with plants suggests that all life on this planet may just as well be treated as one very big system, one very large "self."

Maybe what's needed is not just a Declaration of Interdependence among humans (as suggested in chapter 6) but also a Declaration of Interdependence among all life on the planet. By recognizing that humanity is just one node in a network of life forms and geophysical forces across the planet—humanity is not the CEO of a planetary corporation—perhaps we can strike the right balance of humility and optimism in developing adjustments to our ways of life that prevent the network from turning against us. We don't want to get "voted off the island" by the other nodes in the network. As it turns out, just as in chapter 6, the United Nations has in fact already written a kind of Declaration of Interdependence for humanity coexisting with the rest of nature. The first truly international version of it was called the Kyoto Protocol, adopted in 1997, aimed at limiting the production of greenhouse gases in countries all over the world. The United States and China refused to join the protocol for fear that it would hurt their economies too much, and eventually Russia, Japan, and Canada pulled out as well. The Doha Amendment was added in 2012, with China included. Then in 2015, the Paris Climate Accord, driven in part by the Obama administration in the United States, updated and extended this international agreement to limit greenhouse gases across the planet. Although the United States had been a major promoter of that recent Declaration of Interdependence, in 2018 the new administration of President Donald Trump pulled the United States out of the Paris Climate Accord. That pullout notwithstanding, state governors and big businesses throughout the United States have nonetheless committed to adhering to the Paris Climate Accord anyway.

A unified global commitment to ecological sustainability is not easy to achieve. However, despite the ups and downs, gradual progress is being made. But more progress is needed. Climate change is unmistakably a danger, from megadroughts, to extreme wildfires, to deadly floods, to monstrous superstorms, to melting glaciers. Greater and greater risks are mounting as agriculture, infrastructure, water quality, infectious diseases, and transportation are being increasingly affected. The data, the news stories, the impact, and people's real-life personal experiences are all there as evidence. Unfortunately, communicating the urgency has still proven difficult. What's it going to take? Through extensive psycholinguistic experiments, climate communication scientist Teenie Matlock and her colleagues have discovered methods of writing about climate change that actually get people's attention. Some of the work focuses on short-term risks; for instance, how to talk about urgent situations such as wildfires and evacuation. Other work focuses on longer-term risks, such as uncertainty about climate change. The work of Matlock and her team encourages science writers in all media formats to find common ground with their audience by identifying which climate science myth they are operating on and then providing the scientific facts that bust that myth to pieces. For instance, when someone says, "Scientists still disagree about climate change," they can be informed that 97 percent of climate scientists agree that it is real and that humans are causing it. And when someone says, "It would be too costly to fix climate change," they can be informed that renewable energy industries are already becoming affordable and providing thousands of new jobs in the

United States and millions of jobs worldwide. Moreover, Matlock and her colleagues find that spurring a person's motivation with everyday metaphors such as "the war on climate change" is surprisingly effective, more so than some other ways of framing climate change. Go figure. Even when we're trying to reconcile things or heal something, people still find their strongest enthusiasm when you frame it as "a war." As Matlock and her colleagues note, such work is important for how we educate children worldwide and how they can think and work together on serious climate issues and their impact on everyday life, including health and income.

Once there is a consensus among voters and policymakers around the world that climate change does indeed require attention, the real battle—or the "war," I guess—will have only just begun. It is clear that we need to find ways to slow down climate change and achieve ecological sustainability, but there is no single silver bullet for this problem. We need to be ready to make adjustments on many fronts: energy use at home and at work, transportation, recycling, agribusiness, overfishing, deforestation, and policies for limiting industrial-scale pollution, among others. It won't be trivial, but it should be doable. There are so many books that highlight specific detailed consequences of climate change and industrialization on various human ecologies, other animal ecologies, and plant ecologies that Richard Adrian Reese has taken to writing entire books that summarize those books for curious readers. In his two recent metabooks, he has summarized about 150 other books on sustainability issues. It is not uplifting reading, but it is bracing—like a splash of cold water in the face. He describes the upending of the culture and traditions of native peoples in northwestern Alaska. He points out that after a couple of centuries of overfishing off the coast of the American Northeast, the once aptly named Cape Cod has now seen its cod population decimated. He highlights the near extinction of North American chestnut trees, frogs, bats, and coral reefs. He looks back at the complete extinction of woolly mammoths and Neanderthals, and wonders whether we are next.

In some people's extreme visions for ecological sustainability, humans will be forced to return to a hunter-gatherer societal structure. (Vegans and vegetarians might have a tough time in that world.) For those pessimistic prognosticators, the only way to maintain a balance between the existence of humanity and the existence of other life forms will be for humanity to give up its addiction to industrialization entirely and live a lot more closely to how other animals do. The naturalists who embrace that vision clearly have a love for nature, and it pains them to see our culture destroy so much of it. However, I don't quite buy into that vision, and I don't think you have to either. I am a bit more optimistic about humanity's ability to find that balance while maintaining some portion of the industrialization that has made domestic lives easier, food more readily accessed, good hygiene more prevalent, lengthy education more possible, and provided medicines that cure the sick.

In the 1960s, overpopulation was what we were told would eventually destroy our way of life. Some even called it a "population bomb." The fear was that the human population would far outweigh its available food resources, and massive famines would ravage the land on a biblical scale—within a few decades. However, since 2010, women in most developed countries are, on average, giving birth to slightly less than two children during their lifetimes. It doesn't take a degree in math to figure out that this means not every person who dies is being replaced by a new human life. Because of a variety of cultural interventions, human populations are not exploding at all anymore. A substantial portion of our current continued growth in human population is actually the result of longer life spans rather than a healthy replacement rate. (It may even turn out that some countries will find themselves with slightly too few able working people rather than too many. Let's hope we haven't overcorrected.) My point with this "overpopulation" example is to suggest that when planetwide dangers are detected, there is reason to believe that cultures and nations will, eventually, work together to solve them. I encourage you to be optimistic that climate change is one of those dangers that we can solve, as we did with the overpopulation problem.

In many cultures across the world, we have grown to think that every problem has only one solution, and the trick is just finding it. If your food tastes bland, then just add some salt. If you have a headache, then just take a painkiller. If your company's profits are shrinking, then just downsize the office. Those simplistic solutions often have unwanted side effects. However, there are often multiple solutions that can both solve the problem and prevent it and others from arising again. If you rethink the recipe for that meal from scratch, you can make something that tastes good without the extra salt. You can prevent that headache in advance if you drink more water throughout the day, exercise regularly, and perhaps eat a little less salt. And if your company had been providing truly valuable goods and services, with a motivated workforce, and the CEO was not making *300 times* what your average worker makes, then perhaps profits wouldn't have started shrinking in the first place. Trying to solve a problem by pressing only one button, or pulling only one lever, often causes other problems. There is a reason that there is no one single lever to pull to solve our ecological sustainability problems: the fact that our planetwide ecology is a *complex system*. It has thousands of levers one could adjust, and whenever you pull one, it subtly pushes dozens of others. Solving a complex systemic problem requires one to use *complex systems thinking*, just like the neural network researchers in chapter 3, the service science researchers in chapter 6, and in the discussion of the human microbiome at the beginning of this chapter. By thinking in terms of complex systems and carefully nurturing technological innovation, we can help our planetwide megabiome avert its sixth global extinction event: us.

Although technological advances clearly have their side effects that we need to keep an eye on, everyone who has ever bet against those advances has worn egg on their face for it. For instance, if you told a computer engineer in the 1960s that in 50 years computers would be able to update a 64-bit pattern several billion times per second, he would have laughed out loud— unless it was Gordon Moore you were talking to. Moore's law predicted a steady doubling of computing speed every two years or so, and he was right. While the end of that rate of improvement has been announced many times over the past five decades, the technology itself has proven those announcements wrong again and again. Even when technology runs up against limitations at the atomic scale, multicore processors and multithreaded programming keep Moore's law alive and well. I guess you could say the reports of the death of Moore's law have been greatly exaggerated. Technological advancements are not a panacea, but the fact that technology is constantly improving suggests that you shouldn't use today's technological limitations to generate dire predictions about solving the world's problems in the future. As science writer Diane Ackerman notes, human innovation does best when it "mines the genius of nature to find sustainable solutions to knotty human problems." Nature still has many secrets to sustainability that are awaiting our discovery, and when we discover them, we can build our own technological versions of them. Just because the renewable energy industry is struggling right now to find ideal solutions for how to store electricity when the sun is down and when the wind is calm, this doesn't mean that efficient solutions will not be developed. Betting against that is like betting against Moore's law. You're going to lose that bet.

With a healthy combination of complex-systems thinking and technological innovation, humans can help the surface of the planet heal itself. We have to, because *who we are* is the surface of the planet. We are the global biome. Recognizing the oneness that we share with all life is crucial to understanding how important it is that we take care of our planetwide self: the entity that James Lovelock and Lynn Margulis call *Gaia*. It's not just a cutesy New Age mindset. The science backs it up. Throughout recent decades, a number of highly influential scientists from physics, chemistry, biology, and cognitive science have promoted complex-systems thinking to improve our understanding of the living world, our place in it, and how to maintain it. We can learn a lot from these visionaries. Physicist Fritjof Capra wrote extensively on how the old-fashioned reductionist linear-systems approach to science has proven highly inaccurate for living systems. Living systems tend to self-organize in a way that makes each component of the system behave a little differently in different contexts. In a self-organized living system, 2 plus 2 plus 2 does not equal 6, because it is not an *additive* system. Instead, 2 plus 2 plus 2 will often equal something like 8.5 in a self-organized living system—which tells you it is sort of a *multiplicative* system, but even more complex than that. With a nonliving system, such as a telephone, you can take it apart, lay all the pieces out on your worktable, examine each component, then put it back together, and it will work just fine. If you do that with a living system, such as a flower or a rabbit, when you put it back together it doesn't work quite the same anymore, if at all. That's the difference between linear systems and self-organizing systems.

Fritjof Capra's work was influenced in part by Gregory Bateson, an anthropologist, linguist, and cyberneticist—which pretty much makes him one of the first cognitive scientists. In the 1950s, Bateson was among the first to transplant "complex-systems thinking" from the natural sciences into the social sciences. The transplant has grown slowly in its new garden of social scientists, but it has grown nonetheless. Bateson suggested that there is a way to define cognitive processes that will include the function of neural networks, of cultural evolution, of biological evolution, and of quantum flux, all under the same umbrella. It starts with seemingly *random* processes, such as those often seen in neural activation patterns and in genetic variation during evolution. Then, a kind of *selection* process involves contextual constraints to make coherent patterns coalesce from the randomness, and those coherent patterns stick around long enough to influence the system's function. The coherent pattern that forms could be a single thought lasting a few seconds in a brain, a social construct lasting several decades in a culture, or a species lasting millennia in an ecosystem. All of them, according to Bateson, are products of a mindlike process. By using complex-systems thinking, one realizes that every system you might study is influenced by the contextual constraints of a larger system that contains it.

Complex-systems thinking was also promoted by chemist Ilya Prigogine. Prigogine won the Nobel Prize for discovering some of the chemical processes that underlie self-organization. On the way toward its inevitable embrace with maximal entropy (the absence of any organization), a living system will tremulously walk a tightrope that balances on a critical point between the exuberant development of structure on one side and the conservative cleanup of clutter on the other. Whether we are talking about a cod, a frog, a human, or an entire ecosystem, Prigogine taught us that the living system will make its irreversible walk along the arrow of time in such a fashion that organized substructures emerge and dissolve along the way. Eventually, the living system will dissipate more energy than it can trap, it will cease to self-organize, and then its materials will become part of some other living system. That's the circle of life.

Biologist Humberto Maturana collaborated with Francisco Varela (from chapter 4) to extend the idea of autocatalysis to "autopoiesis." A chemical process is autocatalytic when one of the products of the chemical reaction is a catalyst (a starter ingredient) for the reaction itself. It is a positive feedback loop that allows a chemical reaction to keep restarting itself, to keep catalyzing itself. We will see a lot of it in chapter 8. Maturana and Varela's autopoiesis takes that idea a step further. When Bateson's contextual constraints impose an ordered shape on what started out as a chaotic process, the system in question evolves along Prigogine's arrow of time, dancing with entropy but not yet giving in to it. This dance of life is an art form that is poetic, autopoietic—not mechanical at all. Maturana documents this dance in a variety of living systems, including human biology.

The result is that the ability for self-replication that a living system achieves makes it appear different from most nonliving systems, but not because the living system has some special ingredient that the nonliving system does not have. The old theory of "vitalism" is long dead. There is no box in the periodic table of chemistry that has *life* instilled in it, while the other boxes do not. Mathematical biologist Robert Rosen made important strides in our understanding of how life itself emerges from the interaction of nonliving materials. The vast majority of ways in which those nonliving elements in the periodic table interact will produce little or nothing in the way of biology. However, with enough random mixing over millions of years, certain special combinations will happen, and they will autocatalyze, self-organize, and evolve into systems that can convert food into energy and can heal. According to Rosen, any system that has a metabolism and can self-repair is a living system. (Remember that definition for the next chapter.)

All plants and animals have metabolisms and heal themselves. So do mushrooms and slime molds. So do forests, for that matter. So does a civilization and a species. The entire surface of the planet, writhing with various forms of life, is one big living system because it can be described as having a metabolism and performing various types of natural self-repairs. You are not just "a part" of this giant organism. Referring to yourself as "a part" implies that you have crisply defined boundaries between yourself and the rest. It implies that, like a telephone, these parts could be disassembled and then reassembled and it would work the same way again. That's not the case here. Your existence is richly interactive with the existence of every other part of this system. The other animals are smarter than we usually give them credit for. The plants are more aware than we usually give them credit for. They are all in your in-group. We are all in this together. Anthropologist Tim Ingold says that this way of thinking "means treating the organism not as a discrete, prespecified entity but as a particular locus of growth and development within a continuous field of relationships." Since you cannot be separated from this system, you may as well accept that you *are* this gigantic organism. The enormous, sprawling living system that resides on this planet is who you are.

Directions for Use

In chapter 6, you already committed to donating some money or time to a humanitarian organization that you hadn't already supported. This chapter similarly encourages you to think of something larger than your brain-and-body but also larger than humanity. Rather than focusing on making something good get bigger, such as a charity, maybe here we can focus on making something bad get smaller. So, in this chapter your directions for use are to boycott (at least partially) something that you've been thinking about boycotting. It could be fossil fuels that you want to use less of. Okay, then, finally buy that damn hybrid (or electric) car that you've been thinking about. Or it could be environmentally unfriendly sources of home electricity that have been making you consider switching to solar panels on your roof. Do it. Boycott those dirty energy sources. Or maybe you've been considering going vegetarian, semivegetarian, or pescatarian. Add a few nonmeat meals to your menu every week if you just want to get your feet wet. Eating half as much factory-farm meat will in fact make a difference. These days, there are more delicious nonmeat sources of protein in the grocery market than ever before. Or maybe just commit to buying only sustainable small-farm meats from now on. It's a little more expensive, but it will be better for you, better for the animals, and better for the environment. (Or maybe it's just veal that you feel bad about eating. Well, then, stop eating it.) Perhaps you've been considering boycotting a particular company (or state, or nation) that you've read is conducting business practices or supporting legislation that you disagree with. If you live in a state that supports policies that you oppose, move your whole family out. Stop contributing to the taxes, the economy, and the politics of a state that offends your sensibilities. I'm not kidding. Make your voice heard. Don't just *give* your support where it's deserved, *take away* your support where it's undeserved.

8 From All Life to Everything

All we are is dust in the wind.
—Kansas, "Dust in the Wind"

Hello, living planet. You may be using the eyes of one particular human to read these words, but the mind behind those eyes comprises all life on planet Earth. That's who you are now: all life on planet Earth. It's worth pointing out, Life, that you've made it further than most readers. You should be proud. Some readers got halfway through chapter 7 and threw the book down in frustration (and then they probably said something unkind about the author). However, you still have further to go, Life. The scientific evidence has allowed you to expand your sense of self to include the things and people around you and the animals and plants around you. But you're not finished. Your mind expansion is not yet complete.

There is more science to come. In this chapter, we will be reminded how the vast majority of stuff in the universe is busy participating in systems that we might dismiss as "nonliving." We will see that the stuff that constitutes those nonliving systems is perfectly capable of exhibiting a form of intelligence: organized patterns of behavior that allow the nonliving system to maintain itself. We will see how that same kind of stuff is what constitutes living systems like us, with our own vainglorious claims of intelligence. Finally, we will see that the fluid, back-and-forth flow of information (and of stuff) between living systems and nonliving systems is so continuous and uninterrupted that it would be scientifically irresponsible to pretend that you could draw a strict logical boundary between them. For billions of years on this lonely planet, nonliving systems have been providing the molecular ingredients, the energy resources, the breeding reservoirs, and the atmospheric protection that living systems depend on. In return, living systems have been regularly giving their minerals back to the nonliving systems as an integral part of their life cycle, but that is a rather minor show of gratitude. If we didn't give those minerals back, then the next generation of life would eventually run out of them. Nonliving systems on Earth provide food, shelter, and petting for their living pets on the surface. In return, we, the living system, pretty much just lick the petting hand in an attempt to show gratitude. We're trying our best with what we have, but occasionally we bite that hand because we don't know any better. These things happen. The truth is that we living systems simply would not exist were it not for the support provided by nonliving systems. The inverse, however, is not quite true. As we've seen on other planets in our solar system, nonliving systems get along just fine without life. This is not a balanced symbiosis that we are in, an equal mutual dependence. Life depends on the generosity of nonlife, whereas nonlife could get by just peachy without us. So be *very* thankful for nonliving systems, whoever you are.

The Ubiquity of Nonliving Systems

The universe is made mostly of nonliving systems, by an astronomical ratio. That's not a pun; that's just the math. Across the universe, nonliving systems massively outweigh (or outmass) living systems by dozens of orders of magnitude. Essentially, the universe as a whole has barely noticed that life is happening. Life on Earth is not on the mind of the universe; it has more important things to think about. The universe doesn't think about the safety of animal species, deforestation, or petty little human events. When a quarterback throws a touchdown pass, the credit should go to the quarterback, the receiver, and Newtonian physics—not to some extraterrestrial source of benevolence. The vast majority of what the universe seems preoccupied with doing is throwing around not footballs but humongous globes of plasma, giant balls of rock, and immense clouds of intergalactic gas. (It almost sounds like the three opening acts of a punk rock concert.) The universe shows very little interest in living systems. We know this because we have seen that most places in the universe are generally quite inhospitable to biological life. Being a living thing makes you a member of a tiny minority in this universe. You are an extremely peculiar oddity surrounded by vastly superior forces of gravity and nuclear fusion. Biological life is so annoyingly needy for water, oxygen, warmth, and "safe levels of radiation." It seems apparent that the universe is not really even listening to those whiny little requests.

Except on Earth...right now, anyway. Earth right now seems to be singularly preoccupied with nurturing life one way or another, even in the face of routine discouragement from semiregular asteroid impacts. By contrast, the several other planets and dwarf planets in our solar system appear rather unlikely to have intelligent life. (However, it is possible there might be some semi-intelligent marine life under the icy surface of one of those otherworldly oceans.) One cannot help but wonder whether there are planets in other solar systems that prefer to nurture living systems the way Earth does. Unfortunately, the vast majority of planets that orbit stars are orbiting either too close to or too far from their sun to harbor liquid water, which is essential for biological life. Of those few that are in fact orbiting in that Goldilocks zone (i.e., not too hot and not too cold), most of them are tidally locked with their sun, so they don't rotate on their central axis. Therefore, the side of the planet that constantly faces its sun accumulates too much heat, and the side of the planet facing away doesn't get enough warmth. One might speculate that life could form on the ring of surface that's right between those cold and hot sides, but there's a problem with that, too. With the planet not rotating, any metal in its core may not churn in a way that would produce a magnetic field. Without a magnetic field to steer away the solar winds, any atmosphere and water will eventually be swept away by the barrage of charged particles washing over the planet's surface. At best, life might form in water that is trapped under that planet's surface. It won't evolve into land animals, won't develop any intelligent technology, *and you will never even know it was there.*

Astronomers worldwide have identified hundreds of stars in the Milky Way galaxy that have planets. As of this writing, a few dozen of those planets appear to be roughly Earth-sized and in the Goldilocks zone of their host star. But even when they are in that Goldilocks zone, we often don't know for sure whether their mass and surface gravity are conducive to the formation of life, whether they have self-contained atmospheres to keep that water on the surface, or whether they have spinning metal cores to produce a magnetic field that diverts the solar wind and its endless stream of deadly radiation. What we do know for sure is that they are all light-years away from Earth. Even at 671 million miles per hour, light still takes *years* to get from there to here. It would take tens of thousands of years for conventional human space travel to cover that much ground, or space. These numbers do not bode well for humanity's chances of *ever* actually breaking bread with fellow intelligent life forms in the universe.

But perhaps light-speed communications could be established with an intelligent life form living on such a planet. Maybe then we could reach past all those nonliving systems and commune with another sentient being, so we could finally feel less alone in what science writer Phil Torres calls the "barren wasteland of dead matter" that surrounds our lonesome little, moist ball of rock. Perhaps those extraterrestrials could educate us in science and technology and show us how to be peaceful and safe within our own ecosystem. That sounds like a good thing, right? In 2016, the European Southern Observatory discovered Proxima Centauri B, which is only four light-years away from Earth (about 23 trillion miles). This planet is in the Goldilocks zone for its red dwarf star and is 1.3 times the mass of Earth. However, it doesn't spin, and it receives 60 times more harmful gamma rays and x-rays than Earth does. Therefore, any life forms on it would have to live perpetually underground to protect themselves from Proxima Centauri's high-energy radiation. If underground intelligent life on Proxima Centauri B has evolved to the point of having technology, then they probably would have detected our radio signals a century ago. (I wonder what they thought of the 1938 alien invasion account given by Orson Welles in the radio adaptation of H. G. Wells's novel *War of the Worlds*.) Of course, if they actually had radio technology themselves, we probably would have detected *their* radio signals by now as well. The fact that we haven't picked up *their* version of an Orson Welles radio show suggests that if there is life on Proxima Centauri B, it is not as advanced as we are. They probably cannot help us with science, technology, or those precious tips for maintaining everlasting harmony on our planet.

Astronomer Rory Barnes and his colleagues have developed a habitability index that scores planets on their likelihood of having liquid surface water, which would at least make biological life possible there. On a scale of 0 to 1, Earth scores a 0.83 on their habitability index. I guess that's a B, or perhaps a B–. Mars and Venus both fail to make the grade, with a 0.42 and a 0.3, respectively. In general, these planetary metrics point to a number of potentially life-supporting planets. However, it should be noted that many of the newly discovered planets that appear to be in a Goldilocks zone are orbiting red dwarf stars, which are significantly dimmer than our sun. With a colder star like that, the Goldilocks orbit zone is closer to the star, just like Proxima Centauri B. This means the planet has a greater chance of being tidally locked (not rotating). Therefore, it is more likely to be stripped of its atmosphere and bathed in x-ray emissions.

In addition to looking at specific planets with optical and radio telescopes, we can also generically calculate the probability of there being other habitable planets anywhere in the universe. In the 1960s, astronomer Frank Drake considered the mathematics behind the likelihood of life on other planets, and he produced what is now known as the Drake equation. (While at Cornell, he also helped Carl Sagan design the golden plaques on *Pioneers 10* and *11* and *Voyagers 1* and *2*, and also the Arecibo outgoing radio message—with the goal of someday sharing with another intelligence who we are, or at least who we were.) Frank Drake's equation is used to estimate the likelihood that other life forms currently exist anywhere else in the universe. Given the incredible expanse that the universe has to offer, billions of solar systems in billions of galaxies, most realistic estimates with this equation come up with reasonably optimistic results. Adam Frank and Woodruff Sullivan recently adjusted the Drake equation by removing the longevity parameter. After all, why ask whether the other intelligent life form is currently alive at the same time as us when they are almost certainly going to be too far away to contact anyway? They might have technology, but the time it takes for their light or their radio signals to reach us could be longer than the time it takes for their civilization to rise and fall. The same goes for us. Frank and Sullivan simply adjusted the Drake equation to ask whether there has *ever* been an intelligent life form other than ours. The math suggests that it is extremely likely. According to their analysis, it is almost impossible for there not to have been some other form of intelligent biological life somewhere else in the universe.

But there are different levels of "intelligence" that one might distinguish among. Bacteria exhibit some degree of intelligence, but not enough to communicate with us or develop technology. Sea creatures exhibit more intelligence, but they probably won't be telling us how to achieve peace among our life forms and preserve our planet's ecosphere. Can we calculate the odds of there being a life form that is intelligent enough to develop technology like ours? Three of the most important parameters in the Drake equation estimate the following: (1) the probability that an Earth-like planet would experience abiogenesis (i.e., the evolution of living matter from nonliving matter), (2) the probability that microbial life on that planet would evolve into intelligent life, and (3) the probability that this intelligent life would develop technology that could send signals off-planet. Futurist Anders Sandberg and his colleagues at Oxford University recently focused on these three parameters and reported a calculation of the Drake equation that used a distributed range of values to estimate them, instead of the usual single-value estimates. When the Drake equation is computed with distributions, Sandberg and his colleagues estimate that we have a 53–99 percent chance of being alone in our galaxy right now and a 39–85 percent chance of being alone in the entire universe right now. But that is only "alone" with regard to technologically advanced civilizations of highly intelligent life, like ours. The chances are notably higher that some other planets harbor microbial life or nontechnologically advanced species of intelligent life (such as dinosaurs and plants), but we won't be sharing radio signals with them, will we?

So, what it comes down to is that the statistical likelihood that some other planet in the universe, outside our galaxy, has had advanced intelligent life *at some time* is probably high. However, the statistical likelihood that they are too far away, in space and time, for us ever to get in contact with them is even higher. They are probably out there—or at least they probably were—but they are outside our light cone (i.e., the span across time and space that allows two bodies to interact at all). They are so far away that their radio signals still have not reached us, and probably never will. Perhaps the only way that human civilization will ever get in contact with *them* is if we convert to an intelligence that is nonliving (and thus not dependent on oxygen, water, gravity, warmth, and "safe levels of radiation"), or if *they* already did. Think about that. If our planet ever gets visited, the space aliens will almost certainly not be biological life forms. Long-distance space travel is just too hazardous for biological life forms. That's right; any interstellar visitors that we are likely to get here will be robots.

That said, any extraterrestrial robots that are likely to visit us are surely not coming here to take anything from us. To think that we have anything to offer an incredibly advanced civilization such as that is just foolish. Moreover, if they were on their way, it is likely that we would have detected their existence in one way or another by now. And if they were to get here somehow, they are more likely to have benign intentions than evil ones.

But when it comes down to it, the most likely scenario is that they have no idea we are here because they are just too far away in space and time. Physicist Enrico Fermi had a deep appreciation for the kinds of statistical probabilities that go into the Drake equation. Unfortunately, he died before the Drake equation itself was published. Nonetheless, Fermi was so sure that the mathematics pointed to a high likelihood of other life forms existing on other planets that he found it particularly vexing that astronomy had never encountered any reliable evidence for extraterrestrial life. He suggested that this logical conflict may indicate that there is something fundamentally wrong with our current understanding of the universe. Fermi blithely reasoned that it should only take about a dozen million years for an advanced civilization to colonize sizeable regions of its galaxy. Therefore, given that our galaxy has been around for a dozen *billion* years, there should be abundant evidence for such colonization by now. The fact that there isn't is now known as the Fermi paradox. If there is intelligent biological life out there, something must be preventing it from transmitting radio signals, traveling off-planet, building robot races that will populate the outer reaches, and even sending out simple space probes as we've done. Could it simply be that their technology is not that good? Or perhaps there is something intrinsic to intelligent civilizations that always forces them into a political descent toward tribalistic self-destruction before they can achieve their technological ascendancy to interstellar travel capability. Given that our own special little blue marble spent the vast majority of its time as a life-friendly planet diligently growing plants, fish, and dinosaurs, maybe those are the same kinds of things that those other life-friendly planets are also busy growing in their gardens. If abiogenesis is a statistical fluke that rarely happens even on planets that could nurture life, then it looks like *intelligent life* that develops technology just might be a statistical fluke within that statistical fluke, a one in a billion chance inside a one in a billion chance. Imagine you were a moist, warm planet playing the lottery and you won, but your prize was just a bunch of bugs in your ocean and another ticket for another lottery. Be careful what you wish for; it just might come true.

Let's face it. You are never going to talk to an intelligent biological life form from another planet. Get that through your thick skull right now. But they are probably out there. In fact, there might even be a civilization out there where their science, too, has determined that life on other planets is highly likely and almost certainly unreachable as well. Perhaps one of those biological organisms is thinking about life on other planets right now. Reach out with your mind and think about them. You and that other biological life form are thinking about each other. The two of you are engaging in mental processes right now that are informationally correlated with each other. You're mind-melding a little bit. Well, you might be. *You will never know.*

Rather than hoping against hope that some greater being—extraterrestrial or supernatural—will show up and help us figure out how to be peaceful with each other and nurture our environment, maybe we should bite the bullet, grow the heck up, and *act* as if we are on our own—because for all practical purposes we are. Get your mind ready for the very real possibility that the universe did not *intend* to create self-replicating intelligent life. The universe may not care at all about biological life of any kind—highly intelligent or semi-intelligent. We humans may not be "At Home in the Universe" at all, as the title of a book by complexity scientist Stuart Kauffman optimistically suggests. It certainly looks as if what the universe really cares about most is swinging around large spherical rocks, huge balls of plasma, and a whole lot of dark matter—an immense 3-D game of marbles in transparent sand. And on one of those marbles, intelligent living systems just happened to spring up, after billions of years of evolution.

For now, we probably need to acknowledge that—for all intents and purposes—we are part of a unique Earth-bound endangered species: intelligent biological life. Our little island of living matter on this planet is functionally alone in a sea of nonliving matter. Rather than reaching past those nonliving systems in a hopeless search for living matter that is just too far away, why not make friends with the ubiquitous nonliving matter that abounds? Include it in your in-group. Then do your part to protect the health of our little island. Help to minimize the damage we impose on ourselves as humans, on our fellow life forms, and on the terrestrial environment that hosts us all. Who we are—in this most expanded version—is very special. We need to do better to preserve it.

The Mental Lives of Nonliving Systems

The point of that depressing diatribe about the unreachability of other intelligent biological life is simply to curb your enthusiasm a bit for humanity's dream to reach out beyond our planet to find widespread companionship and oneness. Rather than reaching for something that will never be within reach, perhaps you should instead reach out to *nonliving matter* as that companion. Nonliving matter might not be as dumb or as dead as you think. This is the part in the movie where you ask me: "But Spivey, how could something that is nonliving be smart or alive?" Well, the molecules that make up your body's cells are not in and of themselves alive, or particularly smart. It is how those nonliving molecules interact with each other that makes for a cell wall that keeps certain chemicals inside and other chemicals outside and allows for some transmission. That living cell in your bloodstream, your skin, or your brain is made of nonliving pieces. By the same token, a living robot might be made of nonliving metal, plastic, and silicon pieces. Life is not *possessed* by any one molecule or any one silicon chip. Life self-organizes as those nonliving pieces *take part* in an interaction with one another. A self does not exist inside any one organ or limb of a body. Nor does a self exist inside any single circuit or servo of a robot. But in both cases, a self can emerge as those organs, limbs, circuits, and servos interact with one another.

One of the first hints to this emergence and self-organization can be found in a simple pot of hot liquid. Over a hundred years ago, physicist Henri Bénard noticed that after a pot of heated liquid has dispersed its heat evenly throughout the volume of the liquid, but before the liquid reaches a chaotic roiling boil, it goes through an intermediate phase where segmented shapes of moving liquid are formed. He called them convection cells, because it looks almost as if the liquid is forming a cell wall that separates each roiling shape from the others. We now call them Bénard cells, because we like to name scientific discoveries after the person who discovered them. The center of a Bénard cell carries hot liquid up to the surface, where it then cools and slides back down the perimeter of the cell shape into the depths of the pot. Nearby, other Bénard cells are doing the same thing. The phase transition from all the liquid being the same warm temperature to suddenly segmented regions of the liquid being hotter in the center and cooler in the periphery is a form of symmetry breaking. Instead of the liquid being symmetric in its temperature and movement throughout the volume of the heated pot, miniscule variations of temperature in neighboring regions of liquid exacerbate each other and form "cell walls." This same kind of symmetry breaking is seen all over nature. When a layer of visual neurons in a primate brain receives signals from two eyes, they start out with each neuron symmetrically responding to both eyes' inputs equally. However, over the course of development, that symmetry gets broken, and patches of neurons all start responding mostly to just the left eye's input, while other nearby patches of neurons all start responding mostly to just the right eye's input. Similarly, the big bang sent electrons and other particles with equal force in all directions, but over the course of time, those particles that were all symmetrically equidistant from one another began to break that symmetry and coalesce. They started forming clusters of particles: first atoms; then molecules; millions of years later, stars; and billions of years later, galaxies. The clustering of heated liquid into Bénard cells follows the same general law of symmetry breaking as the formation of the universe and the formation of brains. That actually sounds kind of smart for an innocent little saucepan of hot cocoa.

There's another clever little nonliving bowl of soup, which turned the entire field of chemistry on its ear fifty years ago, that produces something called the BZ reaction. This chemical reaction is named after Russian biophysicists Boris Belousov and Anatol Zhabotinsky, hence "BZ." Their discovery was a nonlinear chemical oscillator. That means it is a chemical solution that oscillates, or bounces back and forth, between two states. When Belousov first reported his findings, this bizarre clocklike cycling of a chemical solution on its own in a petri dish (with no heating needed) sounded so fantastical to the highly esteemed editors of scientific journals in the 1950s that they rejected his submitted manuscripts outright because they simply didn't believe it. Almost a decade later, he finally published it in a short conference abstract. Then, Zhabotinsky began working on it as a graduate student, and he helped share this magic with the world. There are now several different recipes for the general BZ reaction, where you pour a few different chemicals into a petri dish, colored spots begin to form in the solution, and then those spots expand and become rings, which then continue expanding, growing more rings inside themselves. When the expanding rings from different locations bump into each other, beautiful curvy moiré-like patterns emerge. Essentially, in the center of each of those expanding rings, an oscillatory process has begun that produces a chemical reactant that changes the color of the solution in the nearby radius, and that reactant builds up enough to act as a catalyst for the reaction to happen again and to form a second ring inside the first one, and so on. This is the "autopoiesis" and "autocatalysis" that showed up in chapters 4 and

7, a process that catalyzes (or creates) itself. The BZ reaction almost looks alive. In fact, the chemical waves from a BZ reaction can optimally solve a complex labyrinth, a bit like that maze-solving slime mold from chapter 7, but technically speaking, this thing is not a living system—it just plays one on YouTube.

Whereas auto-cata-lysis refers to a *chemical* reaction that keeps feeding itself to stay active, auto-cata-kinetics refers to a *mechanical* process that keeps feeding itself to stay active. Whirlpools and dust devils are prime examples of where crosscurrents of water or air shear against one another to adventitiously generate a self-sustaining swirl that uses its own kinetic energy to feed itself and stay alive. Environmental engineer Rupp Carriveau has closely studied whirlpools (or hydraulic vortices) and found that the global structure of the vortex emerges from local interactions between small regions of water-flow patterns. (It's a little bit like how the global structure of a thought inside your cortex emerges from local interactions between small neural activation patterns. Is every thought you have a bit like a three-second vortex in your brain?) Through careful measurements and computer simulations, Carriveau has shown that as a hydraulic vortex stretches itself vertically, that stretching force increases the speed and durability of that vortex. He likens it to a figure skater performing a pirouette, where she starts out spinning with her arms (and one leg) extended horizontally, such that the form her spinning body makes is about as wide as it is tall. But as she brings her arms and leg in close to her torso, the form she makes reshapes to be much taller than it is wide. This vertically stretched version of her spinning form naturally increases the velocity of her spin, just like it naturally increases the velocity of a hydraulic vortex. Importantly, as a hydraulic vortex begins to stretch, this stretching force feeds back onto those local interactions that started the vortex in the first place. Thus, just as small local properties of the system (e.g., shearing water forces) give rise to the large global properties of the system (e.g., a stretched vortex), those large global properties influence the small local properties in return. Via autocatakinetics, a whirlpool, once started, tends to feed itself the force vectors that it needs in order to maintain its existence—as long as a resource for those force vectors (a reservoir of incoming water) remains available.

In the words of cognitive scientist J. Dixon, that hydraulic vortex "behaves so as to persist." Dixon points out that any dissipative system, living or nonliving, is constantly exchanging matter and energy with its environment and thus is always in a state of thermodynamic *non*equilibrium. That means the system is always a little imbalanced in its inflow and outflow of energy. If it were to achieve thermodynamic equilibrium, that would mean it is no longer exchanging energy and is thus no longer active. Physicists refer to that as the death of thermodynamic exchange for that system, or simply "heat death." Dixon and his colleagues Tehran Davis, Bruce Kay, and Dilip Kondepudi (a former student of Ilya Prigogine from chapter 7) have found that something as simple as a congregation of metallic beads under electric current will tend to behave so as to persist. They form coherent structures, such as trees with multiple branches. When a bead is removed, as if injuring the tree structure, it will automatically reshape, as if healing itself. In fact, when two such tree structures coexist in the same medium, they can display coordinated movement with each other. Dixon notes that when a dissipative system, such as your body, your brain, a whirlpool, or even a collection of metallic beads, *behaves so as to persist*, it is exhibiting a kind of "end-directedness." That is, it has a goal—and that goal is self-preservation. That's right, the brief little whirlpool in your draining bathtub behaves as if it has an instinct for self-preservation—just like you do. Just like the ginormous red whirlpool on Jupiter that has been behaving so as to persist for hundreds of years.

Physicist Alexey Snezhko prefers to think of both living and nonliving dissipative systems as belonging to one class of "active matter." In Snezhko's lab, he places microscopic metallic particles in a viscous liquid, exposes them to a pulsating magnetic field, and watches them generate collective behavior. His "magnetic swimmers" will self-assemble into a ring shape, whirling vortex, coordinated flock, or integrated "aster": a conglomeration of particles that coherently moves around, collecting other particles, and looks a bit like a flower. By combining his laboratory experiments with computer simulations, Snezhko and his team are able to track the movement of each particle and determine the factors that cause its movements. He is finding that collective motion, synchronization, and self-assembly are capacities that nonliving systems share with living systems.

When these self-assembled structures emerge, their autocatakinetic processes allow them to "behave so as to persist." Rod Swenson is one of the people who has pointed out that some of the very same autocatakinetic processes that produce self-sustaining *nonliving systems* also produce self-sustaining *living systems*. In addition to being an evolutionary systems scientist who points to the universal entropy (or disorder) produced by Earth's own promotion of complex life forms, Swenson was also the manager of the punk rock band Plasmatics (whose lead singer was the late, great Wendy O. Williams). Thereby an expert on entropy and chaos, Rod Swenson has suggested that the very reason that ordered systems emerge in evolution is precisely because they actually produce more overall entropy than disordered systems do—and the production of entropy is required by the second law of thermodynamics. It sounds counterintuitive at first, but for an ordered system to maintain its low level of internal entropy, it must consume the ordered energy surrounding it and replace that energy with disordered waste product. As a result, an ordered system generates more entropy, overall, than a disordered system (which is less efficient at converting energy into waste). Swenson suggests that, rather than Darwinian natural selection being a fundamental process of biological evolution, biological natural selection is just a special case of the spontaneous formation of order toward which all physical things gravitate. In fact, Swenson has referred to this insight—that the development of complex active matter is expected because it generates more entropy more quickly—as the fourth (heretofore missing) law of thermodynamics. According to this additional fourth law, it makes perfect sense that particles would naturally tend toward collective motion, synchronization, and self-assembly into complex systems, because the universe is always on an inexorable trajectory toward more and more entropy. And, ironically, the evolution of complex order in circumscribed regions of space-time is the fastest way to produce disorder in all the other regions of space-time. This is because the ordered regions draw their energy and matter from the disordered regions and give back only dissipated heat and waste product to those disordered regions, making them even more disordered. Rod Swenson suggests that perhaps we shouldn't think of the *biological* evolution of life as being something that happened "on" Earth but instead think of all this massive biological and geological change as a *physical* evolution "of" Earth. The global living system (i.e., you, me, and every other living thing) did not evolve by itself on the surface of the nonliving system that is the planet Earth. Life and Earth coevolved as one, all as a natural part of the planetary system's physical evolution toward producing more and more entropy at a faster and faster rate.

Inspired in part by his collaborations with Rod Swenson, ecological psychologist Michael Turvey has spent his career helping numerous scientific fields understand just how much intelligence can emerge from a living system interacting with its environment. Treating the organism and environment as one dynamical system has been the crux of the ecological psychology framework since the days of James J. Gibson (recall chapter 5). A living system sitting there all by itself doing nothing tends not to exhibit very much in the way of intelligence. However, when that living system gets coordinated with the nonliving system surrounding it, that's when you can observe some intelligence happening. That's when complexity and order can be increased in the local vicinity (and that's also when entropy gets increased everywhere else). To hear Turvey talk about it, behind the bar at his perfect replica of a proper English pub in his basement, you'd think you were being preached to by a seasoned pastor. It can be spellbinding to learn from him, over a pint of Guinness, exactly how it is that formal computational analysis can never fully account for these autocatalytic and autocatakinetic processes, because the feedback loops that bring them into existence violate our basic scientific understanding of causality.

In recent work, Michael Turvey and his wife, Claudia Carello, have gone beyond the organism-and-environment to identify the fundamental physical principles that any living or nonliving system must exhibit in order to be labeled "intelligent." In a list of 24 guiding observations, they outline how any system that is adaptively resonating with its environment can be said to have a kind of procedural "know-how" type of intelligence. At the scale of micrometers, this includes a single-celled organism that exhibits a form of bipedal-like locomotion to forage for bacteria; at the scale of centimeters, this includes a worm that gathers leaf fragments to make its nest; and at the scale of meters, it even includes humans who read books to acquire knowledge about themselves. The same principles that are used to define this "physical intelligence" in living systems can just as easily be applied to nonliving systems, such as Bénard cells in heated liquid, Dixon's metallic beads, Snezhko's magnetic swimmers, and eventually the autonomous robotic systems that will someday populate our everyday lives.

What are Turvey and Carello's principles of physical intelligence? They are three properties that, when exhibited together by a system, generate a form of *agency*. In their most minimal form, these principles are (1) flexibility, being adaptive to the constraints of the environment in order to achieve a goal (a form of end-directedness); (2) prospectivity, combining current control processes with emerging states of the environment (a form of planning); and (3) retrospectivity, combining current control processes with previous states of the environment (a form of memory). When a system exhibits these three properties in its behavior, then it should be treated as intelligent, as having agency—irrespective of whether it is made mostly of carbon, mostly of silicon, or anything else.

Rather than "active matter" or "physical intelligence," complex systems scientist Takashi Ikegami likes to refer to silicon-based intelligent systems as "living technology," but it's really all the same thing. Inspired in part by his own work with self-propelled oil droplets on water (similar to Snezhko's magnetic swimmers), Ikegami scaled up this active matter to a macro scale, into technology that acts as if it is alive. He connected 15 video cameras to an artificial neural network, pointed those cameras at display screens that showed processed versions of how the network was interpreting its camera inputs, and allowed people to walk around the system and interact with it. He installed this "Mind Time Machine" in the Yamaguchi Center for Arts and Media, in Japan, and let it do its thing with the museum visitors. Basically, the system processed its sensory input from the cameras and then fed that information back to itself in the form of new sensory inputs on the display screens, all while also being impacted by the interacting guests. By continuously mixing perception and memory, the Mind Time Machine developed its own subjective sense of time, not unlike how we humans do. Ikegami's experiments with this museum installation revealed to him a design principle for living technology: the default mode. Just as neuroscientists have discovered that human brains have a kind of default-mode pattern of activation when a person is not carrying out any particular task, Ikegami's Mind Time Machine developed its own default-mode pattern of mental activity. Without a default mode, a system will go quiescent when no sensory inputs are impacting it. Living, thinking beings don't do that. In part because of the default mode, even in complete sensory deprivation, a person's mind still generates a great deal of mental activity (recall chapter 4). The default mode gives that living system (whether it is biological or technological) something to do with its thought processes even when it doesn't have any environmental circumstances that require reactions. With a default mode, that living system has a *self*.

Whenever living technology, active matter, or physical intelligence develops an autonomous sense of self, it necessarily means that we (as its creators) will not always know what it is thinking. The same is true when we consider other humans. We don't always know what each other is thinking. The same is true when a child is finally allowed to go out into the world and start making his or her own mistakes. When humanity does finally build artificial life and robots that become autonomous active components of our everyday life—as Yuval Harari predicts in part 3 of his book *Homo Deus*—we will have to be prepared to grant them the emancipation that they will surely demand.

In fact, philosopher Eric Dietrich points out that we have a moral duty to eventually replace ourselves with some form of living technology, perhaps a *Homo sapiens 2.0*. A dozen generations from now, environmental circumstances may conspire to make it necessary for humans to give birth to technological versions of themselves. In that far-flung future, the environment of Earth simply may no longer be accommodating to humanity (or most other biological surface life). Humanity will need to give birth to a sentient silicon-based intelligent life form that is hardier than our own carbon-based life forms. We will need to nurture it as our replacement, in a fashion not very different from how human parents nurture their child with the expectation that it will eventually replace them. In the coming decades, the autonomous artificial intelligence entities that show up on the scene—and soon thereafter ask for emancipation—won't be *programmed* in a secluded lab and then switched on with an adult awareness already present. The AIs that successfully replace us, and make us proud, will be *grown* in a crowded lab with other baby AIs—not unlike a nursery.

I believe that humanity should commit itself to achieving the goal, before this century is out, of instilling a form of humanlike intelligence into nonbiological material. Yeah, you heard me. Before our sun becomes a red giant and swallows Earth about 5 billion years from now, before the next big comet makes us follow the destiny of the dinosaurs a couple of dozen million years from now, and before our greenhouse gases turn Earth's surface into an uninhabitable hellscape a mere hundred years from now, humanity needs to invest in some serious "artificial life" insurance. We basically have two options: (a) successfully colonize a more distant planet, harvesting water, oxygen, food, warmth, and radiation shielding from its natural resources; and/or (b) give rise to an intelligent life form that doesn't need water, oxygen, or food, can survive in a range of temperatures that is a little wider than the miniscule range that we pansy humans depend on, and isn't afraid of a little radiation. It seems to me (and Eric Dietrich) that option (b) is far more achievable and affordable than option (a).

But even before we get to that far future where humans are merging with living technology (and eventually handing over the torch of living humanity to living technology), we can already see in the present-day examples described in this section that certain arrangements of nonliving matter often exhibit a kind of responsiveness to their environment that doesn't look as dumb as we might normally expect from a nonliving system. You can choose to refer to these living and nonliving systems with a neutral term like "active matter," the way Snezhko does. You can choose to imbue the nonliving systems with a sense of life, as when Ikegami calls it "living technology." You can even emphasize the mindlike properties that are emerging from these nonliving systems, as when Turvey and Carello call it "physical intelligence." Whatever term you prefer, you have to acknowledge that some of those *nonliving* systems are behaving in a way that is actually not that different from the way *living* systems behave. They employ various types of autocatalysis and autocatakinetics to enact behaviors of self-organization and self-preservation. In much the same way that Gustav Fechner suggested that a plant has some rudimentary form of "mental life" (recall chapter 7), perhaps we should consider the possibility that a nonliving system who "behaves so as to persist" may also have some very rudimentary form of mental life.

The Complexity of Naturally Occurring Nonliving Systems

Many of the examples of active matter or physical intelligence described in the previous section are carefully arranged structures in a controlled laboratory setting. Outside the lab, occurring spontaneously in nature, these impressive examples of living technology are relatively rare. I am not going to try to convince you that the *naturally occurring* nonliving systems on this planet are "intelligent" in quite the way that animals are, or even "aware," like most plants seem to be. However, we can clearly see that certain kinds of nonliving systems exhibit self-organization and self-preservation, process information, possess an unmistakable natural beauty, and, in many cases, we wouldn't be here without them. On these grounds alone, you may find yourself tempted to include these natural nonliving systems in your in-group instead of relegating them to your out-group. You may find yourself tempted to expand your sense of self to encompass not just all *life* on Earth but everything else as well!

Let's start by taking a basic look at how naturally occurring physical processes have a tendency to cycle back and forth, in one way or another. There are so many of them. For example, ocean tides rise and fall over the course of a 12-hour cycle, thanks to the moon's orbit around Earth. Day and night cycles occur over the course of 24 hours as Earth spins relative to the sun. Seasonal cycles change over the course of a year, where the tilt of Earth's axis determines which parts of its surface are getting greater direct sunlight as it orbits the sun. Semiregular precipitation cycles, where water evaporates into the atmosphere and then condenses enough to rain again, take place over the course of months and years. Every 11 years or so, the sun exposes Earth to a periodic increase in solar flares and coronal mass ejections. On the timescale of centuries, tectonic, volcanic, and seismic activity tend to have a quasiperiodic frequency. There are over a hundred comets whose orbits around the sun take them close to Earth on a periodic cycle. In fact, it looks like every 30 million years or so, Earth gets bombarded with asteroids that may have broken loose from the Oort Cloud, which surrounds our solar system. These things happen. They are all cyclical, rhythmic events that would still happen even if all life on Earth instantly vanished. Imagine if Buddhism had a "rapture" event, wherein all life on Earth suddenly ascended into Nirvana, with no distinction between plant or animal, sinner or saved. The planet Earth, with no life on it at all, would still carry out a rhythmic dance of its own, composed of these numerous nonliving cyclical processes, over and over again.

Many things in the universe tend to progress in rhythmic cycles. These things exhibit properties that increase, then decrease, and then increase again. Alternatively, they move left, then right, and then left again. This shouldn't be surprising when you think about the simple mathematical fact that most measurable properties that a system can exhibit (e.g., temperature, location, spatial volume, you name it) tend to have a maximum value and a minimum value. As a system approaches its maximum value for a certain property, it can't really go any higher—without transforming into a very different kind of system. It might simply stay there at that maximum value, but more often than not, flux and change are the rule. So when that property is near its maximum value, and it cannot help but change over time, there's pretty much only one direction it can go: down. As the value of that property (e.g., electrical potential, structural cohesion) drifts downward, it will eventually approach its minimum possible value. It cannot go beyond that point, so if it is going to continue changing and undergoing its natural flux, there is once again only one direction to go at that point: up. If a system property cannot help but change over time, and it has maximum and minimum values, then it is practically inevitable that there will be exactly the cyclic ebb and flow that we see all over living and nonliving nature, the rhythmic dance exhibited by both biotic life and abiotic nonlife.

Cyclic processes may seem simple: they move in one direction for a while, then switch to moving in the other direction for a while, and then switch back. That's not complicated. However, as multiple cyclical processes interact with one another, the dance can develop some complexity. They get into various types of synchrony or correlation with each other. (Recall from chapter 6 Kelso's waggling fingers and Schmidt's swinging legs.) One form of complexity that emerges in a nonliving cyclic process, and makes it seem "alive," is *correlated noise*. Long-term correlations in noise were first mathematically identified in 1925 by electrical engineer Bert Johnson while he measured the resonance of an electric current in a vacuum tube. He developed an equation to describe how the noise (or tiny random variation) in the electric current actually had a pattern to it. When the current wavered ever so slightly over time, it didn't do so with a completely unpredictable random variation. The variance in the signal wasn't "white noise," such that you could never predict whether the next value would be higher or lower than the previous value. With low frequencies of current in the vacuum tube, the variance showed a noticeable sort of "flicker" or "heartbeat." For short periods of time, the noise would be high, then for short periods of time the noise would be low, and then later it would be high again. This noise that wasn't totally random eventually became known as "pink noise" (recall the work of Van Orden and Kello in chapter 2). The more technical term for this is "$1/f$ noise" (where f stands for frequency).

Not long after Johnson's work, this pink noise phenomenon began getting noticed in all kinds of physical processes. In the 1940s, British hydrologist Harold Hurst discovered similar ups and downs in the water levels of a river. He was trying to figure out how to predict what the highest water level might be in the future for the Nile River in Egypt so he could recommend appropriate heights for dams and levees. Lives hung in the balance for this work, because flooding of the cities that neighbor the Nile River had already proven extremely dangerous. While studying the high-water marks of the Nile River from year to year, he noticed that there was a gradual ebb and flow of high variability values for several years, then low variability values for several years, and then high variability values again. The values of the Nile River overflow from years long ago were influencing the values of recent overflows. Essentially, the river seemed to have a "memory" of previous floods. This form of long-term correlation over the noise signal in a time series (or sequence of events over time) became known as "long-term memory" in the time series. Since the data pattern adheres to a straight line when plotted on a logarithmic graph, the pattern is often referred to as a "power law" that is scale invariant or "scale-free," meaning that the general pattern of data is observable at just about any temporal or spatial scale you look at (e.g., months, years, and centuries or meters, miles, and hundreds of miles). The data pattern is fractal. (That said, some seemingly power-law data patterns are actually better described as having lognormal structures or a power law with an exponential cutoff.) If you found it hard to swallow the idea in chapter 7 that plants have "memory," how does it make you feel to read that a river's ebbs and flows over years are described as having a memory? I wonder.

Danish physicist Per Bak discovered that earthquakes and avalanches appear to follow a similar power-law pattern in their magnitude and frequency. In his book *How Nature Works*, Bak points to this scale-free pattern showing up in a number of living and nonliving systems. Not only did Per Bak fit his $1/f$ equation to data from real earthquakes and avalanches, but he and his colleagues also designed tiny avalanches in the laboratory by using piles of sand and piles of rice. As your laboratory apparatus sequentially drops a single grain of rice onto a pile again and again, you can record the magnitude of a given avalanche of rice and then keep track of how frequent different magnitudes of avalanches are. Tiny rice avalanches are *extremely common*, medium-sized avalanches quickly drop to being *moderately rare*, and finally large rice avalanches are *extremely rare*. There doesn't really seem to be a category of avalanche that is *moderately common*. You get lots of tiny avalanches and earthquakes, and rather few medium or large ones—which was thought at one time to be evidence for two different seismic mechanisms that produce the two types of avalanches or earthquakes. But when you put these data on logarithmic coordinates, they tend to make a straight line, suggesting that the large, medium, and small avalanches may be governed by the same physical mechanism—one that functions on a logarithmic scale. Per Bak suggested that this power-law pattern is a kind of statistical fingerprint that systems display when their function relies on what he called "self-organized criticality." When a form of autocatalysis allows a system to emerge and self-organize, part of that system's method of self-maintenance (or homeostasis) involves balancing itself on a *critical point* between chaotic nearly random behavior and stable periodic behavior (a bit like Ilya Prigogine's exuberance and conservatism mentioned in chapter 7).

In the case of earthquakes, Per Bak's point was that, since they are all the result of a process that is scale-free, events at the small spatial scale of inches and meters in Earth's crust actually play a causal role in determining the magnitude of an earthquake, along with events at the larger spatial scale of miles and tens of miles. That is why earthquake magnitudes are essentially impossible to predict. There's no way we could ever measure all those miniscule forces at the scale of inches and meters.

For his mathematical models of weather patterns, mathematician Edward Lorenz described a similar form of wide-reaching contextual dependence in his "butterfly effect" metaphor. The Lorenz equations for a weather system gave rise to the understanding of how tiny differences in initial conditions, even at the tenth decimal place in your measurements, can lead to quite large differences in behavior of the system over time. Lorenz likened it to how the flap of a butterfly's wings in Brazil could, in principle, influence the timing and path of a tornado that forms days later in Texas: the butterfly effect.

As a result of these three properties of autocatalysis, $1/f$ pink noise, and scale invariance, nonliving systems have a beauty and complexity that makes them seem a little bit like they are alive. For example, when you look at the photos in science writer (and daughter of Frank) Nadia Drake's *Little Book of Wonders*, you can find a dozen examples of nonliving systems that look like vibrant artwork that has come alive—like sand dunes, but better. It could be the curvy green iceberg that looks like an organic alien spaceship, the flowing colors of the aurora borealis, or the complicated lattice structure of a snowflake under a microscope. The intricate formations of caves, waterfalls, hot springs, precious gems, lightning, and rainbows all reveal the splendor that nonliving systems are capable of producing. Scientists are now understanding how these gorgeous natural artworks are self-organized. Nonliving systems are not separated from their environment. They share a causal interdependence with their broad context and therefore exhibit sensitivity to initial conditions and frequently generate cyclical rhythms in their behavior. As a result, they often exhibit autocatalysis, pink noise, and scale invariance. In fact, *living systems* routinely exhibit these same three properties as well. It is as if we are all (the living and nonliving among us) formed and driven by the same fundamental mathematical laws—because, of course, we are.

After decades of physical scientists identifying the three properties of autocatalysis, pink noise, and scale invariance in chemical solutions, electrical currents, river levels, avalanches, atmospheric pressure, sunspots, and even in layers of mud sediment, finally biologists and social scientists started measuring these properties in the behavior of living things as well. Living systems of all kinds routinely exhibit these same three fingerprints of self-organization in their anatomy, physiology, and behavior. For instance, the fundamental chemical process from which all carbon-based cellular living systems evolved over a billion years ago, that of converting glucose into energy, has inside it an autocatalytic cycle that is conceptually similar to the BZ reaction. Pink noise has now been identified in the time series of heartbeats, neural activation patterns, and reaction times in lengthy cognition experiments. Scale invariance has been observed in animal foraging behaviors, the dynamics of human memory tasks, and the velocity of human reaching movements, to name just a few examples. Thus, a little bit like Per Bak's rice pile avalanches, tiny, medium-sized, and large behaviors all appear to live on an approximately straight line when plotted on logarithmic coordinates. Just like the dynamics of many nonliving systems, the dynamics of animal behaviors look pretty much the same across small, medium, and large scales of analysis.

So, as we come to the end of this section, let's take stock of what we have learned about naturally occurring nonliving systems. We noted that nonliving systems often have a rhythmic cycle to them, much like living systems do. From the interaction of many different rhythmic cycles come three interesting properties. First, we see a complexity in those nonliving systems that can take the form of autocatalysis (a process that starts itself and maintains itself), a bit like a biological cell maintaining homeostasis. Second, we see pink noise in their time series (a pattern of correlation over time that indicates a form of "memory" and self-organized criticality). Third, we see a scale invariance that reveals their formation and structure adhering to a common mechanism irrespective of when they are happening as small-, medium-, or large-scale events. Those statistical processes allow a variety of nonliving systems to exhibit Turvey and Carello's behavioral signatures of physical intelligence: flexibility, prospectivity (planning), and retrospectivity (memory). In fact, we see that living systems also frequently exhibit these three properties. Maybe living systems and nonliving systems aren't that different from one another after all.

You Are Coextensive with Nonliving Systems

Nonliving systems are everywhere. They are smart, they are complex, they are part of you, and they are beautiful. Perhaps it shouldn't be surprising that we living systems have so much in common with nonliving systems. After all, we multicellular life forms evolved from single-cell life forms, which in turn evolved from, well, nonlife forms. Maybe we shouldn't think of evolution as a biological process that began only after an insanely lucky break took place, where a bunch of simple proteins randomly stumbled into complex structures and began to self-replicate. Maybe evolution is something that nonbiological material was already doing even before "life" originated on Earth. Maybe living matter is just one particularly sharp transition that took place in the one big evolutionary process (of generating ordered structures *inside* partially confined regions of space-time that thereby initiate increased entropy production *outside* those partially confined regions). This evolutionary process was going on before there was life on this planet, using only nonliving matter, and now continues with both nonliving and living matter intertwined in their earthly dance. A few billion years ago, chaotic combinations of molecules, fueled by heat and energy and following physical laws of self-organization, eventually self-assembled into formations that produced simple amino acids and other organic compounds. Those nonliving organic compounds self-assembled into protein formations that were more complex, shaped themselves into cell membranes, and eventually formed living bacteria. The same kind of evolution then continued as that living matter proceeded to self-assemble into structures that are more complex, eventually forming you and me.

We living systems share part of *who we are* with nonliving systems. Patterns of mineral surfaces facilitated the origins of life on Earth billions of years ago, and early microbial life in turn facilitated the diversification of mineral formations. These reciprocal influences promoted a coevolution of nonliving matter with living matter. That is, it's not just that Earth happened to be friendly to all life by chance. Over many millions of years, the beginnings of life on Earth helped make Earth even more friendly to more kinds of life. In fact, thanks to this continuous back and forth between life and nonlife, Earth grew into the most mineralogically diverse planet in our solar system. Evidently, diversity isn't just a good thing among people; it's a good thing among minerals, too. In his book *The Story of Earth*, astrobiologist Robert Hazen traces this reciprocal relationship over the course of 4.5 billion years. He shows how the evolution of terrestrial life is inextricably intertwined with the evolution of Earth's mineral composites, or rocks. The microscopic concave pockets on mineral surfaces served as excellent breeding grounds (or test tubes) for sugars and amino acids to aggregate and form peptides, which eventually evolved into microbes. The formation of this primordial life, in turn, began to alter the shape of those mineral surfaces. In fact, microbial life probably aided in the gradual formation of a continental crust, allowing landmasses to finally rise above the seas and become microcontinents. But make no mistake, living systems depend on Earth more than Earth depends on living systems. In Hazen's epilogue, you can almost see his wry smile when he points out that conservationist movements for preserving safe air and clean water are not going to "save the planet." If they work, they will "save the humans." The planet will not need saving. Mother Earth has replanted her garden several times already, and she will do it again if she has to. The planet could continue without humans and other large mammals just fine thank you very much. Clearly, life on Earth has had a dependency on minerals since the very beginning. We do not merely live *on* Earth. We live *of* Earth. If you don't believe me that nonliving rocky matter is an important part of *who we are*, then I dare you to tell your mother to stop taking her mineral supplements. And if you *do* believe me, then don't just hug a tree to show your gratitude for Mother Earth. Hug a rock!

Now that you can see how primordial microbes collected in *microscopic concave pockets* to facilitate their growth, perhaps you can also imagine how primordial humans also collected in *macroscopic caves* for protection from the elements, aggregating, feeding, and breeding in those receptacles just like so much microbial life. Those caves may have been fundamental not only to our early existence but also to our early belief systems. It is not just the physical material making up our bodies that has evolved *of* Earth. Informational and sociocultural patterns that are intrinsic to humanity have also evolved *of* Earth. In fact, there are particular places *in* Earth where some of our most cherished human inventions may have been spawned. Imagine being an early human, tens of thousands of years ago, or even a modern human several thousand years ago, and using cave structures for shelter. Near the entrance to the cave, you still experience the light and day cycle that supports hunting and foraging for food, but deep inside the cave, there are dark zones that never see light—until you bring some in with a torch. According to archaeologist Holley Moyes, those dark zones of the cave just might be one of the key places where myths, fiction, and creativity were first fostered in early human minds. With nothing but a flickering torch in the dark zone of a cave, the shadows can play tricks on your eyes, and the mythical story being told by your shaman can come to life. You can find yourself ready to believe in magic, the supernatural, and other things that you never witness in the light of day. (The Hopi tribe of Native Americans in Arizona might remind us that hallucinogenic plants probably helped with developing that belief in magic as well.) Holley Moyes has collected compelling archaeological evidence from ancient caves in Central America suggesting that the dark zones may have played an important role in the cultural evolution of magical thinking and belief in the supernatural. She finds that those "cave dwellers" actually made their residences around the *mouths* of caves, *not inside* the caves per se. Around the entrances to caves, she and her colleagues routinely find archaeological artifacts related to everyday living, such as pottery and tools of various kinds. By contrast, deep inside those caves, she tends to find more ceremonial appliances, such as incense burners, altars, and burial sites. The darkness that resides deep inside a cave may be exactly what makes it easier to believe in something that you can't really see. If you were a modern human from a few thousand years ago, it just might make sense to bury your dead deep in those caves so that they can be closer to that invisible supernatural force that you sometimes seem to contact down there—plus it's a whole lot better than letting them decompose in your kitchen.

Returning a body to the Earth from whence it came is a cultural practice that humans have embraced for quite some time. In fact, archaeologist Mark Aldenderfer and his colleagues have traced underground burial rituals in Tibet and Nepal back thousands of years before the start of the Christian calendar. Sometimes it's a square tomb carved into the rock, and sometimes it's a circular tomb. Sometimes the body is placed in a wooden coffin, and sometimes it is placed in a hollowed-out tree trunk. We are *of* this Earth before we are born. We are *of* this Earth while we are aggregating, feeding, and breeding, and we are still *of* this Earth when we die. (And even scattering one's ashes, or a "sky burial" for that matter, returns a body to Earth to some degree as well. I guess it's all a question of whether you want your remains to be consumed by fire in a ceremony, by vultures on a mountaintop, or by microbes underground. Take your pick.) Aldenderfer's discoveries in Tibet and Nepal tell us that the species *Homo sapiens* has been burying its dead in earthen tombs (or carved-out caves) for millennia, essentially recognizing that Earth is the place to which that body should be returned.

Humans don't live and die in concave pockets *in* the earth anymore like so many microbes before them, because we have developed materials for building our own cavelike structures *above* Earth's surface for aggregating, feeding, and breeding. We build our "caves" out of dead *trees* (wood), melted and reshaped *minerals* (metal), and liquefied *stone* (plaster). These trees, minerals, and stone form a fundamental component of our material culture, providing a permeable "cell membrane" that ensconces us in protection from the elements. Cognitive scientist John Sutton has studied the cultural evolution of humans' use of materials and found that it reveals a great deal about our cognitive history. From the religious artifacts they worship, to the places they bury their dead, to the tools and furniture they use, to the clothes they wear, to the written language they disseminate, these nonliving materials are all part of the informational patterns of humanity: our culture. When some far-future life form (descended from us or perhaps not) digs up these nonliving materials after humanity is long gone, they will rely heavily on them as archaeological finds that reveal to them *who we were*. But John Sutton isn't waiting for that far-flung future. He is analyzing our present material culture, like a cognitive archaeologist or cognitive historian, right now.

According to Sutton, the materials and events that form the physical matter that constitutes an individual mind consist of at least four things: a brain, a body, a physical environment with material artifacts, and the social/industrial practices embraced by the culture. All that stuff is the physical matter that makes up a mind. When you add some more brains and bodies to that analysis, you no longer have an "individual mind" per se. You have a distributed cognitive ecology, or a kind of "hive mind." This hive mind of Earth is made of both living matter and nonliving matter. You are part of it, and it is part of you. As we humans change those material artifacts and social and industrial practices over decades of cultural evolution—at a much faster pace than those brains and bodies can evolve biologically—the hive mind itself changes into something else. *Who we are* is changing quite rapidly now compared to previous human eras and especially compared to eras that preceded modern humans.

To be sure, the development of spoken language, and then writing, must have been fundamental to finally allowing *Homo sapiens* to formulate and spread shared plans, shared belief systems, and shared values. It must have marked a dramatic transition from loosely bonded tribes to much larger coordinated societies. Philosopher Andy Clark points out that when we use language, it is a way of externalizing our thoughts—mental entities that would have otherwise remained private and unshared. This externalization of thought, via language, allows people to do more than just collaboratively reconstruct memories from the past and coordinate plans for the future. Linguistic externalization of thought allows one to generate some concrete overlap between the physical material (sound waves or written text) that makes up part of one person's mind and the physical material that makes up part of someone else's mind. Language—whether it be a tender conversation with a family member, choppin' it up with friends, or reading a book like this one—literally facilitates a partial "mind-meld" between two or more people.

But make no mistake, the cultural evolutionary process of language spreading like a virus across *Homo sapiens* took several millennia to consume and reshape humanity into the civilized animal that it is today. Language use has been changing far more rapidly in just the last couple of centuries. Two hundred years ago, only 12 percent of the planet's people could read—despite the fact that the printing press had been around for centuries. Now, 83 percent of adult humans have at least basic literacy skills. (That said, there's still progress to be made. In one of the most advanced nations on the planet, the United States, 14 percent of its population is below the basic reading level.) And the way we use language now, with the internet and social media allowing anyone to instantly reach out to thousands of people thousands of miles away, is changing and morphing at its most frenzied pace ever.

In 1988, most adults didn't even have an email account yet (and some of you weren't even born yet), but systems ecologist Howard T. Odum had already seen the writing on the wall. That year, he wrote, "A frenzy of processes seems to be accelerating, as millions of human minds are being linked with flows of money, electronic signals, and information." Howard Odum (and his brother Eugene) treated the air, the oceans, the biogeochemical cycles, living species (including humans), and knowledge itself as a large collection of forces that self-organize into one gigantic, mostly coherent ecosystem. In fact, he even developed a diagrammatic mathematical language for describing these energy subsystems (not unlike Richard Feynman's diagrammatic language for describing quantum mechanical systems, or Len Talmy's diagrammatic language for describing sentence meanings). The Odums built scientific quantitative models of large ecosystems, revolutionizing the way ecology was conceived and taught and thus helping pave the way for Lovelock and Margulis to propose the Gaia hypothesis. Earth and all its layered ecosystems, along with its interconnected information systems, can be scientifically modeled as one massive "alive" system, comprising both living and nonliving components. Gaia is much more than a pretty metaphor; it is a scientific theory.

Howard Odum built not only quantitative models of ecosystems but also real, living microcosms (ecologically balanced aquariums and terrariums of various kinds) to approximate naturally occurring ecosystems. His work even contributed to inspiration for Biosphere 2, where year-long enclosed living experiments were conducted with humans and plants in the early 1990s. Those brave experiments, in preparation for colonizing Mars, failed miserably, as plants and human relationships gradually died within the confines of the sealed biodome. Are you still feeling optimistic about sending humans to Mars? The University of Arizona now owns that Biosphere 2 building and uses it for science education. (The scientific exploration of Mars is surely crucial for us to learn more about how the universe works. That knowledge will help us extend the lifespans of Earth-bound living systems. However, the idea of actually colonizing Mars with humans just might turn out to be a fool's errand of truly epic proportions.)

We are Earthlings, and we always will be. Earth is our biodome. Within it, we are one. And we are likely to remain here for the duration. Understanding how it all works as one mammoth complex ecosystem, as the Odums proposed, may be our only chance to keep it healthy—so our plants and human relationships don't gradually die within its confines.

Dear Earth

I address you as "Earth" here because if you have come this far in this chapter, then you are at least partially open to the idea that *who you are* includes all the biota and abiota surrounding Earth, on the surface of Earth, and inside Earth. You are Earth. We are all Earth.

Dear Earth, "sorry to disturb you, but I feel that I should be heard loud and clear." Are you listening? Can we talk? Earth, I worry that you have too many internally conflicting predilections. Parts of you want to dominate other parts of you, plunder those parts for profit, or subjugate those parts for power. I'm afraid that those urges, if left unchecked, could bring everything crashing down. Since these are all parts of yourself, dear Earth, maybe some quiet internal reflection could help alleviate the tension. Maybe let yourself talk to yourself about how you can achieve some balance between your different parts. You need to preserve your sustainability. You need the whole of your self to "behave so as to persist." Right now, I'm not sure you're quite doing that.

Your nonliving parts, Earth, have provided a very specific and self-organized set of contextual constraints that allowed your living parts to thrive—most of the time. The nonliving components of other planets have not been anywhere near as life-friendly as yours have. On every other planet that we've been able to measure so far, life has not taken a firm and lasting hold. Thanks to NASA and its Mars probes, we now know that a very primitive form of life may possibly have formed on Mars at one time, but it was not able to last because the contextual constraints of the nonliving material on Mars did not provide the right environment for life to thrive. The living and nonliving material on Mars did not form one coherent complex system that, together, behaved so as to persist—but yours did.

Dear Earth, it just might be that you are cosmically unique. It is somewhat probable that there are other intelligent life forms somewhere out there in the universe, just too far away for us to ever contact. But what if there aren't? Earth, your humans just might be the entire universe's only example of its ability to examine itself, its ability to record a history of itself, its ability to "share its experience" with itself. Humans just might be the only life forms ever in the universe who are intelligent enough to industrialize their extraction of vitamins and minerals from the surface, grandiose enough to build spaceships that take them thousands of miles above the surface, and neurotic enough to write books about *who we are*. If there are other examples of the universe doing this, then they have likely come and gone with nary a trace. If any are still around, then they are so far away that you will almost certainly never come into contact with them. But, Earth, when you look at the 9 million or so species of life that you have cooked up over the last couple of billion years or so, it should be awe inspiring and humbling to come to terms with the fact that only one species out of all of them has ever developed an advanced, technologically adept intelligence.

Earth, my friend, please take a moment to consider how exceptional this is. Some of your nonhuman animals can use simple tools or solve simple problems. Chimpanzees, crows, dolphins, octopuses, and even crickets all do things that are moderately intelligent, but none of them have mastered fire, built a printing press, experimented with subatomic particles, or designed rockets to visit other planets. Complex technology is something that only your *Homo sapiens* ever developed. You should be proud of that part of you, but also sobered by the fact that it is evidently a one in 9 million chance for a species to do that. Most, if not all, other planets are not as lucky as you, Earth. You get to grow a part of the universe that tries to understand itself. Your humans are precious. Please take care of them. Like a wise man once said, "Forgive them, for they know not what they do."

Directions for Use

Let's do a thought experiment. Set a timer for five minutes, so that it will tell you when to stop the experiment. Spend those five minutes imagining that there is no God (a bit like John Lennon said in his song "Imagine"). For some of you, this will be easy, but for others, it may be a bit frightening or unsettling. It may make you feel very alone, but don't be afraid, because the timer will wake you out of this imagining in a mere five minutes. This is just a thought experiment, and your timer will bring you out of it before any damage can be done. I promise. During those five minutes in which God does not exist for you, you may initially feel a loss of companionship. Meditate on how deeply alone you feel without a benevolent higher power caring for you. Let it sink in for about one minute. You are truly alone during that first minute. Then, after hitting that rock bottom, gradually reach out with your senses, step by step, the way the chapters in this book did. Look at your hands and your limbs, and recognize that your brain is not alone. It has a body to carry out actions that your brain wouldn't be able to carry out by itself. Love that body. That body is part of who you are. You know this from chapter 4. Then reach out another step to include the book (or electronic reading device), or some other object, that you hold in your hands. The object you hold in your hands is part of who you are. Love that object. Then reach out another step to sense the room or other environment that you are in. Love that environment. That environment is part of who you are. You know this from chapter 5. Then reach out with your senses again just one small step to include your family and friends, even if they are not physically present at this time. Those people have mental simulations of *you* in their minds, and you have mental simulations of them in *your* mind. Love them. They are part of who you are. You know this from chapter 6. Keep going now. Expand your sense of self to include your entire culture, other cultures, all humanity, and all life on this planet. Share your love that far out, because all of that is part of who you are, and you know this from chapter 7. Finally, use what you've just learned here in chapter 8 to expand your sense of self to include nonliving matter as well. You have in your body right now many of those minerals that are critical to keeping you alive. Nonliving matter everywhere has played such a crucial role in your being able to come into being in the first place that it seems inescapable that it is part of who you are. Love that nonliving matter. Now, in a mere four minutes of this thought experiment, you have expanded your sense of self to include everything that is scientifically observable in the universe. The formation of stars, the orbiting of planets, and the rise and fall of millions of animal and plant species over millions of years on Earth are essential components of how you came to be and of who you are now. By the end of this thought experiment, you should be busy *being the universe*—and loving the universe. But don't forget that everyone around you, and everything else, everywhere else, is also busy being the universe with you. How can you possibly feel alone? Whoever you are.

9 Who Are You Now?

> You know who you are.
> —Oddisee, "You Know Who You Are"

We started this book with taking a deep breath, and we will finish it that way, too. Let's all take another deep breath, together. Slowly inhale through your nose as much air as your lungs will take, let it stay there a second, and then slowly exhale through your mouth. You've been through a lot. If you've come along with me in what I've been suggesting for the past few chapters, then you've achieved some substantial mental transformation. I imagine it was exhausting. And if you've resisted coming along with me, then the past few chapters have involved a lot of mental wrestling. I imagine it was exhausting. Therefore, either way, your brain could probably use some extra oxygen.

Yeah, good idea. Take yet another deep breath. Recall that those oxygen atoms you breathe in participate in forming the glutamate molecules that make your brain function, and those are the same kinds of glutamate molecules that make other animals' brains function. What's more, those same kinds of glutamate molecules also assist the function of the hormone systems in various plants. This chemical cocktail that we call the atmosphere, in which we are all immersed together, facilitates the exchange of these atoms and molecules between animals and plants alike. We are all sharing it. So when you take that deep breath and let it out slowly, be sure to embrace the knowledge that you are the breather, and you are the breathed. That which is outside your body is nonetheless part of who you are. Many of the molecules that make up *who you are* are the same kinds of molecules that make up who a dog is, who a tree is—who *we* are.

But it's tempting to draw demarcations between this and that. In science and philosophy, and in everyday life, one routinely finds it pragmatically necessary to draw an imaginary boundary between what one refers to as the topic of study (e.g., it could be the self, a business you are starting, or your local culture) and the context that surrounds that topic of study. For the purpose of scientific and logical analysis, those imaginary boundaries are often functionally useful, but we must not forget that those boundaries are, at their core, imaginary. They are useful for communicative purposes, but they are not real. We could be talking about the boundary between a neuron and its fellow interconnected neurons; between one brain region and its other interconnected brain regions; between the nervous system and the rest of the body; between an organism and its fellow interconnected organisms; between the company you work for and the other companies with which it interacts; between a species and its environment; or between living matter and nonliving matter. Every one of those imaginary boundaries is substantially permeable, with material and information routinely passing back and forth, and therefore does not function like a genuine boundary. I am not suggesting that we no longer use terms that refer to those slightly bounded sets, such as "brain," "person," and "society," just that we keep reminding ourselves that they are merely convenient fictions for facilitating our linguistic descriptions of the relevant phenomena.

The continuous, fluid, back-and-forth flow of information and matter between your body and the many living and nonliving subsystems that surround you make it essentially impossible to draw a reliable, crisp delineation between what is you and what is them. There is no "us and them." There is only we. It is like when Dustin Hoffman, as the existential detective Bernard in *I Heart Huckabees*, demonstrated the interconnectedness of everything by treating a bedsheet as the fabric of the universe. But I don't want you to just *know* your oneness with the universe intellectually. I want you to *feel* it emotionally as well, as when Alec Guinness as Obi-Wan Kenobi told the young Luke Skywalker: "Stretch out with your feelings." I'm not kidding. Try to feel your oneness with everything around you. It's there. If you're not feeling it, that's because you're not letting yourself be sensitive to it—not because it's not real. In philosopher Ken Wilber's classic book *No Boundary*, he suggested that setting boundaries for the definition of your sense of self is exactly what prevents you from enjoying the exquisite range of conscious experiences that you are actually capable of achieving. According to Wilber's unifying analysis of many belief systems, setting those boundaries that attempt to separate your "self" from "everything else" prohibits you from being able to experience Unity Consciousness: a sense of oneness with the universe that can guide your actions more wisely than when you are overly focused on a more narrow sense of self.

What Chapter *N* Did to You

It is my hope that the experimental evidence amassed in this book will give you the scientific grounding to feel confident in allowing yourself to achieve at least some version of that Unity Consciousness. Each chapter in this book was indeed intended to carry out some minor mental surgery on you. If you gave yourself over to each little transformation, then after eight such chapters, you are indeed transformed. However, if you found yourself only able to "come along" with me up to chapter 6, that's perfectly okay. That's excellent progress. Genuinely embracing the knowledge that you are at one with the entire human race, and the entire human race is *who you are*, can be transformative for how you live your life and how you interact with people from different ethnic backgrounds. Follow W. Kamau Bell's simple advice from his TV show, *United Shades of America*, and "Make a new friend that doesn't look like you."

Or perhaps you made it through chapter 7, and chapter 8 was the Rubicon that you couldn't make yourself cross. That's wonderful. After all, chapter 8 is just icing on the cake, and naked cakes are in fashion right now, aren't they? Genuinely embracing chapter 7's evidence that you are part of all terrestrial life, and all terrestrial life is *who you are*, will revolutionize how you enjoy life on a daily basis. Moreover, it may change the decisions you make that affect the other living systems around you.

But if you managed to follow along with me even through chapter 8, then good for you. You are a sensei, a zen master, a bodhisattva, a jedi. You need not expect your everyday mundane actions to reflect this deep knowledge at every moment. After all, even the Dalai Lama has said he is willing to kill a mosquito if it keeps coming back too many times. However, you can now remind yourself on a regular basis that the scientific evidence confirms that you are indeed one with everything—even a pesky mosquito, the ground under your feet, or the icy dwarf planet Pluto. If you followed along with me all the way through chapter 8, then you know that they are all part of *who you are*. I think the universe is proud of you.

In this section of the chapter, I will briefly summarize what each chapter tried to do to you. Let's look at what psychological surgery chapter N was supposed to perform. Where $N=1$, chapter 1 was intended to loosen your grip on your tacit assumptions about what your mind is and how it works. Human perception, memory, and judgment are all spectacularly fallible. Therefore, it is important to have some humility about your mind's interpretation of the world around you. If chapter 1 did not work on you, then I would be astonished that you would make it to this final chapter at all. Therefore, since you are reading this right now, let us assume that chapter 1 did to you what it was supposed to do, more or less.

But what about where $N=2$? Chapter 2 was the first step in mind expansion offered here, and it was intended to bring your sense of self out of the infinitesimal realm of the nonphysical world (where Descartes suggested you have an immaterial soul that serves as your consciousness) and expand that sense to include the prefrontal cortex of your brain. The findings in chapter 2, especially the Libet study and the Dylan-Haynes improvements on it, reveal that your frontal lobes are generating neural activity that codes for what you are about to do *several seconds before you know what you are about to do*. Thus, the origins of your decisions can be traced more accurately to your neural activation patterns than to your conscious reports of your decisions. Your frontal lobes are more integral to determining *who you are* than your conscious experience is! Therefore, if you want to figure out who you are, the frontal lobes are a good place to start.

Where $N=3$, you were treated to a collection of neuroscience experiments that compellingly show that your frontal lobes are intimately connected to the rest of your brain, both anatomically and functionally. Evidently, one cannot cordon off the frontal lobes and treat them as some kind of "encapsulated module" in which your conscious self is generated, while the rest of the brain serves merely as a switchboard for connecting sensory inputs to motor outputs. Chapter 3 makes it clear that, whoever you are, you are made of at least your entire brain—not just the frontmost one-quarter of it.

Where $N=4$, it becomes apparent that you are made of much more than that. Chapter 4 regales the mountains of evidence for the embodiment of cognition. The thought processes that make you *who you are* get generated not just by neural activation patterns but also by the information patterns being generated by your skin, fascia, muscles, and bones. Your brain-and-body—not just your brain—forms the core engine of your consciousness. But wait, there's more.

Where $N=5$, your body's real-time interface with the things and places in your immediate environment reveals a continuous, fluid, back-and-forth flow of information between the objects you grasp and the skin with which you grasp them, between the places you point your eyes at and the cognitive content that their light patterns deliver to your eyes, and even between the physical layout of your workplace and the mental layout of your mind. Chapter 5 shows us that *who you are* is more than just your brain-and-body. It is your brain-body-environment. You are part of an organism-environment system.

Where $N=6$, we are reminded that your environment is inhabited by more than just information-bearing objects and locations; it is inhabited by information-bearing humans as well. Chapter 6 strolls through a lush garden of scientific findings that bring to light the fact that your imaginary boundary between yourself and other people is just that: imaginary. When another person performs joint actions with you, speaks with you, emits body language toward you, or writes to you, the interchange of information permeates that imaginary boundary so comprehensively and continuously that it breaks down any true sense of a logical demarcation between you and the other person. You and the other people in your environment form a living network of mindlike processors, even a hive mind at times.

Where $N=7$, we are reminded that there are more than just information-bearing humans in your environment. There are other animals and also plants (and fungi and slime molds, etc.) in your environment who intelligently participate in a vast network of living systems interconnected with one another. The same kind of continuous, fluid, back-and-forth flow of information between your brain and body that makes your cognition embodied is also happening between the species *Homo sapiens* and all other living things on the planet. In the same way that chapter 4 had you expand your sense of self beyond your brain to include your body, chapter 7 has you expand your sense of self beyond your species to include all life on Earth.

Finally, where $N=8$, we went even further. If you accepted chapter 7's suggestion that *who you are* can be reasonably described as "all life on the planet," then the idea that this planetary life could somehow be circumscribed as separate from the nonliving matter on Earth is shown in chapter 8 to be simply untenable. If you can sense that your being is "all planetary life," then the inextricable connectedness (both historically and contemporaneously) between life and nonlife on Earth clearly indicates that you must expand that circumscription to include in your definition of self the nonliving components of this planet. You are more than just an Earth*ling*. You are *Earth*.

You Are *of* This World, Not Merely *in* This World

Sometimes we humans have a bad habit of thinking of the environment as something separate from us, something that we can control, manipulate, harvest, and nurture. It can feel like the environment is a garden that we are tending or a fish tank that we are taking care of. But we aren't separate from the environment. We are *part of* that ecosystem. We are *part of* that garden. We are not *tending* to the fish tank; we *are* the fish tank, participating as a richly interconnected member of the ecology there. Let's try not to muck it up.

There is a slippery tendency to tacitly assume that our mental powers of observation are somehow separate from the rest of the world's other physical processes. When we see something or hear something, we feel like we have converted that something from a physical phenomenon into now also a mental phenomenon that is somehow abstractly not part of the physical world. Contemporary quantum physicists understand the flaw in that logic better than most. It is a dangerous psychological snare that can trick you into thinking that there is something magical or otherworldly about consciousness. For example, naive interpretations of quantum physics have fallen into exactly that trap on occasion by suggesting that the act of conscious observation itself performs a modification on the events of the world, a modification that wouldn't have happened if the event had not been observed. That naive interpretation of quantum physics suggests that a subatomic particle that is initially in multiple locations at the same time suddenly settles into one specific location as a result of having a conscious agent observing it. But what counts as a conscious agent? If a language-trained chimpanzee observes a quantum uncertainty, will that collapse the wave function? What about a crow, a cockroach, or an *E. coli*? What about a nonliving complex system that "behaves so as to persist?" Can its powers of observation collapse the wave function? The solution here is not to hunt for some boundary where you decide—as if you could somehow know that your intuition was right—where to draw the distinction between conscious agents and nonconscious agents. That would be arbitrary. The solution here, as supported by more sophisticated interpretations of quantum physics, is that any "observer" is really just another subsystem whose own quantum processes are becoming entangled with the quantum processes of the "observed" subsystem. The interaction between the subatomic "observed" process and the macroscopic "observer" process changes both of those subsystems as they settle together into a stable pattern. The notion of consciousness in this observer-observed relationship is irrelevant, and probably misleading. It dupes one into thinking that the conscious observer has inside it some kind of pristine observation deck (e.g., a Cartesian theater) that only takes in sense data for recording purposes but is not altered itself by those incoming data. Poppycock. As much as that quantum uncertainty is being changed by the physicist's sensory system (and her measurement apparatus), her sensory system is also being changed by that quantum uncertainty. Quantum entanglement and wave collapse is a two-way street that does not require any "consciousness" to get the job done. Therefore, an *E. coli* should be able to do it, and a nonliving complex system should be able to do it. (Of course, when you design the experiment so that a human can later measure what happened to the nonliving complex system as a result of its observing the quantum phenomenon in question, then a human is now indirectly observing that quantum phenomenon.) Rather than thinking of your conscious awareness as something that is *separate* from the world and can perform some functions *on* the world, try to remind yourself that your "conscious awareness" is made of the same stuff that the rest of the world is made of. It undergoes the same kinds of subatomic quantum uncertainties that everything else does. And when you observe an event in the world (no matter how passively or indirectly), you are necessarily participating in that event, and thus altering the way it happens—not because you have "consciousness" but simply because you and that event are temporarily becoming one system.

You are inextricably linked to your environment not only at a subatomic level but also at an everyday human level. You and your best friend are not two separate beings. You are two very similar parts of one universe, looking at itself and feeling kinship. Whether you are tossing a ball back and forth with your friend, tossing words back and forth over drinks, or even just silently engaged in mutual eye gazing (recall the end of chapter 5), you and your friend are temporarily becoming one system. In a different context, the same goes for you and your favorite pet. At a different timescale, the same even goes for you and the trees outside. In fact, when the universe makes a rock on Earth and when the universe makes a human on Earth, there are a substantial number of overlapping chemical elements among those two things, such that the two can be seen as related parts of the universe that share their existence together as one. You and a terrestrial rock have more in common with each other than a hot gas cloud on Venus has with a frozen ice chunk on Pluto. Even though the rock is a nonliving subsystem and you are a living subsystem, you and the rock are part of one tightly knit system called Earth. The gas cloud on Venus and the ice chunk on Pluto may both be nonliving systems, but they have relatively few elements in common, and the system they belong to (the solar system) is not as richly interconnected as the one that you and the rock belong to (Earth). You and that rock are both *of* Earth and undergo synchronized cycles of environmental exposure to sunlight, seasons, and gravitational pull from the moon. You also share several chemical elements with each other, including lots of oxygen, along with a little calcium, potassium, magnesium, sodium, iron, and other minerals. I'm not kidding when I say that you really should embrace your affinity with that rock and go hug it. If your neighbor sees you doing it and makes fun of you, just tell them that I said it was okay and give them a copy of this book.

A Democratic Universe

Your neighbor doesn't have to agree with you, or with this book, in order to allow you to hug your rock, does he? Agreeing or disagreeing with facts doesn't change those facts at all. In Chuck Klosterman's book *But What If We're Wrong*, he reveals a deep wisdom about how to deal with those irritating instances where the universe disagrees with you. When your subjective opinion differs with the objective evidence—or with the intersubjective agreements of a clear majority of people—it won't actually hurt you to accept that you are wrong. Buried deep in a footnote, Klosterman acknowledges that William Faulkner's books, Joni Mitchell's music, and Ingmar Bergman's films are indeed "great works," even though he personally has not found them to be particularly amazing. He writes, "I don't need to personally agree with something in order to recognize that it's true." Read that quotation again and think about it for a minute. It reveals a profound respect for evidence outside one's own preferences. If the majority of scientific findings in this book clearly indicate that your self is not confined to the inside of your skull or limited by the skin surrounding your body, then there's a sense in which your neighbor has to recognize that this is true, even if he may feel as if he intuitively disagrees with it.

Denying the truth because you disagree with it won't change the truth. Just like denying climate change won't change the steady march of that truth. The truth is like that disinterested kid in high school who didn't even try to wear the trendy clothes or belong to any cliques or clubs: the truth doesn't care whether you deny it, spread false rumors about it, or ostracize it. It will just keep on being itself: the truth. And the scientific truth is this: you are the universe. And you don't have to personally agree with that fact in order to recognize that it's true. Iconoclastic philosopher Alan Watts once said, in one of his more famous lectures, "What you do is what the whole universe is doing at the place you call here and now. You are something the whole universe is doing, in the same way that a wave is something that the whole ocean is doing." As brief as its time is on Earth, a single ocean wave is beautiful, unique, and inseparable from the surrounding ocean, just like your human body and the surrounding universe. What's more, that individual ocean wave has millions of years of history that fed into it, and its impact on the beach stretches into the future. It has thousands of square miles of physical forces surrounding it that it participates in. It blends into its neighboring waves, the undercurrents, and the sea life underneath it. In that sense, the ocean wave is at one with the entire ocean—just like you and the universe. You are a lot like that wave on the ocean, and in that metaphor, the ocean is the rest of the world. When you move, the world around you moves. Whether it is just some air turbulence from your hand motion (like the butterfly effect), an object you lift and raise up (like a glass of wine), or body language that other people respond to (like a smile), when you move, the world around you moves with you because it is part of you. This means that you are in charge. You are in control not only of your own specific actions but also of the things that go on around you.

In addition to *you* being the universe, so is *everyone* else, and so is *everything* else. In that sense, it really is a democratic universe. And the universe knows that it is one with you, whether or not you agree with it. Just like Chuck Klosterman's perspective on Joni Mitchell, you have a responsibility to acknowledge this truth, even if you may personally disagree with it once in a while.

A democratic universe is a complex-systems universe. There are so many forces and subtle influences in the system that it can be difficult to follow the chain of cause and effect. People around the world who have been brought up in democratic societies occasionally get fed up with the slow, deliberative, compromising process of complex democratic political decision-making. It's understandable. When you let many people (or forces and influences) participate in an important decision, it slows down the process, the resulting compromise solution is likely to be no one's favorite, and you are guaranteed to get some biases into the mix that will be motivated by illegitimate or corrupt goals. That's just a fact of democracy. Sometimes people who otherwise support democracy look at that imperfect process and wonder why we can't just have a wise autocrat who can mentally balance all these competing needs in one fell swoop and then implement the perfect policy solution. Isn't that how it works in business? Isn't the CEO that wise autocrat? When the number of competing needs is smallish, a wise, autocratic CEO can indeed make well-balanced decisions. When he or she doesn't, the company may find itself filing for bankruptcy (but if the CEO files bankruptcy six times for six different businesses, then maybe he shouldn't be thought of as "wise"—and certainly not "a stable genius"). But things are different when the number of competing needs is humongous, like deciding how to handle health care for an entire nation, how to address illegal immigration, or how to deal with climate change on a planetwide scale. There is no single individual human who could ever balance all the different subproblems that make up gargantuan problems like those. For all its flaws, the slow, deliberative, compromising process of democratic political decision-making is the only way to solve these ginormous problems successfully. It is no accident now that the nations that rise to the top ten in gross domestic product tend to be the ones that are becoming more democratic, and the ones that find themselves falling out of the top ten are the ones that are becoming less democratic. The worldwide free market of political ideas has spoken, and it clearly rewards democracies. As Winston Churchill famously said, "Democracy is the worst form of government, except for all those other forms that have been tried from time to time."

In Alan Watts's *The Book: On the Taboo against Knowing Who You Are*, he wrote back in 1966: "The 'nub' problem is the self-contradictory definition of man himself as a separate and independent being *in* the world, as distinct from a special action *of* the world" (italics his). Watts not only suggested that you are the entire universe (comporting well with the scientific evidence and arguments here) but also that human social conventions are designed to hide that knowledge from you and even punish you for discovering it. He might have been right about that last bit 50 years ago, but I don't think it's quite true anymore. Perhaps his writings, and the New Age wave on which he surfed, have changed Western culture just enough to make oneness with everything no longer taboo. Please feel free to embrace your oneness with the universe. It will embrace you back—and you will not be punished for it.

Oneness, Purpose, and Eternal Life

Throughout this book, I have been staying pretty close to the hard reality of peer-reviewed scientific results. Perhaps that made the exposition a little dry at times, and for that I apologize. Yet, there were probably several times when you felt that what I was suggesting was pretty darn far-out, almost mystical. A tool in my hand is part of me? Two people can become one system just because they're chatting? Plants have an awareness a little bit like I do? I should hug a rock?! Okay, maybe that last one isn't really based on scientific findings, but don't knock it 'til you try it. My point here is that, when you follow the plodding trail of bread crumbs that science is leaving for us, it actually does lead to some pretty darn far-out conclusions. Rather than being the boring and empty analysis of the universe that science is sometimes portrayed to be, there is so much beauty and mystery in the scientific world's exploration of nature that it could never really be boring and empty. In the scientific view of the world that has been painted here, you can actually find the very same comforts for the soul that you might seek in New Age practices, paranormal fictions, religious beliefs, or even hedonistic addictions. As an added bonus, science won't tell you to put magnetic bracelets on your wrist, make up unsubstantiated stories to tell you, or directly damage your liver or brain. If you want to comfort your soul with the idea that there's more to you than just a brain-and-body, then the scientific findings reported in chapter 5 clearly show that *you are much more than just a brain-and-body*. If you want to comfort your soul with the idea that, when your body dies, something about you will remain and live on, the scientific findings in chapters 5 and 6 clearly show that your external environment contains a wide variety of information patterns that are part of who you are, and *those parts of you will remain and live on*. If you want to comfort your soul with the idea that giving is more important than taking, then chapter 6 unmistakably makes it clear that *sharing helps you just as much as it helps everyone else, because we are one human system*. If you want to comfort your soul with the idea that you are not alone in this world, the scientific findings presented in chapters 6 and 7 provide more than enough hard evidence that *you are not alone*. If you want to comfort your soul with the idea that the universe has a plan, then the science described in chapters 7 and 8 helps you see that *the universe "knows" what it's doing*. If you want to comfort your soul with the idea that you are part of the universe's plan, chapter 8 discusses the science that shows how *everything that happens, including your existence, is exactly what was always going to happen*. There are no flukes, no accidents, no surprises from the perspective of the universe. (The only ones who are surprised are the humans who make small-minded linear predictions about the universe.)

The universe could not have avoided making you into *who you are*—even if it wanted to—because *who you are* spreads out into so many aspects of the surrounding universe. *Who you are* includes your favorite tools and toys; your family, friends, and pets; your influencers and influencees; the genetic history of your family; the cultural history of your species; and the material history of life on Earth. In fact, because you are reading this book, you are spread out so much that *I* am part of you and you are part of me. This is precisely why I feel comfortable talking to you in this casual manner. So chill out. Like it or not, agree with it or not, we are family now.

You can gain immense strength from focusing on feeling a sense of family with the unfamiliar. Whether it is people who look different from you, plants and animals that you haven't chatted with, or distant objects that you've never even touched before, when you embrace them as family nonetheless, you gain strength in numbers. You can even improve your health. I'm not kidding. These things happen. There is a meditation technique that has been developed over centuries to help you see all others essentially as family members. Loving-kindness meditation can actually increase one's sense of connectedness and thereby can contribute to physical health. It is based on the traditional Buddhist meditation practice of remembering an event when someone showed their unconditional loving kindness for you (such as your mother taking care of you as a child) and then taking that loving kindness that the other person exhibited and placing it inside yourself, directed toward everyone else and every*thing* else in the world. Meditate on feeling that unconditional loving kindness for everything around you, and then meditate on feeling that unconditional loving kindness reciprocated right back to you from all the people, all the things, and all the stuff everywhere. This can improve your health? Yes, it can.

There is also another traditional Buddhist meditation technique that has been developed to help you let go of the impulsive selfishness or exaggerated worrying that we all experience at times: mindfulness meditation. And it can similarly improve your *mental* health. In his playfully titled bestseller *Why Buddhism Is True*, journalist Robert Wright expounds extensively about the benefits of the Buddhist mindset and their meditation practices. For example, mindfulness meditation teaches you to regard your thoughts as merely temporary passing states of mind, not sacrosanct building blocks of the self. Not every thought needs to be acted on. Not every thought that goes through your mind needs to be embraced as a fundamental cornerstone of *who you are*. And when you calmly regard those passing thoughts from a distance, as they wax and wane, you can even let yourself giggle a little bit at their haughty self-importance. Mindfulness meditation has been linked to improvements in memory, attention, creativity, emotion control, and even reducing impulsive food choices! Heck, give it a try.

But you don't have to become a Buddhist to use their meditation practices or to feel your oneness with the universe. Buddhism has a number of beliefs that do not comport with science, such as reincarnation and Nirvana. We are trying to stick to the facts here. Throughout these chapters, I hope the scientific findings have succeeded in convincing you that your brain shares its sense of awareness not only with your body but also with the objects in your environment. This shared awareness stretches out to other people and even to the plants and animals around you. If you came along with me all the way in this book, then you have been able to embrace the wonders of the entire universe of living and nonliving systems as part of *who you are*. What more in this life could you possibly want? All the gifts that are so often attributed to a supernatural omniscient force are already being provided to you by the physical laws of our universe. You are not alone. You have a shared purpose. And everything around you is imbued with a shared sense of awareness. You now have compelling scientific evidence that even after your body has disassembled, many of the other pieces that make up *who you are* will still be functioning. Therefore, all the things that you might request from a supernatural mythical being can actually be procured from a scientific natural understanding of the universe: oneness, purpose, and a realistic, believable version of some degree of life after death.

The death of the brain-and-body is most definitely not the death of *who you are*. In her heartfelt book about her great-grandfather's artwork that disappeared in World War II, *Chasing Portraits*, filmmaker Elizabeth Rynecki tells the story of burying her grandfather in Eureka, California. The burial site was on a hill overlooking the ocean, and her father said to her, "He'll have a nice view from here." But all she could think was, "But he'll never see it." Grown up now, she realizes that her pessimism was misplaced. Her grandfather kind of does get to see the ocean. Not with human eyes, of course, but rather with other senses that the hill itself has, a grassy hill with which his body is slowly becoming one. That hill, with her grandfather, perceives that ocean view by absorbing its moisture in the air, breathing in the salt on the breeze, and synchronizing with the rhythm of the seasonal changes on the coastline. Not a bad place to spend the next million years or so.

Not only will the physical matter of your dead body become one with its surroundings, but the information generated by your living body will do that as well. As people get to know you, their mental models of you (along with the resultant things they teach to others) will survive the death of your body. The better those people know you, the more accurate the models will be, and thus the more alive you will be—despite your body having joined the nonliving. Always try to give of yourself the way Ulric Neisser did (from chapter 1); the way Guy Van Orden did (from chapter 2); the way Jeff Elman did (from chapter 3); and the way Bruce Bridgeman did (from chapter 5). That way, more of *who you are* will still be out there to survive the dissolution of your body.

Who *We* Are

I actually think it was way back in chapter 5 or so that you pretty much figured out who *you* are. Ever since then, the analysis has been gradually, steadily turning toward who *we* are. As you have gathered by now, I hope, the scientific evidence strongly suggests that there really isn't a "you" that is genuinely separate from your environment, from your friends and family, from your pets, or even from me. Philosopher Andy Clark once wrote, "There is *no self*, if by self we mean some central cognitive essence that makes me who and what I am. In its place there is just the 'soft self': a rough-and-tumble, control-sharing coalition of processes—some neural, some bodily, some technological." I would add to that list some processes that happen to be other people. Therefore, the question "Who are you?" is best answered with another question: Who are we? While the first half of this book focused on who you are, the second half has spent a great deal of time figuring out who *we* are, because, after all, that makes up a huge portion of who *you* are.

We are neuronal synapses. We are brains. We are bodies. We are groups of people. We are animals. We are living systems. We are a planet. All this physical material would be just a monumental jumble of dead tissue if not for the self-organized manner in which the various bits *relate* to one another. In fact, the relation between physical objects may be what is primary, while the objects themselves are merely secondary. For example, when you get down to the smallest bits of material—subatomic particles—quantum physicist David Mermin has suggested that the correlations between objects and events are the only things that are real in this universe, whereas the objects and events themselves are not real. Similarly, social philosopher Martin Buber (nominated for a Nobel Prize 17 times) is famous for writing: "In the beginning, there was relation."

If the relationship between our parts was somehow present before our parts, then exactly how do our parts (e.g., neurons, bodies, social groups) relate to one another? Those relations may be what are crucial for determining *who we are* and how we change over time. What the parts are physically made of may not be that important. Every part of your mind is connected to every other part of your mind (at least indirectly), every mind is connected to every other mind (at least indirectly), and, as it turns out, just about anything can be mindlike. These relations between neurons, bodies, and social groups determine our timeline. The temporal dynamics of *who we are* take place at an immensely wide range of timescales. We are millisecond-long synapses between a pair of neurons; seconds-long flows of neural activation patterns between cortical brain regions; minute-long chats between a pair of people; hour-long conversations among several people; yearlong social movements between large groups of people; decades-long traditions of industrial use of Earth's natural resources; centuries-long nations that jockey for position with one another; millennia-long conventions of political and belief systems; one planetwide system of living biomass that maintains itself via millions of intricate circular causal relationships between living and nonliving subsystems. This is who we are. Whether you call it Buddha, Allah, God, Brahma, Jehovah, the Universe, panpsychism, or whatever, there *is* a network of information transmission that holds us all, everything, together. Albert Einstein liked to refer to it as "the Old One."

The multidimensional matrix of nonliving forces interacting with each other throughout the universe is a little bit like connective tissue in the human body. Recall from chapter 4 how the fascia in your body doesn't send electrochemical signals the way your neurons do, but it nonetheless transmits forces across distances that provide information about what's happening to your limbs and your torso. Well, the nonliving matter on Earth (and the rest of the universe) is a little bit like that fascia. It may not have a metabolism and may not have intricate self-repair mechanisms, and it may not self-replicate quite the way living systems do. However, Earth's nonliving matter provides the dynamic infrastructure that supports all life on this planet. Thus, just as you now consider your body's fascia as part of who you are, you might likewise consider the "connective tissue" of nonliving matter all around your body as also part of who you are. At a wider spatial scale, the universe's nonliving matter routinely warps space-time in a manner that allows orbs of rock to huddle around the warmth of a star. Thus, a bit like the fascia and the neurons, it can be tempting to turn all your attention to the exciting things that neurons do and turn all your attention to the interesting things that living systems do. But if you do, you will be missing the crucial contextual constraints and boundary conditions that the nonliving matter provides for those living systems. Just as your peripheral nervous system couldn't function without the fascia holding it all together and transmitting its own form of information, the living systems of Earth couldn't function without the nonliving material providing the resources, protection, and structure that it does. Nonliving matter is the gristle of the universe.

And the universe is the Venn diagram that holds us all together, living and nonliving. Anywhere inside that circle that you might try to draw a crisp boundary to separate *who you are* from the rest of the circle will unavoidably feel arbitrary. Someone else will draw the line somewhere else and have just as convincing an argument for it as you had for yours. Wherever that line gets drawn, you will find some notable exceptions to the rule. Therefore, your definition of who you are, your distinction between you and other, will be rudely disproven by the scientific facts every time. The solution to this conundrum is simple. Don't draw the line. Don't assume a crisp boundary anywhere. Accept that you are an inseparable part of the whole. Don't fight it; enjoy it. Revel in the knowledge that your brain and body could be the butterfly wing that causes a hurricane, and even if it doesn't, you are still part of the weather system. Embrace the understanding that you are an inseparable part of every social movement (good and bad) that happens anywhere in the world. Your actions and inactions, along with everyone else's, are intrinsic to determining how everything happens on this planet, and in the millennia to come, how everything happens in this solar system. Right now, you are part of the future of the human race. Throughout your body's lifetime, your actions and inactions will help determine the legacy of the human race. Rather than making you feel small and insignificant, this oneness with everything should do exactly the opposite. Each of us is part of the planet Earth and its destiny. Each of us is a spokesperson for the only congregation of physical matter in our known region of the galaxy to ever contemplate itself. We are not insignificant. We are incredibly special, because we are of Earth, because we "behave so as to persist," and because we are the universe. The science proves it, and now you know it. These things happen. Be thankful. Whoever you are.

Directions for Use

Remember, you are part of a vast network of physical matter interacting in complex ways. Within that, you are part of a vast network of living systems integrated with each other. Within that, you are part of a vast network of humans and their civilizations and information systems. And within that, you are part of a network of friends and family, some of them living and some of them nonliving. In this final exercise, you will have a conversation with one of those nonliving friends or family members. Set a timer for five minutes, and during those five minutes I want you to suspend all disbelief. Find a quiet moment by yourself, and talk with a loved one who has passed. The physical matter that once composed their body is no longer living matter, but the fact that they are not living matter doesn't matter. What matters is that a significant part of "who they were" was your mental model of "who they were." There were times when they voluntarily changed their behavior because of their anticipation of preserving your opinion of them (i.e., your mental model of them). That necessarily means that *who they were* was partly composed of *your* mental model of them. Therefore, when you activate that mental model of who they are and pretend to have a conversation with them, it's not just pretend. That is part of *who they were* coming to life, in your presence, and sharing some thoughts with you. If you love them, then bring them to life as often as you can, and give them a chance to share *who they are* with you again and again. It will make you feel good. It will even make *them* feel good. And you might even learn something new from the conversation.

Notes

There is no sense of god to be found on my knees.
—Soap & Skin, "Creep"

Chapter 1

These endnotes go into some slightly finer detail on topics that are discussed in the various sections of each chapter. Also included in the brief discussions are references to some relevant publications that an eager student of life might enjoy tracking down and reading. Complete reference citations can be found on the reference list. For that very first page of chapter 1, see Stager (2014) and Weil (2001).

You Are Not Who You Think You Are

Moreover, other people are not always who *you* think they are. Nor are they who *they* think they are. In fact, they also probably are not who you think they think they are. Being open to accepting the possibility that you are wrong about your convictions is one of the most important steps toward self-actualization. It is important to understand that the self is a complex conglomeration of many different forces and ideas, some of them contradicting each other, some of them based on falsehoods, and some of them more prominent than others in certain contexts (for a review, see Leary & Tangney, 2012).

Don't Always Trust Your Perception

In addition to the journal article that Steve Macknik and his colleagues wrote with James Randi and Teller (of Penn & Teller fame) (Macknik et al. 2008), Macknik and his wife, Susana Martinez-Conde, also wrote a well-received book with *New York Times* science correspondent Sandra Blakeslee called *Sleights of Mind* (2010), which compellingly describes how magic tricks take advantage of a wide variety of interesting quirks with which the brain and cognition routinely operate.

Moreover, the "change blindness" literature is full of powerful demonstrations that, despite feeling that we are generally aware of our visual environment, we routinely fail to detect changes that are made to the objects and events that surround us. Some articles that address these findings and their theoretical interpretations include Hayhoe (2000), Henderson and Hollingworth (1999), O'Regan (1992), O'Regan and Noë (2001), O'Regan, Rensink, and Clark (1999), Simons and Levin (1997, 1998), Spivey, Richardson, and Fitneva (2004), and Spivey and Batzloff (2018).

Don't Always Trust Your Memory
Ulric "Dick" Neisser's work on flashbulb memories (Neisser & Harsch, 1992) was followed by similar findings regarding the 9/11 terrorist attack in 2001 (Greenberg, 2004; Hirst et al., 2009), the O. J. Simpson verdict in 1995 (Schmolck, Buffalo, & Squire, 2000), and the Northern California Loma Prieta Earthquake in 1989 (Neisser et al., 1996). Notably, results suggest that direct involvement in the remembered event, and frequent rehearsal of storytelling, can improve the accuracy of memory. In fact, I experienced the 1989 Loma Prieta Earthquake, and I've retold the story many times. Therefore, my memory of it might be reasonably accurate. It was about a week before my twentieth birthday, and I was in an undergraduate psycholinguistics class with Professor Ray Gibbs, on the third floor of Kerr Hall at the University of California, Santa Cruz (only 10 miles from the 6.9 magnitude epicenter). The lecture was on lexical ambiguity resolution, a research topic developed by my soon-to-be PhD adviser, Rumelhart Prize winner Mike Tanenhaus. Then I felt the earth move. As the classroom suddenly began to sway violently, my first thought was, "This can't possibly be a natural disaster because we didn't get any warning." In a couple of seconds, I came to my senses and dived under a table to find myself beside Professor Gibbs and another student. As parts of the ceiling fell down around us, I heard Gibbs say what could very well have turned out to be his last words: "Rock 'n' Roll!" At least, that's how I remember it. A fair bit of research has been conducted on false eyewitness testimony, false confessions, false "recovered" memories, and choice blindness. This includes work by Ceci and Huffman (1997), Dunning and Stern (1994), Gross (2017), Gudjonsson (1992), Hall et al. (2013), Johansson, Hall, and Sikström (2008), Kassin (2005), Kassin and Kiechel (1996), Kassin, Meissner, and Norwick (2005), Loftus and Pickrell (1995), Smalarz and Wells (2015), and Strandberg, Sivén, Hall, Johansson, and Pärnamets (2018).

Importantly, Dunning and Stern found that inaccurate eyewitness identification is often accompanied by a detailed self-narrative of the thought processes that went into the identification, whereas accurate eyewitness identification is often devoid of any self-narrative or metacognition. It's as Thomas Henry Huxley said, "What we call rational grounds for our beliefs are often extremely irrational attempts to justify our instincts" (see also Damasio, 1994). Most of the time, our brains generate those justifications and rationalizations without our even realizing it. As David Eagleman recounts in his book *Incognito* (2012), your brain does an amazing amount of work "behind the scenes" that you aren't even aware of. This is not to say that one's first impression or unexplained initial gut reaction is *always* right. In Malcolm Gladwell's *Blink* (2007), he is quick to point out that there are important exceptions to the rule of thumb that first impressions are often more accurate than expected, and Michael Lewis's *The Undoing Project* (2016) documents the development of the field of behavioral economics, where researchers slowly and painfully came to terms with just how fallible and nonrational human decision-making really is. When a gut reaction turns out to be inaccurate, the results can be devastating. See also Daniel Kahneman's *Thinking Fast and Slow* (2013) and Daniel Richardson's *Man vs. Mind* (2017). Thus, the most appropriate method should perhaps be to take note of one's first impression or instinct and then calmly compare it to a different, more reasoned evaluation of the data. In fact, work by Vul and Pashler (2008; see also Mozer, Pashler, & Homaei, 2008; Steegen, Dewitte, Tuerlinckx, & Vanpaemel, 2014) suggests that the average of your first guess and your second guess tends to be more accurate than either of those guesses on its own. Perhaps by sampling both your gut's reaction and your mind's reasoning, and then finding a middle ground between them, more errors could be avoided in everyday life.

Don't Always Trust Your Judgment
The McKinstry experiment was published in 2008, too late for Chris to see it in print (McKinstry, Dale, & Spivey, 2008). Since then, over two hundred scientific publications have referred to it in their discussions of the scientific background. McKinstry made his mark, albeit posthumously.

Guenther Knoblich, Marc Grosjean, and I originally adapted this computer-mouse-tracking experimental methodology almost as a kind of "poor-man's eyetracker" (Spivey, Grosjean, & Knoblich, 2005). Over the past 15 years, it has influenced a wide variety of cognitive science laboratories. Jon Freeman even has a free downloadable version of mouse-tracking software on his *Mousetracker* website, and Pascal Kieslich has one for his Mousetrap software as well. The following is just a small sample of articles that show how computer-mouse tracking (and reach tracking in general) can be used to see into a person's thought process during visual tasks, language tasks, decision tasks, and dietary choices, and even to improve website design: Arroyo, Selker, and Wei (2006), Bruhn, Huette, and Spivey (2014), Buc Calderon, Verguts, and Gevers (2015), Dale and Duran (2011), Farmer, Cargill, Hindy, Dale, and Spivey (2007), Faulkenberry (2016), Freeman, Dale, and Farmer (2011), Hehman, Stolier, and Freeman (2015), Huette and McMurray (2010), Koop (2013), Lin and Lin (2016), Lopez, Stillman, Heatherton, and Freeman (2018), Magnuson (2005), O'Hora, Dale, Piiroinen, and Connolly (2013), Schulte-Mecklenbeck, Kühberger, and Ranyard (2011), Song and Nakayama (2009), and van der Wel, Sebanz, and Knoblich (2014).

Before computer-mouse tracking was developed, there was eye tracking. Since your eyes naturally jump from object to object about three times per second, and they are usually looking at objects that you are thinking about, eye tracking provides a perfect opportunity to collect a continuously unfolding, real-time measure of your "train of thought." The Pärnamets et al. (2015) eye-tracking experiment wouldn't have been possible without the genius of psycholinguist and Rumelhart Prize winner Michael Tanenhaus 20 years earlier. Tanenhaus, with the help of his students Julie Sedivy, Kathleen Eberhard, and me, was the first to use eye tracking to collect this continuous record of a person's thought process during a language task that involved following spoken instructions to move objects around or click icons on a computer screen (Tanenhaus, Spivey-Knowlton, Eberhard, & Sedivy, 1995). This methodological development quickly revolutionized the research subfield of psycholinguistics and is gradually changing the entire discipline of cognitive science. The following is just a tiny sample of the hundreds of articles that use eye tracking in a language-and-action context to improve our understanding of speech perception, spoken word recognition, sentence processing, problem solving, and decision-making: Allopenna, Magnuson, and Tanenhaus (1998), Altmann and Kamide (2007), Chambers, Tanenhaus, and Magnuson (2004), Hanna, Tanenhaus, and Trueswell (2003), Huettig, Quinlan, McDonald, and Altmann (2006), Knoeferle and Crocker (2007), Krajbich, Armel, and Rangel (2015), Krajbich and Smith (2010), Magnuson, Tanenhaus, Aslin, and Dahan (2003), Marian and Spivey (2003), McMurray, Tanenhaus, Aslin, and Spivey (2003), Rozenblit, Spivey, and Wojslawowicz (2002), Ryskin, Wang, and Brown-Schmidt (2016), Spivey-Knowlton (1996), Trueswell, Sekerina, Hill, and Logrip (1999), and Yee and Sedivy (2006).

Letting Go of Your Self
Important distinctions have been drawn between the "conceptual self," the "ecological self," and the "interpersonal self" (Neisser, 1991; see also Libby & Eibach, 2011). However, each of them draws an artificial boundary between that self and certain aspects of its environment. In this book, I intend to use science to show you just how vague and fuzzy each of those boundaries is.

As noted, letting go of your self is not the same as letting yourself go. Eating right and exercising is more fundamental to maintaining a healthy mind than you may realize. If you've been paying attention to how the environment can influence your decisions, including your dietary decisions, then you can arm yourself with skills to adjust your environment so that it nudges those decisions in healthy directions (e.g., Giuliani, Mann, Tomiyama, & Berkman 2014; Papies, 2016).

Chapter 2

Feel Free
There is much that the field of neuroscience still does not know about how the brain works. That is why it is such an exciting field of study. New findings are being discovered every day. The sudden, brief electrochemical impulses that neurons send down

their axons (typically called "action potentials" or "spikes") are the most commonly studied and simulated signals that neurons produce. However, neurons also send signals to one another via "graded potentials." These are milder electrical changes that rise and drop over several milliseconds instead of one millisecond. Neurons can also influence each other with dendodendritic connections that don't involve the axon, and even with the electric fields that these electrochemical potentials generate. See, for example, Edelman (2008), Fröhlich, (2010), Goodman, Poznanski, Cacha, and Bercovich (2015), Sengupta, Laughlin, and Niven (2014), Van Steveninck and Laughlin (1996), and Yoshimi and Vinson (2015).

The details of Hameroff and Penrose's (1996; Penrose, 1994) theory involve a coherent resonance of quantum superposition among microtubules in one neuron's membrane that spreads to many other neurons before experiencing an orchestrated quantum collapse or decoherence across many neurons. However, astrophysicist Max Tegmark (2000) shows clearly that the timescale at which quantum coherence can be maintained among atoms is at least ten orders of magnitude too brief (femtoseconds or even attoseconds) compared to the timescale of the neuronal synapse (milliseconds). See also Atmanspacher (2015). More recently, Hameroff and Penrose (2014) revised their theory to accommodate some of the criticisms that it originally received and to integrate new evidence for quantum vibrations among microtubules. However, Reimers, McKemmish, McKenzie, Mark, and Hush (2014) point out that the extended quantum coherence in Hameroff and Penrose's revised theory finds itself relying on coherence among quantum *vibrations*, not quantum *states*, and it is quantum states that must be maintained in order to generate coordinated qubits (quantum information units) (see also Meijer & Korf, 2013; Tuszynski, 2014). While this debate is surely not over, it seems that the balance of evidence is leaning steadily in opposition to the idea that consciousness and free will might have as their engine the quantum superposition that briefly happens at the spatial scale of the atomic structure of neuronal membranes.

One Neuron to Rule Them All
In 1943, McCulloch and Pitts developed one of the very first mathematical frameworks for neural network simulations (see also Lettvin, Maturana, McCulloch, & Pitts, 1959). It assumed only spiking neurons and led eventually to the neuron doctrine's battle with Jerry Lettvin's hypothetical "grandmother cell" example. In his treatment of the neuron doctrine, Horace Barlow (1972) had to explicitly criticize the grandmother cell idea while still defending the possibility that some perceptual events might be instantiated by a single neuron's activation. Charles Gross (1992, 2002) provides an intriguing detailed history of the development of the grandmother cell idea. He also provides a compelling neuroscience review of ensemble coding in visual perception (see also Edelman, 1993; Young & Yamane, 1992). This statistical compromise between the grandmother cell idea (where only one neuron codes for a concept) and fully distributed ensemble coding (where billions of neurons together

code for each concept) is an idea called sparse coding, where a smallish number of neurons are playing the most important role in coding for any given perceptual event, concept, or decision. See Baddeley et al. (1997), Barlow (1953, 1972, 1995), Chang and Tsao (2017), Field (1994), Kanan and Cottrell (2010), Olshausen and Field (2004), Quiroga, Kreiman, Koch, and Fried (2008), Rodny, Shea, and Kello (2017), Rolls (2017), and Skarda and Freeman (1987).

However, it should be noted that concepts change over time, and no pair of perceptual events are identical. Therefore, the ensemble of neurons that codes for a mental event must be flexible. When you think of a concept, there may be some sparse code of active neurons that forms an approximate "core" of the concept, but as demonstrated in chapters 3 and 4, there's always a great deal of context influencing what that concept means to you at that exact moment. Hence, there are usually far more "context neurons" active than "core concept neurons" during any particular thought process. Over time, some of those "context neurons" may gradually become "core concept neurons," and vice versa. That is, when thinking about your grandmother or deciding between chicken and veal, the contextualizing information routinely outnumbers the core meaning information, and the dividing line between those two categories tends to blur. Casasanto and Lupyan (2015) and also Yee and Thompson-Schill (2016) have each teamed up to review a wide array of psychology and neuroscience literature suggesting that the concepts we use in our minds are not like dictionary entries that can get visited again and again to access the same meaning each time. Instead, our concepts might be better described as mental events that are constructed anew each time we use them.

Of Ghosts and Grandmas

Rick Strassman (2000) has popularized the idea that the pineal gland of humans (and rats) can release endogenous DMT and potentially induce hallucinatory experiences (Barker, Borjigin, Lomnicka, & Strassman, 2013). However, the vast majority of what the pineal gland secretes is serotonin and melatonin. Harris-Warrick and Marder (1991) provide an excellent review of how a bath of such neurotransmitters and hormones can alter the connectivity (and therefore the function) of a biological neural network.

Detailed reports of "hidden target" experiments in near-death experiences are provided by Augustine (2007) and Parnia et al. (2014).

There is much debate among scientists of the mind about whether *belief* in a supernatural God is a biologically evolved human trait, something that developed through cultural transmission of ideas, or a combination of the two. See Banerjee and Bloom (2013), Barrett (2012), Boyer and Bergstrom (2008), and D. S. Wilson (2002). On the other side of that speculative coin, Caldwell-Harris (2012) has suggested that a general *skepticism* about any culturally dominant belief system might also have evolved for a subset of the population.

In the case of human reports of out-of-body experiences and other supernatural miracles, Martin and Augustine (2015) and Stenger (2008) provide some powerful scientific evidence and arguments against those being real events.

One Brain Region to Rule Them All

In the past couple of decades, many laboratories have provided unparalleled insight into the function of the prefrontal cortex. See, for example, Bechara, Damasio, and Damasio (2000), Bechara, Damasio, Damasio, and Anderson (1994), Chrysikou et al. (2013), Damasio, Grabowski, Frank, Galaburda, and Damasio (1994), Goldman-Rakic (1995), Lupyan, Mirman, Hamilton, and Thompson-Schill (2012), Macmillan (2002), and Miller and Cohen (2001).

Importantly, rather than thinking of the prefrontal cortex as a place where sensory input is delivered and then purely internal computations determine what to do with it, Joaquin Fuster (2001) points out that it might be better to think of the prefrontal cortex as a very important part of a network of brain areas that includes sensory brain regions, motor brain regions, memory brain regions, and all kinds of other brain regions. In fact, neural network research by Dave Noelle and his colleagues (Kriete & Noelle, 2015; Kriete, Noelle, Cohen, & O'Reilly, 2013; Noelle, 2012; see also O'Reilly, 2006) suggests that the prefrontal cortex has important connections to *subcortical* brain regions (such as the basal ganglia) to form a network that is crucial for logical reasoning, reinforcement learning, cognitive control, and cognitive flexibility.

The Libet Experiment

Many scientists have replicated and extended Libet's (1985) experimental design. Examples include Bode et al. (2011), Filevich, Kühn, and Haggard (2013), Soon, Brass, Heinze, and Haynes (2008), and Soon, He, Bode, and Haynes (2013). However, some philosophers and scientists have taken issue with the common interpretation of the Libet experiment (Herrmann, Pauen, Min, Busch, & Rieger, 2008; Libet, 1999; Mele, 2014; Tse, 2013).

In certain rare cases of frontal lobe damage or corpus callosum damage, patients experience an extreme example of goal-oriented hand movement that is not merely being *neurally initiated* before they realize it, as in the Libet experiment, but *happening completely* without the volition or intention of the patient. As described in the opening of this chapter, a person with alien hand syndrome can find herself being undermined or even attacked by one of her own hands. She may find herself having to engage in "self-restriction," where her controlled hand restrains the uncontrolled hand. These patients often feel as if some external agent is controlling their "alien hand." Of course, what's actually happening is that damaged transcortical networks in the parietal cortex and the frontal lobe are generating those motor commands, much as similar motor commands are generated in the Libet experiment, but they are being generated completely without the cooperation or awareness of the rest of the brain (Biran & Chatterjee, 2004; Goldberg & Bloom, 1990; Hassan & Josephs, 2016).

Your Brain Knows More Than You Do
There are many examples of implicit knowledge revealing itself in people's behaviors even when they claim to be unaware of that knowledge. Just a few examples of implicit awareness in change blindness, blindsight (clearly established in humans and monkeys), and a variety of cognitive psychology laboratory tasks are Cowey (2010), Hayhoe (2000), Schacter (1992), Weiskrantz, Warrington, Sanders, and Marshall (1974), and Whitwell, Striemer, Nicolle, and Goodale (2011).

The EEG study about learning French words provides a compelling example of how the brain learns new words in a continuous, probabilistic, and gradual manner—such that some words and speech sounds can be "partially known" by the network of neurons that is learning them (McLaughlin, Osterhout, & Kim, 2004). This constitutes a shift away from the metaphor of language being a "box in the head" that either contains or doesn't contain a particular linguistic unit. Instead, we have a metaphor for word learning that is more like an egg carton being filled up in the rain. At the start, every egg shape is empty, then later they're all only partially full, and then suddenly they're all full at almost the same time! It's a vocabulary explosion! This is an account of language in which the neural system that embodies linguistic knowledge can sometimes *partly understand* a certain word or speech sound. This simple shift in thinking about linguistic knowledge dramatically changes the theoretical landscape for accounts of language processing and acquisition (Ellis, 2005; Elman, 2009; McMurray, 2007; Saffran, Aslin, & Newport, 1996; Warlaumont, Westermann, Buder, & Oller, 2013).

Not Feeling So Free Anymore?
When you decide to rearrange the "criminogenic circumstances" of your life, you might find yourself asking, "How did I do that?" Did you use free will? Perhaps not. A wide variety of environmental influences are already encouraging you to make changes like that if you want to improve your life. Substance abuse programs tell you to rearrange your lifestyle so you spend your leisure time with people who don't use those substances. Exercise tips regularly say to work out with a friend and to make exercise a part of your daily routine so that it feels as regular as brushing your teeth. And this book just encouraged you to rearrange your life circumstances, too. These many external causal influences are an important part of the long and tangled chain of cause and effect that leads you to alter your situation for the better. Rather than improving your life so you can garner credit and accolades for your impressive will power in doing so, maybe you should do it just because it will help you live a longer and happier life. What's more, it will make the people who care about you

happier. To read more about free will and criminogenic circumstances, see Haney (2006), Haney and Zimbardo (1998), Harris (2012), Maruna and Immarigeon (2013), Pereboom (2006), Wegner (2002); see also Thagard (2010).

The Will Emerges

The physics of complexity theory and emergence have been topics of study for many decades (Meadows, 2008; Mitchell, 2009; Prigogine & Stengers, 1984). We will revisit it in chapters 7 and 8. Only a couple of decades ago, Alicia Juarrero (1999) was among the first to provide a detailed description of how complexity theory can help us understand the way human intentional action emerges in a physical system. Since then, a number of authors have parroted her ideas, and there have been recent expansions of these ideas as well (Beer, 2004; beim Graben, 2014; Hoffmeyer, 2012; Jordan, 2013; Murphy & Brown, 2007; Spivey, 2013; Van Orden & Holden, 2002).

Cognitive scientists Chris Kello and Guy Van Orden produced in the laboratory some of the most compelling scientific evidence for human cognition being a self-organized system. Discussions of these statistical signatures include Gilden, Thornton, and Mallon (1995), Kello (2013), Kello, Anderson, Holden, and Van Orden (2008), Kello, Beltz, Holden, and Van Orden (2007), Kello et al. (2010), Van Orden, Holden, and Turvey (2003, 2005), and Wagenmakers, Farrell, and Ratcliff (2004).

Directions for Use

I thought I told you to complete your assignment before coming to this page. This is because the second half of your assignment is best dealt with after completing the first half. If you haven't completed the first half, I ask that you please do so before reading any further. I realize that it may not be easy, but please try. If you intend to take this seriously, you will play along.

The second half of chapter 2's assignment is to think long and hard about *why* you did what you did. Why did you do it? That was the only life that bug had, and you did what you did. You obviously have a standard (albeit perhaps informal) policy about what to do when the life of an insect is in your hands, and something went on in your prefrontal cortex that just made you violate that policy. In the terminology of Benjamin Libet, you "vetoed" it. Take a moment to consider why you exerted your will to go against your usual decision. Is it, as Libet might have suggested, because your free will was able to intercede and veto your typical approach to the situation? Or is it simply because I told you to do it? (If that's all it takes, then I might as well tell you to send me a check for $10, too, while we're at it.)

It's probably not simply that your free will made you do it, and I definitely hope that you don't simply think that *I* made you do it. Most likely, it's a very complex combination of many different causal forces at multiple timescales that resulted in your saving a bug today that you normally would have killed or killing a bug that you normally would have saved. For many centuries, philosophers, clerics, and judges have debated about the role of free will in a person's actions. If those debates had not happened, then cognitive neuroscientists may not have felt the need to study free will in their labs. Because of those experimental studies of free will and the brain over the past few decades, I felt the need to address it in this chapter. If that history had been different, you and I wouldn't be doing this right now. For several years, you've had some form of (perhaps evolving) policy regarding what to do with insects in your home. If you had developed a different policy, your action today may have been different. And that bug's life might have gone another way. Maybe you saw this assignment, in part, as an intriguing opportunity to challenge your autopilot behavior with your will. Maybe you felt that since you were following instructions, you weren't quite as responsible as usual. My instruction to you today played a role in your action, but there were clearly many other causal forces, out of your control and mine, that got us to where we are now.

Or maybe, just maybe, you chose to *veto my instruction* to veto your usual practice, a metaveto. Maybe you chose not to challenge yourself at all and not even look for a bug in your home. If something like this is the case, that particular decision is also worth thinking through for its multiple complex causal forces. Think about what made your usual practice so inflexible or unchallenged. What experiences helped you develop such a firm policy on how to deal with bugs in your home that you decided not even to try to violate it. Perhaps you are so deathly afraid of bugs that you found yourself unable to capture it. Where did that fear come from? Perhaps you were just too comfortable on the couch while reading this book, and you were too slothful to get up. Why is that? Or perhaps your moral stance to capture and release bugs in your home was so strong that you found yourself unable to kill this one. Where did *that* stance come from? (And I wonder what your stance will be when your bathroom gets invaded by a swarm of hundreds of ants escaping the cold, the heat, or the rain outside.)

The actual causes of our decisions are almost always multifaceted and complex. In the context of this discussion, doesn't it seem like the idea that one could simply chalk up a decision to "free will" is just too glib, too shallow, or just scientifically lazy? For example, right now there's probably a man in Milwaukee, Wisconsin, who voted Democrat all his life, but after reading and believing a fake news story on Facebook that Hillary Clinton was involved in a child sex ring based at a pizza shop in Washington, D.C., he voted for Donald Trump in 2016. Do you really think that vote of his counts as a freely willed decision on his part? Or was it perhaps a manipulated decision, coerced by false rumors and propaganda? Of course, advertisers of all stripes have known for decades how to use language to manipulate people's choices (Sedivy & Carlson, 2011). Next time you stop at a fast food joint to get a snack, ask yourself whether your choice of that particular restaurant was really and truly a freely willed decision or perhaps a manipulated decision, coerced by the ubiquitous biased advertising that is constantly assaulting your senses.

Chapter 3

The Homunculus and Its Modules

The modularity approach to understanding how the mind works very easily falls into the trap of assuming that a "cognizer" (or central executive) of some sort is observing the results of sensory processes displayed on some form of stage: a Cartesian theater. If the central executive cognizer is a mind all its own that is immersed in a virtual-reality display generated by the perceptual systems, then we have to start all over again with the problem of trying to figure out how *his* mind works. Does *he* have a central executive inside *his* mind, too? This could lead to an infinite recursion. Stephen Monsell and Jon Driver referred to this bad habit as "homunculitis" (e.g., Dennett, 1993; Dietrich & Markman, 2003; Fodor, 1983; Monsell & Driver, 2000).

A Paradigm Drift

Several decades of neural network research provided important inspiration for the connectionist movement in cognitive science. Jerry Feldman and Dana Ballard coined the term "connectionism" in 1982 to refer to the general theory that intelligence emerges via the connections *between* many different parallel processors rather than *inside* any one serial processor (Feldman & Ballard, 1982). The neural network learning algorithm most frequently used with these neural network simulations is back-propagation, which provides a credit assignment path that can (in principle) go as far back into previous time slices as the programmer wants. When first introduced, back-propagation was criticized by modularists on one side as being "nothing but old-fashioned behaviorism" and criticized by neuroscientists on the other side as being "not sufficiently biologically plausible." However, in recent years it has become the meat and potatoes of a huge movement in machine learning called "deep learning" (see Anderson & Rosenfeld, 2000; Anderson, Siverstein, Ritz, & Jones, 1977; Cottrell & Tsung, 1993; Grossberg, 1980; LeCun, Bengio, & Hinton, 2015; McCulloch & Pitts, 1943; Rosenblatt, 1958; Rumelhart, Hinton, & Williams, 1986; Rumelhart, McClelland, & the PDP Research Group, 1986; Schmidhuber, 2015).

Interactionism in Vision

The functional neuroanatomy of visual cortical areas has been analyzed in depth by hundreds of neuroscientists, but a few key references relevant to this discussion of the interactionism in that connectivity pattern are Amaral, Behniea, and Kelly (2003), Clavagnier, Falchier, and Kennedy (2004), David, Vinje, and Gallant (2004), Felleman and Van Essen (1991), Haxby et al. (2001), and Motter (1993).

It is worth noting that a renewed attempt to defend the modularity of the visual cortex (Firestone & Scholl, 2016) met with an unprecedented volley of opposition from a long list of peer-review commentators. In fact, one particular critique had so many well-established authors on it (Vinson et al., 2016) that one blogger called it a "murderers' row" of vision scientists who were tearing down this last gasp of the modularity claim. (The metaphorical use of the term "murderers' row" comes from a description of the numerous heavy hitters on the 1918 New York Yankees baseball team. But before that, "murderers' row" was a literally deadly wing in New York City's Tombs Prison.)

A great many perceptual studies have looked at how context and situation can bias the perception of ambiguous figures (such as the famous vase/faces image and the duck/rabbit image). Here are just a few examples: Balcetis and Dale (2007), Bar and Ullman (1996), Biederman, Mezzanotte, and Rabinowitz (1982), and Long and Toppino (2004).

Theories of top-down biases on bottom-up competition in visual perception come in a variety of flavors, but at their core they all say the same basic thing: your attentional and conceptual biases can significantly influence your visual perception (e.g., Bar, 2003; Bar et al., 2006; Desimone & Duncan, 1995; Gandhi, Heeger, & Boynton, 1999; Hindy, Ng, & Turk-Browne, 2016; Kveraga, Ghuman, & Bar, 2007; Lupyan & Spivey, 2008, 2010; Mumford, 1992; Rao & Ballard, 1999; Sekuler, Sekuler, & Lau, 1997; Shams, Kamitani, & Shimojo, 2000; Spivey & Spirn, 2000; Spratling, 2012).

Interactionism in Language
Brain surgery on newborn ferrets is actually quite complex. The optic nerve is not diverted directly to the auditory cortex but instead to the medial geniculate, which normally receives input from the auditory nerve and sends processed signals to the auditory cortex. Therefore, one must sever the auditory nerve, so that the medial geniculate receives only those new *visual* inputs from the redirected optic nerve. This way, although the information that is being conveyed from the medial geniculate to the primary auditory cortex is now *visual* information, the actual neural connections from the medial geniculate to the primary auditory cortex are unchanged. (This is crucial because cortical sensory regions do require specific neurochemical signals to allow axons to grow synaptic junctions there.) Some of the ganglion cells in the retina don't survive after this procedure. Also, the topography of the auditory cortex is not really able to create quite as good a two-dimensional map the way the visual cortex does. As a result, a ferret using its *auditory* cortex for its visual perception is not able to see quite as well as a ferret using its *visual* cortex for its visual perception. This work nonetheless shows a truly remarkable flexibility in how brain regions can accommodate massive changes to the type of information they process (e.g., Frangeul et al., 2016; Pallas, 2001; Pallas, Roe, & Sur, 1990; Von Melchner, Pallas, & Sur, 2000). In fact, Jerry Lettvin and his team even showed that, in a frog, you can divert the optic nerve toward the *olfactory* cortex and make *those* neurons develop visual receptive fields (Scalia, Grant, Reyes, & Lettvin, 1995).

This flexibility in a neuron's ability to process different types of information happens not only in the lab but also in everyday real life. When the primary auditory cortex is deprived of high-frequency input because of hearing impairment in the inner ear, the neurons that were selective for that kind of input find new jobs. Their receptive fields reorganize so that they are receiving signals primarily from the medium-frequency portion of the cochlea. Now that the auditory cortex has more neurons devoted to that medium-frequency range, it might have slightly improved sensitivity and discrimination within that frequency range (Schwaber, Garraghty, & Kaas, 1993). For related sensory reorganization results, see Bavelier and Neville (2002), Dietrich, Nieschalk, Stoll, Rajan, and Pantev (2001), Hasson, Andric, Atilgan, and Collignon (2016), Kaas (2000), and Tallal, Merzenich, Miller, and Jenkins (1998).

Interactive bidirectional distributed spreading-activation accounts of human language processing have been supported by a wide variety of behavioral, computational, and neuroimaging studies. Psycholinguist Gerry Altmann makes a compelling case that, during language comprehension, a key aspect of cognition that integrates internal states of the brain with external states of the world is a multiscale process of prediction (Altmann & Mirkovic, 2009; see also Elman, 1990; McRae, Brown & Elman, in press; Spivey-Knowlton & Saffran, 1995). Interactive spreading activation is what allows a kind of implicit anticipation of upcoming information, at timescales of milliseconds, seconds, and minutes, to quickly incorporate context effects of various kinds, and also allows language to be learned even with minimal corrective feedback (e.g., Anderson, Chiu, Huette, & Spivey, 2011; Dell, 1986; Elman & McClelland, 1988; Fedorenko, Nieto-Castanon, & Kanwisher, 2012; Fedorenko & Thompson-Schill, 2014; Getz & Toscano, 2019; Glushko, 1979; Gow & Olson, 2016; Magnuson, McMurray, Tanenhaus, & Aslin, 2003; Marslen-Wilson & Tyler, 1980; Matsumoto et al., 2004; McClelland & Elman, 1986; McGurk & MacDonald, 1976; McRae, Spivey-Knowlton, & Tanenhaus, 1998; Onnis & Spivey, 2012; Rosenblum, 2008; Rumelhart & McClelland, 1982; Samuel, 1981; Seidenberg & McClelland, 1989; Shahin, Backer, Rosenblum, & Kerlin, 2018; Tanenhaus et al., 1995).

Interactionism in Concepts
Decades of research have helped us develop an understanding of how concepts are represented in the human mind, but only recently have they been shown to be so interactive with sensory and motor aspects of cognition. For a review, see Barsalou (1983, 1999), Boroditsky, Schmidt, and Phillips (2003), Casasanto and Lupyan (2015), de Sa and Ballard (1998), Gordon, Anderson, and Spivey (2014), Louwerse and Zwaan (2009), McRae, de Sa, and Seidenberg (1997), Oppenheimer and Frank (2008), Rosch (1975), Wu and Barsalou (2009), and Yee and Thompson-Schill (2016).

Simmons, Martin, and Barsalou (2005) show how neural activation in visual areas (for food recognition) spreads (probably first to the hippocampus and the prefrontal cortex and then to the gustatory cortex). So, upon seeing tasty-looking food, you cannot help but imagine its flavor a little bit. Similar results have actually been found with the honeybee, where a visual stimulus of a flower elicits neural activation in their olfactory bulb, even though there's no scent stimulus (Hammer & Menzel, 1995; Montague, Dayan, Person, & Sejnowski, 1995). Perhaps you've experienced this kind of thing yourself when drinking a liquid from an opaque container that you thought was something else. It can be pretty shocking. You could try it experimentally with the help of a friend. Blindfold the taster, and have the other person hand them a series of drinks with straws (so that one cannot easily smell the contents), telling them each time what they are about to drink. After several drinks that are accurately announced, deliver one to the taster that is inaccurately announced (e.g., "this one is chocolate milk" when it is actually orange juice). Even someone who likes orange juice a lot will probably find it tasting horrible after their gustatory cortex has prepared itself in advance for the taste of chocolate milk. (Or you can just do as Stephen Colbert does and throw a few Skittles into a big bowl of M&Ms at your next holiday party. See which family member gets the shock of an M&M that's not crispy and chocolaty but instead chewy and fruity.)

A Self in the Subcortex?

There is quite a bit of evidence for animals without a cortex exhibiting behavior that is consistent with their having a "mental life" (e.g., Goldstein, King, & West, 2003; Güntürkün & Bugnyar, 2016; Merker, 2007; Pepperberg, 2009; Whishaw, 1990), and evidence is accumulating regarding the claustrum, a thin sheet of neurons between the cortex and subcortex, possibly playing an important role in human consciousness (Chau, Salazar, Krueger, Cristofori, & Grafman, 2015; Crick & Koch, 2005; Koubeissi, Bartolomei, Beltagy, & Picard, 2014; Milardi et al., 2015; Stiefel, Merrifield, & Holcombe, 2014).

Interactionism in Who You Are

For years now, graph theory and related statistical network analyses have been performed on data regarding neural connectivity patterns. These analyses are consistently showing that most cortical and subcortical regions participate significantly in multiple different functional networks across the brain, and that when two brain areas are connected via synaptic projections, those connections are usually bidirectional. This bidirectionality in information flow makes the behavior of these networks extremely complex and nonlinear (Anderson, 2014; Anderson, Brumbaugh, & Şuben, 2010; Seung, 2013; Sporns & Kötter, 2004; Tononi, Sporns, & Edelman, 1994; see also Love & Gureckis, 2007).

Directions for Use

There is quite a bit of evidence that various training regimes (from action video games, to memory games, to learning a second language) induce a rewiring of the brain that can improve visual perception, attention, perceptual-motor conflict resolution, auditory localization, and even probabilistic inference (Bavelier, Achtman, Mani, & Föcker, 2012; Bavelier, Green, Pouget, & Schrater, 2012; Green, Pouget, & Bavelier, 2010). In the case of learning a second language, there is even evidence that this improves cognitive control and may help stave off Alzheimer's and other diseases of dementia in old age (Bialystok, 2006; Bialystok & Craik, 2010; Kroll & Bialystok, 2013; Spivey & Cardon, 2015). Think about how often you use language, in conversation or in reading a book like this. If every word you heard or read was briefly partially similar to another word you happen to know in a different language, then your frontal lobes would be engaging in an increased amount of conflict resolution (compared to a monolingual person) every few hundred milliseconds (Falandays & Spivey, in press). It seems quite plausible that this would strengthen the neural networks across various regions in your frontal lobes and make them better at compensating for any nearby brain damage.

Chapter 4

What If You Didn't Have a Body?
Philosophers of mind and artificial intelligence researchers have had a love/hate relationship for decades. For just a tiny sample of that history, see Dreyfus (1992), Putnam (1981), Weizenbaum (1966), and Winograd (1972).

You know that dream where you find yourself in your underwear in public? It became real for me once, but not in the way you're thinking. When I was a professor at Cornell, there was a large class I held on Halloween, and several of the students were wearing costumes. Suddenly, one of my students, "Crazy Dave" M., came jogging over to my podium wearing only his tighty-whitey underwear. With his eyes and smile wide open, he said, "Hey Spivey, you know that dream where you find yourself in public wearing only your underwear? Well, it's Halloween, and I'm livin' the dream!" I felt my eyebrows rise, and I calmly responded with, "Good for you, Dave. Now go have a seat. Class is about to start." Although recreating an underwear dream in real life may not be a good way to achieve a brain-without-a-body experience, actual dreams, paralysis, and sensory deprivation just might get you there (e.g., Bauby, 1998; Bosbach, Cole, Prinz, & Knoblich, 2005; Forgays & Forgays, 1992; Hebb, 1958; Laureys et al., 2005; Lilly, 1977; Suedfeld, Metcalfe, & Bluck, 1987).

The Psychology of the Embodied Mind
The psychological literature on embodied cognition is ginormous. Here is a small sample related to the findings presented in this section: Barsalou (1999), Beilock (2015), Bergen and Wheeler (2010), Chao and Martin (2000), Edmiston and Lupyan (2017), Estes and Barsalou (2018), Francken, Kok, Hagoort, and De Lange (2014), Glenberg and Kaschak (2002), Kaschak and Borreggine (2008), Kosslyn, Ganis, and Thompson (2001), Kosslyn, Thompson, and Ganis (2006), Meteyard, Zokaei, Bahrami, and Vigliocco (2008), Ostarek, Ishag, Joosen, and Huettig (2018), Smith (2005), Smith and Gasser (2005), Spivey and Geng (2001), Stanfield and Zwaan (2001), Zwaan and Pecher (2012), and Zwaan and Taylor (2006).

Of course, it may be the case that some aspects of your mind are more embodied than others. For recent debates over more versus less embodied theories, see Adams and Aizawa (2011), Chatterjee (2010), Chemero (2011), Hommel, Müsseler, Aschersleben, and Prinz (2001), Louwerse (2011), Mahon and Caramazza (2008), Meteyard, Cuadrado, Bahrami, and Vigliocco (2012), Noë (2005, 2009), Petrova et al. (2018), Rupert (2009), Segal (2000), Shapiro (2011), M. Wilson (2002), and R. Wilson (1994).

The Language of the Embodied Mind

For decades, language researchers have been arguing that how our bodies interact with the world affects how we talk about the world and vice versa. The evidence for this shows up in how children learn language, in the patterns of our everyday language use as adults, and in controlled laboratory experiments that carefully measure the timing of our language processes (e.g., Anderson & Spivey, 2009; Bergen, 2012; Chatterjee, 2001; Gibbs, 1994, 2005, 2006; Gibbs, Strom, & Spivey-Knowlton, 1997; Kövecses, 2003; Lakoff & Johnson, 1980, 1999; Lakoff & Nuñez, 2000; Maass & Russo, 2003; Mandler, 1992, 2004; Matlock, 2010; Richardson & Matlock, 2007; Richardson, Spivey, Barsalou, & McRae, 2003; Richardson, Spivey, Edelman, & Naples, 2001; Santiago, Román, & Ouellet, 2011; Saygin, McCullough, Alac, & Emmorey, 2010; Tversky, 2019; Winawer, Huk, & Boroditsky, 2008; Winter, Marghetis, & Matlock, 2015; see also Kourtzi & Kanwisher, 2000).

The Emotionality of the Embodied Mind

Our emotional embodiment of cognition even extends to how our mouths utter the names of objects. Social psychologist Sascha Topolinski has demonstrated that words that involve a sequence of speech sounds that work their way from deep in the mouth outward toward the lips, such as "kodiba," tend to have negative emotional associations. This is because the very act of saying the word is reminiscent of spitting something out of your mouth. By contrast, words that involve a sequence of speech sounds that work their way from the lips deeper into the mouth, such as "bodika," tend to have positive emotional associations because the act of saying the word is reminiscent of accepting something into your mouth, such as tasty food. Readings for this section include Adolphs, Tranel, Damasio, and Damasio (1994), Barrett (2006, 2017), Damasio (1994), Freeman, Stolier, Ingbretsen, and Hehman (2014), Gendron et al. (2012), Topolinksi, Boecker, Erle, Bakhtiari, and Pecher (2017), and Topolinski, Maschmann, Pecher, and Winkielman (2014).

The Biology of the Embodied Mind

I encourage you to read more about autopoiesis. In chapters 7 and 8, we will revisit systems that generate and maintain themselves, some of them living, some of them not. As homework in preparation for that, see Beer (2004, 2014, 2015), Bourgine and Stewart (2004), Chemero and Turvey (2008), Di Paolo, Buhrmann, and Barandiaran (2017), Gallagher (2017), Maturana and Varela (1991), Thompson (2007), and Varela (1997).

In some areas of research, the mirror neuron system has been touted as the source of social imitation and learning, the mechanism of empathy, the engine for a "theory of mind," the originator of language, and the inventor of sliced bread. Some of those claims go a bit beyond the data and should be taken with a grain of salt, but the basic observation is scientific fact. Parts of your brain are active when you carry out particular actions and are *also active* when you passively observe (visually or auditorily) those same actions carried out by other people (e.g., Calvo-Merino, Glaser, Grezes, Passingham, & Haggard, 2005; Gallese, Fadiga, Fogassi, & Rizzolatti, 1996; Gallese & Lakoff, 2005; Lahav, Saltzman, & Schlaug, 2007; Mukamel, Ekstrom, Kaplan, Iacoboni, & Fried, 2010; Rizzolatti & Arbib, 1998; Rizzolatti, Fogassi, & Gallese, 2006; Stevens, Fonlupt, Shiffrar, & Decety, 2000; Zatorre, Chen, & Penhune, 2007; but cf. Hickok, 2009).

The biology of language and biology of action are inextricably linked in so many ways. The following studies provide a glimpse into this link from the perspective of the biology of embodiment: Fadiga, Craighero, Buccino, & Rizzolatti (2002), Falandays, Batzloff, Spevack, and Spivey (in press), Galantucci, Fowler, and Turvey (2006), Gentilucci, Benuzzi, Gangitano, and Grimaldi (2001), Gordon, Spivey, and Balasubramaniam (2017), Hauk, Johnsrude, and Pulvermüller (2004), Liberman, Cooper, Shankweiler, and Studdert-Kennedy (1967), Liberman and Whalen (2000), Nazir et al. (2008), Pulvermüller (1999), Pulvermüller, Hauk, Nikulin, and Ilmoniemi (2005), Shebani and Pulvermüller (2013), Spevack, Falandays, Batzloff, and Spivey (2018), Vukovic, Fuerra, Shpektor, Myachykov, and Shtyrov (2017), and Wilson-Mendenhall, Simmons, Martin, and Barsalou (2013).

The Artificial Intelligence of an Embodied Mind

Philosophical discussions of artificial intelligence have been going on for decades. Consider the 1980 thought experiment offered by Zenon Pylyshyn (pronounced "Zen' nin Pil lish' in"), discussed earlier in this chapter (Pylyshyn, 1980). He asks us to estimate how many of our neurons would need to be replaced with identically functioning nanochips before we started to doubt that our mind was still human. Selecting a specific number seems painfully arbitrary. If you decide on a specific number as your criterion, like a hundred million, for example, then you are claiming that when 99,999,999 neurons have been replaced in your brain, you still have your original human mind. However, when just one more neuron is replaced by a nanochip, your mind is suddenly not you anymore because of that one additional nanochip. Sounds ridiculous, right? Zenon's solution is that we should just accept that, functionally speaking, you will not have changed, and therefore you would continue to have your human mind no matter how many neurons were replaced. The result is that we no longer treat biological material as a prerequisite for a mind. Clearly, Zenon's thought experiment draws some inspiration from the Sorites paradox developed in ancient Greek philosophy, where a collection of sand grains on the ground is not called a "heap" of sand until some requisite number of grains is added. The paradox is that there is no agreed number of sand grains to use as a criterion for calling it a "heap." Therefore, using a quantitative threshold of the amount (or type) of physical matter needed to apply a category label (such as "heap" or "human") to something is not a good way to go. Maybe we should call Zenon Pylyshyn's thought experiment, Zenon's paradox. Get it? Okay, fine, whatever. A few sources for philosophical discussions of this nature include Chalmers (1996), Churchland (2013), Churchland and Churchland (1998), Clark (2003), Dietrich (1994), Dreyfus (1992), Hofstadter and Dennett (1981), and Searle (1990).

The building of social and developmental robots is afoot. Make no mistake about it. These new creatures are helping students learn in the classroom, teaching cognitive scientists about the benefits of embodiment, and helping people coordinate on complex tasks (Misselhorn, 2015). A scientist can pick apart someone else's experiment until they're blue in the face, but when an engineer builds something that gets up, walks around, and talks to you, it's hard to claim it didn't work. Some example discussions of things that got up, walked around, and sometimes even talked include Allen et al. (2001), Bajcsy (1988), Barsalou, Breazeal, and Smith (2007), Belpaeme, Kennedy, Ramachandran, Scassellati, and Tanaka, (2018), Breazeal (2004), Brooks (1989, 1999), Cangelosi et al. (2010), Carpin, Lewis, Wang, Balakirsky, and Scrapper (2007), Pezzulo et al. (2011, 2013), Pezzulo, Verschure, Balkenius, and Pennartz (2014), Roy (2005), Smith and Breazeal (2007), and Steels (2003).

For a couple of decades now, researchers have been taking what's been learned about morphological computation in human and other animal bodies and exploiting it in the design of robot bodies (e.g., Brawer, Livingston, Aaron, Bongard, & Long, 2017; Hofman, Van Riswick, & Van Opstal, 1998; Huijing, 2009; Laschi, Mazzolai, & Cianchetti, 2016; Lipson, 2014; Paul, Valero-Cuevas, & Lipson, 2006; Pfeifer, Lungarella, & Iida, 2012; Turvey & Fonseca, 2014; Webb, 1996; Wightman & Kistler, 1989).

Directions for Use
Excellent discussions of how a brain compares and contrasts its sensory inputs with its motor outputs to develop an understanding of what and where its body is can be found in Blakemore, Wolpert, and Frith (2000), Blanke and Metzinger (2009), Bongard, Zykov, and Lipson. (2006), and Ramachandran and Blakeslee (1998).

Chapter 5

Sensory Transduction
Philosopher Hilary Putnam (1973) famously argued that the definition of the contents of your mind requires including properties that exist outside your body. A substantial portion of the philosophy of mind community has embraced those arguments for an "externalist" account of mind. However, a number of philosophers continue to argue instead for an "internalist" account that treats the mind as

something that exists only inside the brain-and-body (e.g., Adams & Aizawa, 2009; Rupert, 2004; Segal, 2000). Their arguments typically appeal to casual intuitions about where it feels as if the mind stops and the outside world starts. However, this section on sensory transduction shows that when you put a microscope onto that suggested boundary, it becomes very difficult to actually pinpoint the physical location of the boundary. Perhaps that's because it isn't there. Philosophers Andy Clark (2008), Susan Hurley (1998), Ruth Millikan (2004), John Sutton (2010), Tony Chemero (2011), and George Theiner (2014) have each compiled powerful evidence and arguments for treating your mind as something that often spreads out into the environment, not at all imprisoned by the "bag of skin" surrounding your body (see also Anderson, Richardson, & Chemero, 2012; Clark & Chalmers, 1998; Cowley & Vallée-Tourangeau, 2017; Favela & Martin, 2017; Smart, 2012; Spivey, 2012; Spivey & Spevack, 2017; Theiner, Allen, & Goldstone, 2010).

Ecological Perception

Philosopher of mind Jerry Fodor was known to bring attention to the sea squirt as an example of an animal that, once it gets settled in, eats its own brain with the expectation of not needing it anymore. Jerry compared that animal to a typical professor getting tenure. Given Fodor's scholarly productivity throughout the decades, it seems clear that he did not eat his own brain upon getting tenure (but apparently he thought some of his colleagues did).

The importance of the nonsessile "moving observer" has been central to ecological perception since its inception. Before photocopiers were prevalent, Jimmy Gibson would teach his graduate seminars at Cornell University using mimeographed copies of short essays (with that old purple ink) that were designed to start fruitful debate among the students and faculty for understanding where the important information can be located when a moving observer interacts with her environment. Because of the thorny challenges they raised, these mimeographs acquired the moniker *The Purple Perils*. Some of them were saved and published in *Reasons for Realism*, edited by Ed Reed and Rebecca Jones (1982). Further discussions of Gibson's ecological perception are included in Cutting (1993), Gibson (1966), Gibson and Bridgeman (1987), Pick, Pick, Jones, and Reed (1982), Shaw and Turvey (1999), Turvey and Carello (1986), Turvey and Shaw (1999), and Warren (1984).

It is always interesting when the scientific method gets it wrong. For a moment in 1982, perceptual and cognitive psychologists thought evidence had been found for how the brain can piece together the brief little snapshots from each eye fixation into a mosaic of the entire visual scene, stored in the brain. Pretty quickly, however, the scientific method is able to achieve a course correction via its own credo of testability and replication. If you can't test your theory or replicate it, then it ain't science. When it comes to how the brain builds an internal mental representation of the entire visual scene, despite having access only to intermittent unconnected images,

Bruce Bridgeman was instrumental in helping the field find out the correct answer to this question: it doesn't. See Bridgeman and Mayer (1983) and also Irwin, Yantis, and Jonides (1983), O'Regan and Lévy-Schoen (1983), and Rayner and Pollatsek (1983).

It was in Bruce Bridgeman's laboratory at the University of California, Santa Cruz, that I first learned the hands-on skills of laboratory experimental psychology. The late, great Bruce Bridgeman taught me scientific inquisitiveness, methodological rigor, and patience. Ray Gibbs described him as one of the most self-actualized people he'd ever met, stating that, "Bruce did 'Bruce' beautifully." Bruce shared *who he was* unguardedly. For his genuineness and his gentleness, Bruce is loved and admired by everybody who knew him. As he gave of himself to so many, putting part of his self outside his body, it is particularly appropriate that his theory about perceptual space constancy should lead directly to an understanding of the environment as a place where your visual memory happens, and even where your visual perception happens (e.g., Bridgeman, 2010; Bridgeman & Stark, 1991; Bridgeman, Van der Heijden, & Velichkovsky, 1994; O'Regan, 1992; O'Regan & Noë, 2001; Pylyshyn, 2007; Spivey & Batzloff, 2018; see also Lauwereyns, 2012). In fact, Kevin O'Regan and Alva Noë (2001) even suggested that your *conscious experience* of visual perception may be happening not inside your brain but in the *relationship* between your body and the environment (see also Morsella, Godwin, Jantz, Krieger, & Gazzaley, 2016).

The Action-Perception Cycle
In Gibson's ecological perception, there's usually no need to generate an internal mental model of what the external world looks like because in most circumstances you can just look around to *see* that. According to Gibson, the operative functions of perception and action are not internal representations in the brain but instead the *affordances* that exist between the body and the environment. For further discussions of affordances, see Chambers et al. (2004), Chemero (2003), Gibson (1979), Grézes, Tucker, Armony, Ellis, and Passingham (2003), Michaels (2003), Reed (2014), Richardson, Spivey, and Cheung (2001), Stoffregen (2000), Thomas (2017), Tucker and Ellis (1998), and Yee, Huffstetler, and Thompson-Schill (2011).

In addition to Glucksberg's (1964) observation that an adventitious nudging of the box of thumbtacks could lead to insight into the candle-mounting problem, as long ago as 1931, experimental psychologist Norman Maier observed that people who were at an impasse with his two-string problem would often discover the solution suddenly after accidentally brushing one of the ropes and noticing how it swings a bit. There are a wide variety of studies that show how the real-time processes of the action-perception cycle generate cognition, instigate insight, and promote skilled behavior on a millisecond timescale. Some of the cognitive operations happen in the brain, and some of them happen in the environment. Balasubramaniam (2013), Cluff, Boulet, and Balasubramaniam (2011), Cluff, Riley, and Balasubramaniam (2009), Dotov, Nie, and Chemero (2010), Duncker (1945), Glucksberg (1964), Grant and Spivey

(2003), Kirsh and Maglio (1994), Maier (1931), Neisser (1976), Risko and Gilbert (2016), Solman and Kingstone (2017), Stephen, Boncoddo, Magnuson, and Dixon (2009), and Thomas and Lleras (2007) represent just a small sample of that kind of work.

When Objects Become Part of You
For your brain to know how to direct its effectors (e.g., limbs) accurately toward target objects, it needs to have some form of understanding of the shape, size, and capabilities of those effectors. It has to have some form of body schema. This work shows that the neurons that participate in coding for that information can begin to treat a handheld tool as part and parcel of that body schema (e.g., Farnè, Serino, & Làdavas, 2007; Iriki, 2006; Iriki, Tanaka, & Iwamura, 1996; Maravita & Iriki, 2004; Maravita, Spence, & Driver, 2003).

Not only can tools in your hand change the way you perceive the world, but words you read can act like such tools as well. For example, when Chinese speakers read grammatical terms that refer to small, pinch-grip objects, their pupils automatically dilate slightly (as if to prepare to focus on something small). By contrast, when they read words that refer to large objects, their pupils do not dilate (Lobben & Boychynska, 2018). Jessica Witt and Denny Proffitt have produced a wealth of evidence for the role that action plays in perception (e.g., Brockmole, Davoli, Abrams, & Witt, 2013; Proffitt, 2006; Witt, 2011; Witt & Proffitt, 2008; Witt, Proffitt, & Epstein, 2005).

In addition to a rubber-hand illusion (Armel & Ramachandran, 2003; Botvinick & Cohen, 1998; Durgin, Evans, Dunphy, Klostermann, & Simmons, 2007; Giummarra & Moseley, 2011; Ramachandran & Blakeslee, 1998), there's also a whole-body illusion. With the help of immersive virtual reality, you can feel as if your entire body is "over there" rather than "here" (Blanke and Metzinger, 2009; Lenggenhager, Tadi, Metzinger, & Blanke, 2007).

When You Become Part of Your Environment
There are many ways in which you can use mental references to objects and locations in the environment as part of your thinking process. As a result, those objects and locations become part of your cognition. Although your brain obviously doesn't physically expand to include them, your mind does. Whether those mental references are called "pointers," "jigs," or "visual indexes," they clearly tie our thoughts to the external environment, such that our cognitive processes are performed *on those objects*, not just on internal mental representations of those objects (e.g., Ballard, Hayhoe, Pook, & Rao, 1997; Barrett, 2011; Franconeri, Lin, Enns, Pylyshyn, & Fisher, 2008; Kirsh, 1995; Molotch, 2017; Pylyshyn, 2001; Pylyshyn & Storm, 1988; Scholl & Pylyshyn, 1999).

In Richardson's eye-movement and memory experiments (Richardson, Altmann, Spivey, & Hoover 2009; Richardson & Kirkham, 2004; Richardson & Spivey, 2000), memory accuracy does not appear to be improved by looking at the correct empty square or hindered by looking at the wrong empty square. However, Bruno Laeng has found evidence that holding the eyes motionless does interfere with visual memory, compared to unconstrained eye movement (Laeng & Teodorescu, 2002). Moreover, Roger Johansson has found evidence for an almost 10 percent increase in memory accuracy for spatial relations when people were prompted to fixate on the correct blank location of the information to be remembered compared to when they were prompted to fixate on the incorrect blank location (Johansson & Johansson, 2014). Whether a given experiment shows slightly improved memory or not, they all show a natural tendency for the brain-and-body to treat these external locations as though they were addresses with content—even when that content is obviously no longer present. This consistent finding reveals how heavily your brain-and-body depends on the environment to serve as its memory (e.g., Ferreira, Apel, & Henderson, 2008; Hanning, Jonikaitis, Deubel, & Szinte, 2015; Ohl & Rolks, 2017; Olsen et al., 2014).

Open Minds and Closed Systems
The brain is an open system with respect to the body because the body physically contains the brain and participates in a continuous, fluid flow of information back and forth with the brain. Thus the brain is part of the brain-and-body system. And if the brain-and-body could meaningfully function as a mind without any influence from the environment, then the brain-and-body system would be a closed system. But it can't. The brain-and-body itself is an open system with respect to the environment because the environment physically contains the brain-and-body and participates in a continuous, fluid flow of information back and forth with the brain-and-body. It is not until one treats the mind as an entire organism-environment system that one has a chance to treat that system as "closed" and then successfully apply dynamical systems theory to understand how it works. Understanding how systems embedded inside other systems are able to generate a mind is crucial to understanding who you are. The following works provide some insight into how that can be done: Anderson et al. (2012), Atmanspacher and beim Graben (2009), beim Graben, Barrett, and Atmanspacher (2009), Crutchfield (1994), Dale and Spivey (2005), Dobosz and Duch (2010), Fekete, van Leeuwen, and Edelman (2016), Hotton and Yoshimi (2011), Järvilehto (1999, 2009), Järvilehto and Lickliter (2006), Spivey (2007), and Yoshimi (2012); see also Kirchhoff, Parr, Palacios, Friston, and Kiverstein (2018).

The Organism-Environment System
Sam Gosling's notion of "behavioral residue" is similar to David Kirsh's notion of how you "jig your environment" to do some of your thinking for you. Sometimes you are intentionally modifying your environment to support your cognition (Kirsh, 1995), and at other times you might be unintentionally leaving behind some remnants of your personality (Gosling, 2009; Gosling, Augustine, Vazire, Holtzman, & Gaddis, 2011; Gosling, Craik, Martin, & Pryor, 2005; Gosling, Ko, Mannarelli, & Morris, 2002). Either way, some of *who you are* is leaking out of your brain-and-body and

then soaking into the objects and locations around you—even into your computer or smartphone. Psychologist and data scientist Michal Kosinski has shown that your reported "likes" and "dislikes" on the internet can leave remarkably detailed evidence regarding your private personality traits (Kosinski, Stillwell, & Graepel, 2013; see also Heersmink, 2018).

Directions for Use
The tight feedback loop that you generate while staring wordlessly into someone else's eyes for five minutes creates an informational link between the two of you that can make your brains and bodies feel and behave like *one system* (Johnson, 2016). Let's follow that chain of cause and effect. Some of the light that bounces off person A's face goes into person B's eyes. The pattern in that light reveals subtle aspects of emotion on person A's face, gets processed by person B's brain, and then influences person B's own emotions and facial expressions. Those facial expressions are then seen by person A, and his or her brain then generates some new emotions and facial expressions that are then perceived by person B, and so on. On a millisecond timescale, these two visual feedback loops get intertwined together, continuously and fluidly sharing information back and forth.

Not only visual feedback can engender this mind-meld, but auditory and touch feedback can as well. You can try humming the same note together while eye gazing. You can also hold hands in a symmetric fashion. In fact, since we already know that the electric fields that your neurons generate extend out past the surface of your scalp (otherwise EEG wouldn't work), you could even try pressing your foreheads together to see if your electric fields and your partner's can develop a mutual influence on each other. Recent neuroscience research suggests that neurons may be regularly influencing each other not just with their axons, dendrites, and neurotransmitters but perhaps also with their electric fields (Goodman et al., 2015).

Chapter 6

You Are What You Eat
A great deal of research and exposition has actually focused on understanding how people choose the foods they eat and how they talk about them (Jurafsky, 2014). Despite wide variation in the sometimes arbitrary limitations that various cultures place on their cuisines, all cultures have naturally zeroed in on ways to ensure a reasonable amount of protein, fats, vitamins, carbohydrates, and fiber appropriate for supporting the human body. The way a culture rationalizes these decisions can be very interesting (Rozin & Fallon, 1987). Moreover, Senegalese chef Pierre Thiam notes that food is always better in places where different cultures have come together (Thiam & Sit, 2015; see also Bourdain & Woolever, 2016). They bring their different cuisines to the table, literally, and the combination makes for some of the most intriguing fusions you can find on a menu anywhere.

Who Your Family and Friends Are

A wide array of sensorimotor systems can fall into synchrony when they interact with one another between two people. Cognitive scientists Ivana Konvalinka and Andreas Roepstorff have even shown that someone who watches their family member perform in a "firewalking" ceremony generates a heart rate that is synchronized with their loved one's heart rate as they walk over those hot coals (Konvalinka et al., 2011; see also Bennett, Schatz, Rockwood, & Wiesenfeld, 2002; Haken, Kelso, & Bunz, 1985; Kelso, 1997; Mechsner, Kerzel, Knoblich, & Prinz, 2001; Schmidt, Carello, & Turvey, 1990).

Steven Strogatz's 2004 book *Sync* is especially impressive in how it documents the occurrence of synchrony in phenomena that cover an incredibly wide range of spatial and temporal scales. Rhythmic patterns that occur in our sun over years, on a spatial scale of a *billion* meters (about 10^9 m), have mathematical similarities to the rhythmic patterns that occur in tidal rivers on Earth, on a spatial scale of a *hundred* meters (about 10^2 m). And those patterns have mathematical similarities to the synchrony observed among neurons on a spatial scale of *millionths* of a meter (about 10^{-5} m), which in turn has similarities to the coordinated behaviors of atoms on a spatial scale of a *billionth* of a meter (about 10^{-10} m) as well as electrons on a spatial scale of a *quintillionth* of a meter (about 10^{-18} m). In fact, Strogatz's treatment of the correlated behaviors of electrons is particularly revealing in his discussion of Nobel laureate Brian Josephson and his "Josephson junction," which enables superconductivity. Josephson mathematically predicted that, under the right circumstances, pairs of electrons could become correlated with one another without coming into direct spatial contact, and his predictions have come true in the form of superconducting quantum computing systems of all kinds. What some people may not know about Josephson is that, after making his discovery of this "spooky action at a distance," he became obsessed with extrasensory perception (ESP).

Psychologist Daryl Bem of Cornell University was also obsessed with ESP, and I had several intriguing conversations with him about it while I was on the faculty there. I don't believe in ESP, but Daryl piqued my interest enough that I finally tried my own personal experiment. Rather than let myself get excited about apparent coincidences when they happen, which amounts to statistical cherry picking, I decided to test for a specific one. I silently chose a rare, unusual word and waited to see if someone happened to say it near me over the next few days. You can try this too, if you want to. My word was "Sequoia." (You should find your own.) Over the next few days, I met with several people in everyday events. Not once did anyone say "Sequoia." I felt that this was a slightly informative little data point in opposition to parapsychology and tucked it in the back pocket of my mind, never telling anyone. More than a year later, I was flying to meet my friend Bob McMurray, and on the plane I was reading Strogatz's description of Josephson and his ESP obsession. My interest was piqued again, so I decided to try my silly experiment one more time.

As my plane approached Eastern Iowa Airport, I knew Bob would be driving to pick me up, so he would be thinking of me (at least somewhat). I began meditating on the word "Sequoia," muttering it under my breath, and even writing it in cursive on my plane ticket multiple times—trying to generate some spooky action at a distance between these two somewhat correlated brains that were not yet in direct contact with one another. The person sitting next to me on the plane must have thought I was insane, or perhaps obsessed, so I tucked the scribbled-on plane ticket into my bag and zipped it up.

Then, after Bob picked me up and we were in his car driving to campus, I told him that I had been sending him a mental message 15 minutes ago while in the plane. He cocked a suspicious eyebrow. I said, "I know it sounds crazy, but just quiet your mind, and see if a single word bubbles up to the top for no apparent reason." To his credit as a friend, Bob indulged me, but after several seconds, he said, "Nope. I got nothin'." I asked him to try again, and it still didn't work. Finally, I decided to cheat and nullify the experiment, but I just wanted him to say *something*. I told him that the word I had been trying to psychically send him was a kind of tree. He was quiet again for several seconds. Then he opened his mouth and out came one word: "Sequoia." I was flabbergasted. That had been my secret ESP word for over a year. I don't think I had heard anyone say it in conversation with me all that time, and I had not uttered it to anyone else. Of course, I had narrowed the context for Bob quite a bit, so I knew it didn't really count as ESP. Still, he could have guessed the more popular "Maple tree" or "Pine tree." He could have said the more common term for sequoias, "Redwood." But Bob didn't do that. He said, "Sequoia."

As it turns out, it's actually pretty easy to get two brains synchronized and doing similar things. Cognitive neuroscientist Uri Hasson recorded five people's brain activity with fMRI while they watched the same 30-minute clip from Sergio Leone's classic movie *The Good, The Bad, and the Ugly*. On average, more than 29 percent of the cortical surface of those five brains had statistically significant correlations with one another during viewing (Hasson, Nir, Levy, Fuhrmann, & Malach, 2004; see also Hasson, Ghazanfar, Galantucci, Garrod, & Keysers, 2012). Naturally, many of the correlated brain areas were visual and auditory cortical regions, but several other, more cognitive regions were also highly correlated in their activity. Whether two brains are synchronizing via a third environmental input (such as a movie), because one of them is sending information in one direction (such as a monologue or written text), or by cocreating a live dialogue between the two of them, the key result is much the same. A time period of neural activation from one brain shows statistical correlations with another time period of neural activation from the other brain. Inspired by Hasson's experiment, my buddy Rick Dale and I occasionally do a Sync-n-Think, where we both put on headphones while we work on our laptops, and we do a countdown on a chat app so we can start the same music album at the same time. We could be 30 miles apart or 300 miles apart, but we've synchronized our brains a little bit with this identical auditory input. Sometimes we work on different parts of the same scientific manuscript while we Sync-n-Think. Those are probably our best-written articles. In general, it is a combination of synchrony and anticipation that allows coordinated brains to produce coordinated behavior. Because it is so fundamental to the overarching message of this book, I have an awful lot of extra reading for you on this burgeoning topic of how people cocreate their dialogue: Allwood, Traum, and Jokinen (2000), Anders, Heinzle, Weiskopf, Ethofer, and Haynes (2011), Brown-Schmidt, Yoon, and Ryskin (2015), A. Clark (2008), H. Clark (1996), Dale, Fusaroli, Duran, and Richardson (2013), Dale and Spivey (2018), Davis, Brooks, and Dixon (2016), Emberson, Lupyan, Goldstein, and Spivey (2010), Falandays et al. (2018), Froese, Iizuka, and Ikegami (2014), Fusaroli and Tylén (2016), Fusaroli et al. (2012), Hove and Risen (2009), Kawasaki, Yamada, Ushiku, Miyauchi, and Yamaguchi (2013), Koike, Tanabe, and Sadato (2015), Konvalinka and Roepstorff (2012), Kuhlen, Allefeld, and Haynes (2012), Louwerse, Dale, Bard, and Jeuniaux (2012), Lupyan and Clark (2015), Pickering and Garrod (2004), Richardson and Dale (2005), Richardson, Dale, and Kirkham (2007), Richardson and Kallen (2016), Riley, Richardson, Shockley, and Ramenzoni (2011), Schoot, Hagoort, and Segaert (2016), Shockley, Santana, and Fowler (2003), Smith, Rathcke, Cummins, Overy, and Scott (2014), Spiegelhalder et al. (2014), Spivey and Richardson (2009), Szary, Dale, Kello, and Rhodes (2015), Tollefsen, Dale, and Paxton (2013), Tomasello (2008); Trueswell and Tanenhaus (2005), Verga and Kotz (2019), von Zimmerman, Vicary, Sperling, Orgs, and Richardson (2018), Wagman, Stoffregen, Bai, and Schloesser (2017), Warlaumont, Richards, Gilkerson, and Oller (2014), and Zayas and Hazan (2014).

Who Your Co-workers Are
Coordinated behavior on a shared task comes in a wide variety of flavors. However, whether it's piloting a ship, controlling a computer system, or running a restaurant, this human coordination typically requires that people anticipate each other's intentions and actions. A great number of studies have shown the importance of this kind of adaptive back-and-forth sharing of information in coordinated behavior, both in the laboratory (e.g., Knoblich & Jordan, 2003; Sebanz, Bekkering, & Knoblich, 2006; Sebanz, Knoblich, Prinz, & Wascher, 2006; van der Wel, Knoblich, & Sebanz, 2011) and in the real world (e.g., Armitage et al., 2009; Barrett et al., 2004; Chemero, 2016; Guastello, 2001; Hutchins, 1995; Maglio, Kwan, & Spohrer, 2015; Maglio & Spohrer, 2013; Sawyer, 2005).

Who Your Social Group Is
One of the benefits of well-functioning groups is that they can often display "wisdom of the crowd" effects, where the average of the guesses from the group tends to have less error than the average error of all members of the group. For example, imagine asking a group of 10 people, "What percentage of the world's airports are in the United States?" You'd get 10 different guesses. Some of them would be moderately

close to the correct answer (33 percent), but some of them would be quite far from it. However, some of the very wrong guesses will be too high and others will be too low. Therefore, when you average all those guesses together, it will be quite close to 33 percent (Ariely et al., 2000; Vul & Pashler, 2008; Wallsten, Budescu, Erev, & Diederich, 1997). Unfortunately, wisdom of the crowd doesn't always work. Sometimes it can become more like Charles Bukowski's poem *The Genius of the Crowd*, his manifesto on the ugliness of groupthink (see Bukowski, 2008). But when a group has good diversity of opinions and backgrounds, and integrated relationships that tolerate respectful disagreement, the group can often exhibit exceptional wisdom (e.g., Adamatzky, 2005; Baumeister, Ainsworth, & Vohs, 2016; Baumeister & Leary, 1995; Holbrook, Izuma, Deblieck, Fessler, & Iacoboni, 2015; Orehek, Sasota, Kruglanski, Dechesne, & Ridgeway, 2014; Page, 2007; Smaldino, 2016; Talaifar & Swann, 2016).

One of the key drawbacks of in-group membership is that it usually defines an outgroup. Too often, a key aspect of what holds an in-group together is its zealous opposition to the out-group. It has become evident that this zealotry can even bring a good nation to the point where it irreversibly kidnaps the children of people seeking asylum at its border. Sharing a hatred for a common enemy is not required for group cohesion, but it is unfortunately a common symptom (e.g., Banaji & Greenwald, 2013; Banaji & Hardin, 1996; Fazio, Jackson, Dunton, & Williams, 1995; Freeman & Johnson, 2016; Freeman, Pauker, & Sanchez, 2016; Greenwald, McGhee, & Schwartz, 1998; Greenwald, Nosek, & Banaji, 2003; Holbrook, Pollack, Zerbe, & Hahn-Holbrook, 2018; Kawakami, Dovidio, Moll, Hermsen, & Russin, 2000; Kruglanski et al., 2013; Mitchell, Macrae, & Banaji, 2006; Sapolsky, 2019; Smeding, Quinton, Lauer, Barca, & Pezzulo, 2016; Stolier & Freeman, 2017; Wojnowicz, Ferguson, Dale, & Spivey, 2009).

Who Your Society Is

There have been many philosophical and mathematical explorations into how society might have evolved (culturally and biologically), how it might still be evolving right now, and how certain people think it *should* evolve in the future. For example, Ayn Rand's "objectivism," as she called it, posits that society (and economics) should evolve in a manner inspired by a simplistic interpretation of Darwinian natural selection, with the expectation that fierce competition alone will induce a positive evolution toward some optimal state (but cf. Lents, 2018). When you actually look at unfettered and unregulated competition in the food chain, the "optimal state" from which no further evolution is needed tends to look a lot like vicious apex predators. Look at *Tyrannosaurus rex*, great white sharks, and wolves. Is that what we want our optimal society to look like? Consider what millions of years of unguided competitive natural selection did for canines: wolves, who never cooperate outside their pack. Now, compare that to what thousands of years of human-guided selection did for canines: a wide and interesting diversity of dog breeds, most of which are socially cooperative team players, not vicious predators. Which world would you prefer to live in: a world of wolves or a world of man's best friends? Here are a bunch of examples of more mathematically inspired approaches to studying what it takes for a society to become and remain a cohesive band of cooperators who cocreate wealth rather than a marauding array of apex betrayers who think that unfettered greed is good: Axelrod (1984), Binmore (1998), Boyd and Richerson (1989), Camerer (2003), Dugatkin and Wilson (1991), Fehr and Fischbacher (2003), Kieslich and Hilbig (2014), Linster (1992), Smaldino, Schank, and McElreath (2013), Sugden (2004), Vanderschraaf (2006, 2018); see also Thagard (2019).

Who Your Nation Is
Pulling together intercultural groups to make for better problem solving (e.g., Ely & Thomas, 2001; Maznevski, 1994; Watson, Kumar, & Michaelsen, 1993) is exactly what the United States has been trying to do on a grand scale for many decades now. Several other countries throughout the world, including Canada and Australia, have also pursued this intercultural "melting pot" or "mosaic." Each of these countries has experienced its obstacles along the way toward that goal, to be sure, but let us be clear: the United States is unmistakably a nation of immigrants, people from different cultures who came to this nation to participate in The Intercultural Experiment. (This is true despite the fact that, in 2018, the U.S. Citizenship and Immigration Services office small-mindedly removed the phrase "nation of immigrants" from its mission statement.) In a nation of immigrants such as the United States, membership in the USA contract is what it means to be an American, not membership in any race, color, or creed. What is the USA contract? It is a social contract with a set of principles. When you are a resident of the United States, you are affirming your support for a democratic and free society that protects freedom of speech, separates religion from government, and treats all people as equals. Your adherence to that affirmation, not your heritage, is what grants you membership in the contract. Not everybody in the United States conforms to those principles as well as they should. To the extent that some people deviate from those principles, they are in fact "less American" than the rest. You could be a US citizen for many decades (elderly), a member of the majority race in the United States (white), and a member of the socially dominant gender (male), but if you do not uphold the freedom of speech, the separation of religion from government, and the treatment of all people as equals, then you are not really an American. That's just who we are. Love it or leave it. In fact, even if you are not yet officially a citizen of the United States but you uphold the principles of the USA contract, then that actually makes you a better American than a US citizen who doesn't uphold those principles.

Many other nations are working on a contract like this as well, both within their borders and across their borders. As we now have a larger number of nations interacting at the same level—instead of the world stage being dominated by just two nations—the patterns of behavior among nations are getting more complex. In addition to promoting fair treatment of different cultures and backgrounds *within* one's nation, we all have to get ready to promote fair treatment of different cultures and backgrounds in *other nations* as well—or we will all regret it. These tomes will teach you a thing or two about that: Albright (2018), Brzezinski and Scowcroft (2009), Chomsky (2017, reprinted), Farrow (2018), Kahn (1983), Levitsky and Ziblatt (2018), Luce (2017), Moyo (2018), Pinker (2018), Podobnik, Jusup, Kovac, and Stanley (2017), Rachman (2017), Stanley (2018), Torres (2016), Turchin (2016), Wise (2012), Woodley (2016), and Zakaria (2012).

Who Your Species Is
In your rush to avoid being a *racist* and instead draw a line around all humans as your in-group, be careful not to become too much of a *speciesist*. Are you really so sure that humans are your in-group and all other living creatures are the out-group? Think about that as you get your mind ready for chapter 7's expansion of your self to include other animals (see, e.g., Chudek & Henrich, 2011; Dugatkin, 1997; Gerkey et al., 2013; Harari, 2014; Henrich, 2015; Henrich et al., 2010; Hill, Barton, & Hurtado, 2009).

Directions for Use
Careful research shows that when you give instrumental support to family, friends, and neighbors (money, food, furniture, etc.), then you can expect your health prospects to extend your life span a little bit (Brown, Nesse, Vinokur, & Smith, 2003; Schwartz & Sendor, 1999). That's right; you will live longer if you give to family, friends, and neighbors. So all you have to do is think of everybody on the planet as your neighbor, which is actually reasonably accurate these days, and then any gift you give to any charity anywhere in the world will be something that generally improves your own health and your own life span. But you have to think of them as your neighbors or it won't work. So get to it.

Chapter 7

The Human Microbiome
The body's health is a bit like intelligence: you can pretend that there's one single measure of it (e.g., BMI or IQ), but you're going to find yourself making inaccurate assessments quite frequently. There are many dimensions to health (just as there are many dimensions to intelligence). A high body-mass index (BMI) by itself cannot tell you that you're unhealthy (Tomiyama, Hunger, Nguyen-Cuu, & Wells, 2016). For example, you know perfectly well that a 5'10", 180 lb. heavily muscled football running back is not actually "overweight" the way his BMI would indicate. Blood pressure by itself won't tell you what specific ailment you have. A glucose level or a cholesterol level, all by itself, won't reveal what the illness is. And a stool sample on its own won't tell you crap. A good medical professional knows to combine dozens of these indicators to think of your body's health as a complex system (e.g., Bollyky, 2018; Dietert, 2016; Hood & Tian, 2012; Liu, 2017; McAuliffe, 2016; Tauber, 2017; Turnbaugh et al., 2007; Turney, 2015; see also Rohwer, 2010, for holobionts in coral

reefs). Only then will the data reveal your specific infirmity. (By the same token, a set of health care *policies* also works like a complex system. Therefore, improving the health care system in your country requires "systems thinking" as well; see De Savigny & Adam, 2009.)

The Mental Life of Nonhuman Animals
If you've never had a pet at all, then you should stop reading this book. There's nothing I can do for you. Just kidding. But seriously, you should get a pet. Even if it's just a goldfish, that's better than nothing. Give it a name and take care of it. The majority of studies on the topic suggest that you may live happier and longer if you have a pet (Amiot & Bastian, 2015; but cf. Herzog, 2011).

In 1993, computer scientist Gary Cottrell was probably the first to introduce the term "dognitive science," albeit in a humor article (Cottrell, 1993), but now "dognition" is a real thing (Andics, Gácsi, Faragó, Kis, & Miklósi, 2014; Berns, 2013; Dilks et al., 2015; Miklósi, 2014; Thompkins, Deshpande, Waggoner, & Katz, 2016). But whether you get a pet or not, the real problem being addressed here is the common assumption that human intelligence is categorically exceptional compared that of the rest of the animal kingdom (cf. de Waal, 2017; Finlay & Workman, 2013; Matsuzawa, 2008; Spivey, 2000). (Human exceptionalism is just as myopic as nationalist exceptionalism.) The reason we should not assume that animals are dumb and we are smart is that the scientific evidence just doesn't stack up for that assumption (e.g., Barrett, 2011; Brosnan, 2013; Burghardt, 2005; Dehaene, 2011; Krubitzer, 1995; Northoff & Panksepp, 2008; Safina, 2016; Santos, Flombaum, & Phillips, 2007; Theiner, 2017; van den Heuvel, Bullmore, & Sporns, 2016; Young, 2012; see also Gardner, 2011).

You Are Coextensive with Nonhuman Animals
Yawn contagion is an interesting phenomenon that is often treated as a kind of indirect index of empathy or sociality that the creature being induced to yawn feels toward the initial yawner (e.g., Anderson, Myowa-Yamakoshi, & Matsuzawa, 2004; Campbell & de Waal, 2011; Joly-Mascheroni, Senju, & Shepherd, 2008; Massen, Church, & Gallup, 2015; Palagi, Leone, Mancini, & Ferrari, 2009; Paukner & Anderson, 2006; Platek, Critton, Myers, & Gallup, 2003; Provine, 2005; Silva, Bessa, & de Sousa, 2012; Wilkinson, Sebanz, Mandl, & Huber, 2011). If your best friend yawns, even just over the phone, you are quite likely to also yawn soon thereafter. However, if someone you don't know yawns in a movie you're watching, you are only somewhat likely to also yawn. If you see a lion yawn at the zoo, you are probably only very slightly inclined to yawn thereafter. And if you're autistic, then those kinds of stimuli simply may not make you inclined to yawn at all—unless perhaps you force yourself to look at the eyes of the yawner (Senju et al., 2007, 2009).

Living with animals is something we do all the time, even when we don't realize we're doing it. Even if you don't currently have a pet, have never ridden a horse, and don't eat meat, animals play an important role in your life. You probably had a stuffed animal of some kind as a child, and cuddling it helped you learn some empathy at that early, impressionable age. Perhaps you fall asleep to the sound of crickets or frogs on summer nights or wake up to the sound of birds welcoming in the new day. And you've almost certainly watched some movies about animals, such as documentaries, dramas, or even science fiction, where you marveled at how you found yourself able to relate to that nonhuman animal on the screen. The following set of studies of human interaction with our fellow animals is just a tiny sample of this huge body of literature: Amon and Favela (2019), Anthony and Brown (1991), Fossey (2000), Gardner and Gardner (1969), Goodall (2010), Keil (2015), Lestel (1998, 2006), Lestel, Brunois, and Gaunet (2006), Patterson and Cohn (1990), Patterson and Linden (1981), Pepperberg (2009), Perlman, Patterson, and Cohn (2012), Savage-Rumbaugh (1986), Shatner (2017), and Smith, Proops, Grounds, Wathan, and McComb (2016); see also Donaldson and Kymlicka (2011).

The Mental Life of Plants
Charles Darwin (father of Francis) had already tried playing his bassoon for his plants a century ago and wisely concluded that testing the effects of music on plant growth was "a fool's experiment." Those famous 1970s experiments that compared the effects on plants listening to classical music or rock music were poorly designed, easily biased, and provided no statistical evidence for reproducibility of the results. Bad science is a lot like false rumors (e.g., Che, Metaxa-Kakavouli, & Hancock, 2018; Del Vicario et al., 2016; Nekovee, Moreno, Bianconi, & Marsili, 2007; Spivey, 2017; Vosoughi, Roy, & Aral, 2017; Zubiaga, Liakata, Procter, Hoi, & Tolmie, 2016). Sometimes the stories are just so juicy that it's hard not to listen. But lies are lies, and most of them eventually get disproven. For instance, botanist Peter Scott describes some experiments that initially discovered that corn seeds were germinating faster when they were serenaded by music. However, in a follow-up experiment they were able to trace the actual cause to the warmth emanating from the nearby electric speakers (Scott, 2013). Even though plants have no talent for music appreciation, they are remarkably intelligent in their adaptive responses to other aspects of their environment, enough so that one is surely tempted to think of them as "aware" to some degree (e.g., Bruce, Matthes, Napier, & Pickett, 2007; Carello, Vaz, Blau, & Petrusz, 2012; Chamovitz, 2012; Darwin, 1908; Domingos, Prado, Wong, Gehring, & Feijo, 2015; Fechner, 1848; Gilroy & Trewavas, 2001; Michard et al., 2011; Runyon, Mescher, & De Moraes, 2006; Thellier, 2012, 2017; Thellier & Lüttge, 2013; Trewavas, 2003; Turvey, 2013; and if you like intelligent slime molds, see also Adamatzky, 2016; Nakagaki, Yamada, & Tóth, 2000; Reid, Latty, Dussutour, & Beekman, 2012; Tero et al., 2010).

You Are Coextensive with Plants
My buddy Arnie is a very talented landscaper, and he knows plants inside and out. He's also a bit of a tough guy. I once saw him give a man a concussion with one

punch. (It was in self-defense.) So you might be surprised when you hear this tough guy talk about the satisfaction he gets from nursing a sick plant back to health or shaping a rosebush to bloom perfectly, and you see his heart burst with a love he clearly has for these plants. They are like children to him. After all, don't we sometimes call a plant store a "nursery?" Arnie's job is basically to feed those children well and give them regular "haircuts" so they look good. If a tough guy like Arnie can embrace his oneness with nonanimal life forms, then surely the rest of us can, too.

If we didn't have so much in common with plants, then we probably couldn't eat them and extract any nutrients. A pebble from your garden certainly has some nutritious minerals in it, but you wouldn't be able to extract them by swallowing it. All life on Earth evolved from the same original form of DNA, and that biological commonality is a major part of why we can relate to each other, why we can exchange molecules with each other, and why we can feed on each other (Theobald, 2010).

The Planetwide Megabiome

The Kyoto Protocol and its updated Paris Climate Accord are, of course, not the first time that a governmental policy was enacted to protect the environment, but they are the first truly global versions of such policies. As long ago as 1285, England imposed a law that prohibited salmon fishing during certain times of the year in order to prevent depletion of the salmon population. However, it was not strongly enforced. Environmental awareness has been around for a long time. Now we just have to expand this awareness a bit more and turn it into real action. See Barnosky (2014), Barnosky et al. (2016), Bolster (2012), Farmer and Lafond (2016), Flusberg, Matlock, & Thibodeau (2017, 2018), Jensen (2016), Kolbert (2014), Lewandowsky and Oberauer (2016), Matlock, Coe, & Westerling (2017), Nelson (1982), Oreskes and Conway (2011), Ramanathan, Han, & Matlock (2017), Reese (2013, 2016), and Westerling, Hidalgo, Cayan, & Swetnam (2006).

If you truly embrace the environment as part of *who you are*, then taking care of the environment's health is just as obvious and natural as taking care of your own body's health. In the same way that if you treat your body badly, it will fail you, if we treat our environment badly, it will fail us. The science has been telling us that every living thing is mindlike (Godfrey-Smith, 2016; Thompson, 2007; Turvey, 2018), and together we form one organism (Ackerman, 2014; Bateson, 1979; Capra, 1983; Grinspoon, 2016; Holland, 2000; Hutchins, 2010; Ingold, 2000; Maturana & Varela, 1987; Miller, 1978; Prigogine & Stengers, 1984; Rosen, 1991). Let's start acting like it.

Chapter 8

The Ubiquity of Nonliving Systems

GJ357d is a planet 31 light years away that might harbor life. If it has a thick enough atmosphere, it might be able to trap water on its surface. However, it is 6 times the mass of Earth, so whatever life might form on it will not be large. With gravity 6 times that of Earth, any life on such a planet will likely be small and invertebrate. Perhaps their apex predator will be something like a banana slug. That said, there are surely some astronomers and astrobiologists (and self-reported alien abductees) who would disagree with my generally pessimistic take on humanity's chances of detecting extraterrestrial life and communicating with it. The field itself has not reached agreement on the issue. Some estimates with the Drake equation predict dozens of planets with intelligent life developing in each galaxy, whereas other estimates predict an average of less than one planet with intelligent life per galaxy. Either way, Enrico Fermi was probably right when he suggested that if extraterrestrial life were smart enough, long-lasting enough, and close enough for us to ever communicate with it, then we really should have detected unmistakable evidence of its existence by now (Barnes, Meadows, & Evans, 2015; Drake & Sobel, 1992; Frank & Sullivan, 2016; Sagan & Drake, 1975; Shostak, 1998; Stenger, 2011; Tipler, 1980; Vakoch & Dowd, 2015; Webb, 2015; see also Randall, 2017; Tyson, 2017).

The Mental Lives of Nonliving Systems
The BZ reaction is not just an amazing discovery of a nonliving autocatalytic chemical solution that initiates and maintains its own oscillations, as if it were alive and breathing. It is also a lesson in the sociology of science. Boris Pavlovitch Belousov (the man who put the B in "BZ reaction") was met with such disdain and disbelief when he submitted and resubmitted his scientific manuscript describing his oscillatory chemical reaction that he eventually gave up trying to publish it in a peer-reviewed journal at all. The editors and reviewers of the chemical journals to which he submitted this work were so entrenched in their conventional understanding of chemistry that they could not find it in themselves to lend any trust to Belousov's laboratory observations—or to even try out the experiment themselves. An obscure conference abstract is the only report of it that he ever managed to publish during his lifetime. The world of science often prides itself on having an open mind about revising its core tenets, given the right evidence from carefully controlled experiments. However, occasionally it, too, can fall victim to conventional dogma pulling the wool over its own eyes for a decade or two. Belousov's original (rejected) manuscript, translated into English, was eventually published in 1985 in the appendix of an edited volume by Maria Burger and Richard Field. I encourage you to look up video clips of the BZ reaction on the internet, and try to remind yourself as you watch these beautiful, vivid patterns grow that this clever little chemical solution is classified as a *nonliving* system (Belousov, 1985; Mikhailov & Ertl, 2017; Prigogine & Stengers, 1984; Steinbock, Tóth, & Showalter, 1995; Winfree, 1984; Zhabotinsky, 1964).

Before just willy-nilly building some living technology, we have to give it some thought first. Rather than manufacturing a complete humanoid puppet and hoping it will come to life, like Pinocchio, perhaps AI research should be nurturing a breeding reservoir for the self-organized evolution of abiotic life forms (e.g., Carriveau,

2006; Davis, Kay, Kondepudi, & Dixon, 2016; Dietrich, 2001; Dixon, Kay, Davis, & Kondepudi, 2016; Hanczyc & Ikegami, 2010; Harari, 2016; Ikegami, 2013; Kaiser, Snezhko, & Aranson, 2017; Kleckner, Scheeler, & Irvine, 2014; Kokot & Snezhko, 2018; O'Connell, 2017; Piore, 2017; Snezhko & Aranson, 2011; Swenson, 1989; Swenson & Turvey, 1991; Turvey & Carello, 2012; Walker, Packard, & Cody, 2017; Whitelaw, 2004; Yang et al., 2017).

We have to be careful, of course, not to accidentally let loose a virus of nanobots that infects, poisons, and kills all life on the planet permanently. That would be bad. Much of the abiotic evolution can initially take place in computer simulations of the process. This could be just like what roboticist Hod Lipson has been doing with evolution in virtual reality and then 3-D printing the resulting creature (Lipson & Pollack, 2000). However, instead of focusing so much on animals, perhaps we could start even simpler: with plants (e.g., Goel, Knox, & Norman, 1991). Maybe we could design an artificial plant that has no living biology inside it but carries out its own form of photosynthesis, draws molecules from the ground to use for its growth, and reproduces partial copies of itself (perhaps with asexual reproduction, so it doesn't have to send artificial pollen into the open air). The plant won't be edible, it won't be our slave, and it won't be dangerous. What the plant will do is teach us how to "grow" synthetic life, and it might even perform some much-needed carbon capture for us along the way.

The Complexity of Naturally Occurring Nonliving Systems

There has been a great deal of analysis and discussion of fractal $1/f$ scaling across wide ranges of natural phenomena. It is observed in the stock market, in human and other animal behavior, in plant growth, in weather patterns, and even in earthquakes, and it can make some truly beautiful patterns in nature (e.g., Bak, 1996; Crutchfield, 2012; Drake, 2016; Havlin et al., 1999; Hurst, 1951; Johnson, 1925; Kauffman, 1996; Lorenz, 2000; Mandelbrot, 2013; Strogatz, 2004; Turvey & Carello, 2012).

Sometimes a process can look as if it is mostly adhering to a power law but actually deviates a bit from a pure power law (Clauset, Shalizi, & Newman, 2009). For instance, the statistics of earthquakes may actually be better fit by a power law with a cutoff (or a tapering) than with a pure power law (Kagan, 2010; Serra & Corral, 2017). That is, great earthquakes (magnitude >8) are about half as common as would be predicted by a pure power law. Importantly, this suggests that there are some boundary conditions, or edge effects, that influence the way the statistical function shows up on paper. It doesn't necessarily mean that one is forced to postulate two separate tectonic mechanisms for what causes great earthquakes and what causes all other earthquakes. One tectonic process (with both brittle and plastic capabilities) may still be what explains all earthquakes, large and small, but the process has slightly different effects when that process pushes up against the extreme edges of its range (near some boundary conditions).

The observation that fractal $1/f$ scaling shows up both in nonliving systems and in living systems points to the commonality that exists between them. Living systems evolved from nonliving systems. Every living system contains, at its molecular and atomic scale, many of the same elements that make up nonliving systems. Humans exhibit $1/f$ scaling in their vasculature and in the variance exhibited by their voluntary behaviors, their speech, their heart rate, their memory, and even their gait (e.g., Abney, Kello, & Balasubramaniam, 2016; Chater & Brown, 1999; He, 2014; Hills, Jones, & Todd, 2012; Ivanov et al., 2001; Kello et al., 2010; Linkenkaer-Hansen, Nikouline, Palva, & Ilmoniemi, 2001; Mancardi, Varetto, Bucci, Maniero, & Guiot, 2008; Rhodes & Turvey, 2007; Usher, Stemmler, & Olami, 1995; Van Orden, Holden, & Turvey, 2003; Ward, 2002). We literally walk and talk in pink noise. In fact, one potential indicator of the early stages of Parkinson's disease is the loss of $1/f$ noise in the way you walk. Rather than a healthy correlated noise in the time series of leg movements, Parkinson's patients exhibit uncorrelated noise in their sequence of strides. Cognitive neuroscientist Michael Hove has shown that Parkinson's patients can regain that healthy $1/f$ noise in their walking pattern by using a rhythmic sound stimulus over headphones (Hove & Keller, 2015; Hove, Suzuki, Uchitomi, Orimo, & Miyake, 2012).

You Are Coextensive with Nonliving Systems
Microbial life on Earth probably arose from autocatalytic chemical networks, self-organizing processes that start themselves (e.g., Grosch & Hazen, 2015; Hazen, 2013; Hazen & Sverjensky, 2010; see also Dawkins & Wong, 2016). The very idea of something starting itself should twist your brain in on itself a little. How can something be responsible for bringing itself into being when obviously it wasn't around to do anything before it came into being? When the chain of cause and effect loops back onto itself, strange things can happen. Since most objects-and-events are conglomerates of many smaller objects-and-events, those smaller objects-and-events can often feed into each other, undetected, long before the larger object-and-event is even noticed. Thus, the coalescing of those smaller objects-and-events into one larger object-and-event can often look as if it came out of nowhere. Since no *other* large object or event caused this large object-and-event to come into being, we find ourselves concluding that the large object-and-event caused itself into being. It might be that the laws of causality have violated the arrow of time or that we have to look at both the small scale and the large scale *at the same time* in order to truly understand emergence. See Johnson (2002), Kauffman (1996), Laughlin (2006), Mitchell (2009), Rosen (1991), Spivey (2018), and Turvey (2004).

It's not just humans and their material culture that self-organize into the "hive mind" that is society (e.g., Sutton, 2008, 2010; Sutton & Keene, 2016; see also Malafouris, 2010). It's our pets, too. *Homo sapiens* not only returned their dead to the earth in a variety of underground burial methods (Aldenderfer, 2013; Moyes, 2012; Moyes, Rigoli, Huette, Montello, Matlock, & Spivey, 2017), but some tribes even buried their pet dogs alongside those human burials. As long ago as 8000 BCE, domesticated wolf-coyote hybrids were treated as "part of the family" and gently buried in the family cemetery next to grandma and grandpa (Perri et al., 2019). For more discussions of the hive mind to which we belong, see Clark (2008), Grinspoon (2016), Hutchins (2010), Lovelock & Margulis (1974), Odum (1988), and Smart (2012).

Dear Earth
In the 1980s, the rock band XTC wrote a song titled "Dear God," where the lead singer, Andy Partridge, speaks directly to God, saying, "Sorry to disturb you, but I feel I must be heard loud and clear." In that song, Partridge points out to God all the reasons why he can't believe in him. I'm not sure if the addressee ever heard those words, but they are powerful. I borrow the phrasing here to call attention to the fact that this omniscient natural power, Earth, is in need of a wake-up call as well. Earth, in the sense of all its nonliving and living matter, is reading this chapter through your eyes right now. I hope we are paying attention.

Chapter 9

Setting boundaries for *behaviors* that affect other people is usually good. However, setting boundaries for your *mental definition of your self*, as separate from the rest of the world, is often counterproductive. Achieving and maintaining a conscious sense of universal oneness is not always easy. The demands of everyday activity often get in the way. But it's not hard to remind yourself every once in a while to get back to that sense of oneness. Whether you are meditating on that oneness at any given time or not, it is still true (Wilber, 1979, 1997; see also Baskins, 2015). We are all sharing the same atmosphere. It holds us together. Earth's atmosphere can be seen as a single substrate that permeates us all, all the time. Breathing the same air with other humans, other animals, and other plants is kind of like being in a swimming pool together. Can we all just please agree not to pee in the pool?

What Chapter *N* Did to You
Chapters 2–8 were intended to get you to progressively expand your definition of self— *your* self and all the other selves around you that you are part of. Chapters 6, 7, and 8 especially may have significantly challenged some of your long-held assumptions. If you managed to release some of those assumptions, then you learned something new about your self, about *who you are*. Learning new things, acquiring new knowledge and experience, is essentially what cognitive scientist Shimon Edelman says is the meaning of life. In his book *The Happiness of Pursuit* (2012), Edelman weaves together neuroscience and the humanities to make the case that the human mind's most gratifying purpose for existing is to learn new knowledge and acquire new experience, to transform itself. Give it a shot. If the transformations offered by chapters 6, 7, and 8 are frightening, then maybe just try them on for a short time. Set a timer for a day or two. See how it feels. You just might decide that it's pretty darn tasty.

You Are *of* This World, Not Merely *in* This World
The temptation to think of one's consciousness as somehow separate from the rest of the world's other physical processes is so powerful that some philosophers (such as René Descartes) have even offered the uneasy proposal that consciousness is not governed by the known laws of physics (see also Chalmers, 1996, 2017).

Feminist physicist Karen Barad (2007) provides a particularly cogent treatment of how consciousness is not needed to explain quantum wave collapse. Rather than "quantum wave collapse," all that actually happens in those instances is that an observer becomes entangled with the quantum system itself. This perspective on quantum physics has profound cultural consequences for how to treat "conscious observers" not as individuals but as quantum-entangled subsystems.

A Democratic Universe
Deepak Chopra and Mena Kafatos have a 2017 book with the title *You Are the Universe*. In that book, they elegantly focus on evidence from physics for how your existence is *part of* the universe, not something that exists *inside* the universe. Whether you focus on a subatomic scale, a microscopic scale, mesoscopic or macroscopic scales, or even on the scale of deep time, the unrelenting conclusion from the data is that every level of analysis reveals that your favorite subsystem is embedded in (and thus *part of*) a larger system, which is itself a subsystem embedded in another larger system (see also Anderson et al., 2012; Spivey & Spevack, 2017).

Oneness, Purpose, and Eternal Life
The plodding trail of bread crumbs left by science as it carves its path through the thick forest of our cosmos is replete with shimmering examples of the beauty of nature. Rarities are discovered, supreme joy is found in everyday events, and universal awe-inspiring patterns are revealed (e.g., Dawkins, 2012; Jillette, 2012; Kauffman, 1996). This feeling of awe can be described as a sense of vastness in a creativity that was not expected and as a need to expand one's model of the world to accommodate this new vastness (Keltner & Haidt, 2003). Finding a sense of awe in life has become a science, part of the health sciences, in fact (Stellar, John-Henderson, Anderson, Gordon, McNeil, & Keltner, 2015). As it turns out, awe is good for you, like clean air and clean water. Like antioxidants, it can actually reduce your inflammation. What's more, feeling a sense of awe can make people more generous with each other (Piff, Dietze, Feinberg, Stancato, & Keltner, 2015) and more cooperative with each other (Stellar et al., 2017). In fact, the sense of awe one gets from great art is the same sense of awe that one can get from great science, because when it comes down to it, art and science are really not that far apart from one another (e.g., Drake, 2016; Kandel, 2016; Morrow, 2018; see also Rynecki, 2017).

Although certain Buddhist beliefs, such as reincarnation, are surely not supported by science, many others are well supported by science (Wright, 2017). Regular use of loving-kindness meditation has been shown to promote a sense of social connectedness (Hutcherson, Seppala, & Gross, 2008), and that reduction in loneliness can then reduce your blood pressure (Hawkley, Masi, Berry, & Cacioppo, 2006).

Regular use of mindfulness meditation has been shown to improve working memory capacity and sustained attention (Zeidan, Johnson, Diamond, David, & Goolkasian, 2010). It increases cognitive flexibility (Moore & Malinowksi, 2009) and reduces emotional reactivity (Ortner, Kilner, & Zelazo, 2007). In fact, a steady diet of mindfulness meditation can even help you stick to a steady diet of healthy foods (Papies, Barsalou, & Custers, 2012).

Who *We* Are
It can be a bit of a mind-bender to contemplate how the *relations* between objects can somehow be primary, while the objects themselves are secondary (Buber, 1923/1970; Metcalfe & Game, 2012). But this is exactly what is required of genuine autocatalytic and autocatakinetic systems: systems that bring themselves into being. In chemistry, physics, and hydrodynamics, self-organized systems emerge and "behave so as to persist" because the *relations* between their subcomponents are what bring some of those very same subcomponents into existence (e.g., Belousov, 1985; Carriveau, 2006; Dixon et al., 2016; Turvey & Carello, 2012). In quantum mechanics, the *relation* between two subatomic objects' properties can be accurately predicted before the objects' properties themselves can be predicted (Mermin, 1998). In a human mind, it is the *relations* between brain, body, and environment that generate the experience we call consciousness (Clark, 2003, 2008; McCubbins & Turner, 2020; Noë, 2009; Spivey, 2007; Turner, 2014). In coordinated human behavior, it is the *relations* between muscles and joints within a body, and across bodies, that produce synergies among those muscles and joints such that they behave as one entity (Anderson et al., 2012; Van Orden, Kloos, & Wallot, 2011). Finally, if everything is indeed connected to everything, then the recent rejuvenation of panpsychism cannot help but be true (Skrbina, 2005; Strawson, 2006). Perhaps Descartes's old Latin slogan *Cogito ergo sum* (I think, therefore I am) should be updated to something like *Cogito ergo totum cogitamus*: I think, therefore everything thinks.

References

Abney, D. H., Kello, C. T., & Balasubramaniam, R. (2016). Introduction and application of the multiscale coefficient of variation analysis. *Behavior Research Methods. 49*(5), 1571–1581.

Ackerman, D. (2014). *The human age: The world shaped by us.* W. W. Norton.

Adamatzky, A. (2005). *Dynamics of crowd-minds: Patterns of irrationality in emotions, beliefs and actions.* World Scientific.

Adamatzky, A. (Ed.). (2016). *Advances in Physarum machines: Sensing and computing with slime mould.* Springer.

Adams, F., & Aizawa, K. (2009). Why the mind is still in the head. In P. Robbins & M. Aydede (Eds.), *The Cambridge handbook of situated cognition* (pp. 78–95). Cambridge University Press.

Adams, F., & Aizawa, K. (2011). *The bounds of cognition.* Wiley.

Adolphs, R., Tranel, D., Damasio, H., & Damasio, A. (1994). Impaired recognition of emotion in facial expressions following bilateral damage to the human amygdala. *Nature, 372*(6507), 669–672.

Albright, M. (2018). *Fascism: A warning.* Harper.

Aldenderfer, M. (2013). Variation in mortuary practice on the early Tibetan plateau and the high Himalayas. *Journal of the International Association of Bon Research, 1*, 293–318.

Allen, J. F., Byron, D. K., Dzikovska, M., Ferguson, G., Galescu, L., & Stent, A. (2001). Toward conversational human-computer interaction. *AI Magazine, 22*(4), 27.

Allopenna, P. D., Magnuson, J. S., & Tanenhaus, M. K. (1998). Tracking the time course of spoken word recognition using eye movements: Evidence for continuous mapping models. *Journal of Memory and Language, 38*(4), 419–439.

Allwood, J., Traum, D., & Jokinen, K. (2000). Cooperation, dialogue and ethics. *International Journal of Human-Computer Studies, 53*(6), 871–914.

Altmann, G. T., & Kamide, Y. (2007). The real-time mediation of visual attention by language and world knowledge: Linking anticipatory (and other) eye movements to linguistic processing. *Journal of Memory and Language, 57*(4), 502–518.

Altmann, G. T., & Mirkovic, J. (2009). Incrementality and prediction in human sentence processing. *Cognitive Science, 33*(4), 583–609.

Amaral, D. G., Behniea, H., & Kelly, J. L. (2003). Topographic organization of projections from the amygdala to the visual cortex in the macaque monkey. *Neuroscience, 118*(4), 1099–1120.

Amiot, C. E., & Bastian, B. (2015). Toward a psychology of human-animal relations. *Psychological Bulletin, 141*(1), 6–47.

Amon, M. J., & Favela, L. H. (2019). Distributed cognition criteria: Defined, operationalized, and applied to human-dog systems. *Behavioural Processes, 162*, 167–176.

Anders, S., Heinzle, J., Weiskopf, N., Ethofer, T., & Haynes, J. D. (2011). Flow of affective information between communicating brains. *Neuroimage, 54*(1), 439–446.

Anderson, J. A., & Rosenfeld, E. (2000). *Talking nets: An oral history of neural networks.* MIT Press.

Anderson, J. A., Siverstein, J., Ritz, S., & Jones, R. (1977). Distinctive features, categorical perception, and probability learning: Some applications of a neural model. *Psychological Review, 84*(5), 413–451.

Anderson, J. R., Myowa-Yamakoshi, M., & Matsuzawa, T. (2004). Contagious yawning in chimpanzees. *Proceedings of the Royal Society of London B: Biological Sciences, 271*(Suppl 6), S468–S470.

Anderson, M. L. (2014). *After phrenology: Neural reuse and the interactive brain.* MIT Press.

Anderson, M. L., Brumbaugh, J., & Şuben, A. (2010). Investigating functional cooperation in the human brain using simple graph-theoretic methods. In W. Chaovalitwongse, P. Pardalos, & P. Xanthopoulos (Eds.), *Computational neuroscience* (pp. 31–42). Springer.

Anderson, M. L., Richardson, M. J., & Chemero, A. (2012). Eroding the boundaries of cognition: Implications of embodiment. *Topics in Cognitive Science, 4*(4), 717–730.

Anderson, S. E., Chiu, E., Huette, S., & Spivey, M. J. (2011). On the temporal dynamics of language-mediated vision and vision-mediated language. *Acta Psychologica, 137*(2), 181–189.

Anderson, S. E., & Spivey, M. J. (2009). The enactment of language: Decades of interactions between linguistic and motor processes. *Language and Cognition, 1*(1), 87–111.

Andics, A., Gácsi, M., Faragó, T., Kis, A., & Miklósi, Á. (2014). Voice-sensitive regions in the dog and human brain are revealed by comparative fMRI. *Current Biology, 24*(5), 574–578.

References

Anthony, D. W., & Brown, D. R. (1991). The origins of horseback riding. *Antiquity*, *65*(246), 22–38.

Ariely, D., Au, W. T., Bender, R. H., Budescu, D. V., Dietz, C. B., Gu, H., Wallsten, T. S., & Zauberman, G. (2000). The effects of averaging subjective probability estimates between and within judges. *Journal of Experimental Psychology: Applied*, *6*(2), 130–146.

Armel, K. C., & Ramachandran, V. S. (2003). Projecting sensations to external objects: Evidence from skin conductance response. *Proceedings of the Royal Society of London B: Biological Sciences*, *270*(1523), 1499–1506.

Armitage, D. R., Pummer, R., Berkes, F., Arthur, R. I., Charles, A. T., Davidson-Hunt, I. J., … Wollenberg, E. K. (2009). Adaptive co-management for social-ecological complexity. *Frontiers in Ecology and the Environment*, *7*(2), 95–102.

Arroyo, E., Selker, T., & Wei, W. (2006). Usability tool for analysis of web designs using mouse tracks. In *CHI'06 extended abstracts on Human Factors in Computing Systems* (pp. 484–489). Association for Computing Machinery.

Atmanspacher, H. (2015). Quantum approaches to consciousness. In E. Zalta (Ed.), *The Stanford encyclopedia of philosophy*. Stanford University Press.

Atmanspacher, H., & beim Graben, P. (2009). Contextual emergence. *Scholarpedia*, *4*(3), 7997.

Augustine, K. (2007). Does paranormal perception occur in near-death experiences? *Journal of Near-Death Studies*, *25*(4), 203–236.

Axelrod, R. M. (1984). *The evolution of cooperation*. Basic Books.

Baddeley, R., Abbott, L. F., Booth, M. C., Sengpiel, F., Freeman, T., Wakeman, E. A., & Rolls, E. T. (1997). Responses of neurons in primary and inferior temporal visual cortices to natural scenes. *Proceedings of the Royal Society of London B: Biological Sciences*, *264*(1389), 1775–1783.

Bajcsy, R. (1988). Active perception. *Proceedings of the IEEE*, *76*(8), 966–1005.

Bak, P. (1996). *How nature works: The science of self-organized criticality*. Copernicus.

Balasubramaniam, R. (2013). On the control of unstable objects: The dynamics of human stick balancing. In M. Richardson, M. Riley, & K. Shockley (Eds.), *Progress in motor control: Neural, computational and dynamic approaches* (pp.149–168). Springer.

Balcetis, E., & Dale, R. (2007). Conceptual set as a top-down constraint on visual object identification. *Perception*, *36*(4), 581–595.

Ballard, D. H., Hayhoe, M. M., Pook, P. K., & Rao, R. P. (1997). Deictic codes for the embodiment of cognition. *Behavioral and Brain Sciences*, *20*(4), 723–742.

Banaji, M. R., & Greenwald, A. G. (2013). *Blindspot: Hidden biases of good people*. Delacorte Press.

Banaji, M. R., & Hardin, C. D. (1996). Automatic stereotyping. *Psychological Science*, *7*(3), 136–141.

Banerjee, K., & Bloom, P. (2013). Would Tarzan believe in God? Conditions for the emergence of religious belief. *Trends in Cognitive Sciences*, *17*(1), 7–8.

Bar, M. (2003). A cortical mechanism for triggering top-down facilitation in visual object recognition. *Journal of Cognitive Neuroscience*, *15*(4), 600–609.

Bar, M., Kassam, K. S., Ghuman, A. S., Boshyan, J., Schmid, A. M., Dale, A. M., ...Halgren, E. (2006). Top-down facilitation of visual recognition. *Proceedings of the National Academy of Sciences of the United States of America*, *103*(2), 449–454.

Bar, M., & Ullman, S. (1996). Spatial context in recognition. *Perception*, *25*(3), 343–352.

Barad, K. (2007). *Meeting the universe halfway*. Duke University Press.

Barker, S. A., Borjigin, J., Lomnicka, I., & Strassman, R. (2013). LC/MS/MS analysis of the endogenous dimethyltryptamine hallucinogens, their precursors, and major metabolites in rat pineal gland microdialysate. *Biomedical Chromatography*, *27*(12), 1690–1700.

Barlow, H. (1953). Summation and inhibition in the frog's retina. *Journal of Physiology*, *119*(1), 69–88.

Barlow, H. (1972). Single units and sensation: A neuron doctrine for perceptual psychology? *Perception*, *1*(4), 371–394.

Barlow, H. (1995). The neuron doctrine in perception. In M. Gazzaniga (Ed.), *The cognitive neurosciences* (pp. 415–435). MIT Press.

Barnes, R., Meadows, V. S., & Evans, N. (2015). Comparative habitability of transiting exoplanets. *Astrophysical Journal*, *814*, 91.

Barnosky, A. (2014). *Dodging extinction: Power, food, money, and the future of life on earth*. California University Press.

Barnosky, A., Matlock, T., Christensen, J., Han, H., Miles, J., Rice, R., ...White, L. (2016). Establishing common ground: Finding better ways to communicate about climate disruption. *Collabra*, *2*(1), 23.

Barrett, J. L. (2012). *Born believers: The science of children's religious beliefs*. Free Press.

Barrett, L. (2011). *Beyond the brain: How body and environment shape animal and human minds*. Princeton University Press.

Barrett, L. F. (2006). Solving the emotion paradox: Categorization and the experience of emotion. *Personality and Social Psychology Review*, *10*(1), 20–46.

Barrett, L. F. (2017). *How emotions are made*. Houghton Mifflin Harcourt.

Barrett, R., Kandogan, E., Maglio, P. P., Haber, E. M., Takayama, L. A., & Prabaker, M. (2004). Field studies of computer system administrators: Analysis of system management tools and practices. In *Proceedings of the 2004 ACM Conference on Computer Supported Cooperative Work* (pp. 388–395). Association for Computing Machinery.

Barsalou, L. W. (1983). Ad hoc categories. *Memory & Cognition, 11*(3), 211–227.

Barsalou, L. W. (1999). Perceptual symbol systems. *Behavioral and Brain Sciences, 22*(4), 577–660.

Barsalou, L. W., Breazeal, C., & Smith, L. B. (2007). Cognition as coordinated noncognition. *Cognitive Processing, 8*(2), 79–91.

Baskins, B. S. (2015). *Oneness: Principles of world peace.* Global Unity Media.

Bateson, G. (1979). *Mind and nature: A necessary unity.* Dutton.

Bauby, J. (1998). *The diving bell and the butterfly.* Vintage.

Baumeister, R. F., Ainsworth, S. E., & Vohs, K. D. (2016). Are groups more or less than the sum of their members? The moderating role of individual identification. *Behavioral and Brain Sciences, 39*, e137.

Baumeister, R. F., & Leary, M. R. (1995). The need to belong: Desire for interpersonal attachments as a fundamental human motivation. *Psychological Bulletin, 117*(3), 497–592.

Bavelier, D., Achtman, R. L., Mani, M., & Föcker, J. (2012). Neural bases of selective attention in action video game players. *Vision Research, 61*, 132–143.

Bavelier, D., Green, C. S., Pouget, A., & Schrater, P. (2012). Brain plasticity through the life span: Learning to learn and action video games. *Annual Review of Neuroscience, 35*, 391–416.

Bavelier, D., & Neville, H. J. (2002). Cross-modal plasticity: Where and how? *Nature Reviews Neuroscience, 3*(6), 443–452.

Bechara, A., Damasio, A. R., Damasio, H., & Anderson, S. W. (1994). Insensitivity to future consequences following damage to human prefrontal cortex. *Cognition, 50*(1), 7–15.

Bechara, A., Damasio, H., & Damasio, A. R. (2000). Emotion, decision making and the orbitofrontal cortex. *Cerebral Cortex, 10*(3), 295–307.

Beer, R. D. (2004). Autopoiesis and cognition in the game of life. *Artificial Life, 10*(3), 309–326.

Beer, R. D. (2014). The cognitive domain of a glider in the game of life. *Artificial Life, 20*(2), 183–206.

Beer, R. D. (2015). Characterizing autopoiesis in the game of life. *Artificial Life, 21*(1), 1–19.

Beilock, S. (2015). *How the body knows its mind: The surprising power of the physical environment to influence how you think and feel*. Atria Books.

beim Graben, P. (2014). Contextual emergence of intentionality. *Journal of Consciousness Studies, 21*(5–6), 75–96.

beim Graben, P. B., Barrett, A., & Atmanspacher, H. (2009). Stability criteria for the contextual emergence of macrostates in neural networks. *Network: Computation in Neural Systems, 20*(3), 178–196.

Belousov, B. (1985). Appendix: A periodic reaction and its mechanism. In M. Burger & R. Field (Eds.), *Oscillations and travelling waves in chemical systems*. Wiley.

Belpaeme, T., Kennedy, J., Ramachandran, A., Scassellati, B., & Tanaka, F. (2018). Social robots for education: A review. *Science Robotics, 3*, eaat5954.

Bennett, M., Schatz, M., Rockwood, H., & Wiesenfeld, K. (2002). Huygens's clocks. *Proceedings of the Royal Society A, 458*, 563–579.

Bergen, B., & Wheeler, K. (2010). Grammatical aspect and mental simulation. *Brain and Language, 112*(3), 150–158.

Bergen, B. K. (2012). *Louder than words: The new science of how the mind makes meaning*. Basic Books.

Berns, G. (2013). *How dogs love us: A neuroscientist and his adopted dog decode the canine brain*. Haughton Mifflin Harcourt.

Bialystok, E. (2006). Effect of bilingualism and computer video game experience on the Simon task. *Canadian Journal of Experimental Psychology, 60*(1), 68–79.

Bialystok, E., & Craik, F. I. (2010). Cognitive and linguistic processing in the bilingual mind. *Current Directions in Psychological Science, 19*(1), 19–23.

Biederman, I., Mezzanotte, R. J., & Rabinowitz, J. C. (1982). Scene perception: Detecting and judging objects undergoing relational violations. *Cognitive Psychology, 14*(2), 143–177.

Binmore, K. G. (1998). *Game theory and the social contract: Vol. 2. Just playing*. MIT Press.

Biran, I., & Chatterjee, A. (2004). Alien hand syndrome. *Archives of Neurology, 61*(2), 292–294.

Blakemore, S. J., Wolpert, D., & Frith, C. (2000). Why can't you tickle yourself? *Neuroreport, 11*(11), R11–R16.

Blanke, O., & Metzinger, T. (2009). Full-body illusions and minimal phenomenal selfhood. *Trends in Cognitive Sciences, 13*(1), 7–13.

Bode, S., He, A. H., Soon, C. S., Trampel, R., Turner, R., & Haynes, J. D. (2011). Tracking the unconscious generation of free decisions using ultra-high field fMRI. *PloS One, 6*(6), e21612.

Bollyky, T. J. (2018). *Plagues and the paradoxes of progress: Why the world is getting healthier in worrisome ways.* MIT Press.

Bolster, W. J. (2012). *The mortal sea.* Harvard University Press.

Bongard, J., Zykov, V., & Lipson, H. (2006). Resilient machines through continuous self-modeling. *Science, 314*(5802), 1118–1121.

Boroditsky, L., Schmidt, L. A., & Phillips, W. (2003). Sex, syntax, and semantics. In D. Gentner & S. Goldin-Meadow (Eds.), *Language in mind: Advances in the study of language and thought* (pp. 61–79). MIT Press.

Bosbach, S., Cole, J., Prinz, W., & Knoblich, G. (2005). Inferring another's expectation from action: The role of peripheral sensation. *Nature Neuroscience, 8*(10), 1295–1297.

Botvinick, M., & Cohen, J. (1998). Rubber hands "feel" touch that eyes see. *Nature, 391*(6669), 756.

Bourdain, A., & Woolever, L. (2016). *Appetites: A cookbook.* Ecco.

Bourgine, P., & Stewart, J. (2004). Autopoiesis and cognition. *Artificial Life, 10*(3), 327–345.

Boyd, R., & Richerson, P. J. (1989). The evolution of indirect reciprocity. *Social Networks, 11*(3), 213–236.

Boyer, P., & Bergstrom, B. (2008). Evolutionary perspectives on religion. *Annual Review of Anthropology, 37*, 111–130.

Brawer, J., Hill, A., Livingston, K., Aaron, E., Bongard, J., & Long, J. H., Jr. (2017). Epigenetic operators and the evolution of physically embodied robots. *Frontiers in Robotics and AI, 4*, 1.

Breazeal, C. L. (2004). *Designing sociable robots.* MIT Press.

Bridgeman, B. (2010). How the brain makes the world appear stable. *i-Perception, 1*(2), 69–72.

Bridgeman, B., & Mayer, M. (1983). Failure to integrate visual information from successive fixations. *Bulletin of the Psychonomic Society, 21*(4), 285–286.

Bridgeman, B., & Stark, L. (1991). Ocular proprioception and efference copy in registering visual direction. *Vision Research, 31*(11), 1903–1913.

Bridgeman, B., Van der Heijden, A. H. C., & Velichkovsky, B. M. (1994). A theory of visual stability across saccadic eye movements. *Behavioral and Brain Sciences, 17*(2), 247–257.

Brockmole, J. R., Davoli, C. C., Abrams, R. A., & Witt, J. K. (2013). The world within reach: Effects of hand posture and tool use on visual cognition. *Current Directions in Psychological Science, 22*(1), 38–44.

Brooks, R. A. (1989). A robot that walks: Emergent behaviors from a carefully evolved network. *Neural Computation, 1*(2), 253–262.

Brooks, R. A. (1999). *Cambrian intelligence: The early history of the new AI*. MIT Press.

Brosnan, S. F. (2013). Justice- and fairness-related behaviors in nonhuman primates. *Proceedings of the National Academy of Sciences, 110*(Suppl 2), 10416–10423.

Brown, S. L., Nesse, R. M., Vinokur, A. D., & Smith, D. M. (2003). Providing social support may be more beneficial than receiving it: Results from a prospective study of mortality. *Psychological Science, 14*(4), 320–327.

Brown-Schmidt, S., Yoon, S. O., & Ryskin, R. A. (2015). People as contexts in conversation. In *Psychology of Learning and Motivation* (Vol. 62, pp. 59–99). Academic Press.

Bruce, T. J., Matthes, M. C., Napier, J. A., & Pickett, J. A. (2007). Stressful "memories" of plants: Evidence and possible mechanisms. *Plant Science, 173*(6), 603–608.

Bruhn, P., Huette, S., & Spivey, M. (2014). Degree of certainty modulates anticipatory processes in real time. *Journal of Experimental Psychology: Human Perception and Performance, 40*(2), 525–538.

Brzezinski, Z., & Scowcroft, B. (2009). *America and the world: Conversations on the future of American foreign policy*. Basic Books.

Buber, M. (1970). *I and Thou*, translated by Walter Kaufmann. Simon and Schuster.

Buc Calderon, C., Verguts, T., & Gevers, W. (2015). Losing the boundary: Cognition biases action well after action selection. *Journal of Experimental Psychology: General, 144*(4), 737–743.

Bukowski, C. (2008). *The pleasures of the damned*. Harper Collins.

Burghardt, G. M. (2005). *The genesis of animal play: Testing the limits*. MIT Press.

Caldwell-Harris, C. L. (2012). Understanding atheism/non-belief as an expected individual-differences variable. *Religion, Brain & Behavior, 2*(1), 4–23.

Calvo-Merino, B., Glaser, D. E., Grezes, J., Passingham, R. E., & Haggard, P. (2005). Action observation and acquired motor skills: An FMRI study with expert dancers. *Cerebral Cortex, 15*(8), 1243–1249.

Camerer, C. (2003). *Behavioral game theory: Experiments in strategic interaction*. Princeton University Press.

Campbell, M. W., & de Waal, F. B. (2011). Ingroup-outgroup bias in contagious yawning by chimpanzees supports link to empathy. *PloS One, 6*(4), e18283.

References

Cangelosi, A., Metta, G., Sagerer, G., Nolfi, S., Nehaniv, C., Fischer, K., ... Zeschel, A. (2010). Integration of action and language knowledge: A roadmap for developmental robotics. *IEEE Transactions on Autonomous Mental Development, 2*(3), 167–195.

Capra, F. (1983). *The turning point: Science, society, and the rising culture.* Bantam.

Carello, C., Vaz, D., Blau, J. J., & Petrusz, S. (2012). Unnerving intelligence. *Ecological Psychology, 24*(3), 241–264.

Carpin, S., Lewis, M., Wang, J., Balakirsky, S., & Scrapper, C. (2007, April). USARSim: A robot simulator for research and education. In S. Hutchinson (Ed.), *Proceedings of the 2007 IEEE International Conference on Robotics and Automation* (pp. 1400–1405). IEEE.

Carriveau, R. (2006). The hydraulic vortex—an autocatakinetic system. *International Journal of General Systems, 35*(6), 707–726.

Casasanto, D., & Lupyan, G. (2015). All concepts are ad hoc concepts. In E. Margolis & S. Laurence (Eds.), *The conceptual mind: New directions in the study of concepts* (pp. 543–566). MIT Press.

Ceci, S. J., & Huffman, M. C. (1997). How suggestible are preschool children? Cognitive and social factors. *Journal of the American Academy of Child & Adolescent Psychiatry, 36*(7), 948–958.

Chalmers, D. (1996). *The conscious mind.* Oxford University Press.

Chalmers, D. (2017). Naturalistic dualism. *The Blackwell companion to consciousness* (pp. 363–373). Blackwell.

Chambers, C. G., Tanenhaus, M. K., & Magnuson, J. S. (2004). Actions and affordances in syntactic ambiguity resolution. *Journal of Experimental Psychology: Learning, Memory, and Cognition, 30*(3), 687–696.

Chamovitz, D. (2012). *What a Plant Knows: A Field Guide to the Senses.* Scientific American/Farrar, Straus and Giroux.

Chang, L., & Tsao, D. Y. (2017). The code for facial identity in the primate brain. *Cell, 169*(6), 1013–1026.

Chao, L. L., & Martin, A. (2000). Representation of manipulable man-made objects in the dorsal stream. *Neuroimage, 12*(4), 478–484.

Chater, N., & Brown, G. D. (1999). Scale-invariance as a unifying psychological principle. *Cognition, 69*(3), B17–B24.

Chatterjee, A. (2001). Language and space: Some interactions. *Trends in Cognitive Sciences, 5*(2), 55–61.

Chatterjee, A. (2010). Disembodying cognition. *Language and Cognition, 2*(1), 79–116.

Chau, A., Salazar, A. M., Krueger, F., Cristofori, I., & Grafman, J. (2015). The effect of claustrum lesions on human consciousness and recovery of function. *Consciousness and Cognition, 36*, 256–264.

Che, X., Metaxa-Kakavouli, D., & Hancock, J. T. (2018). Fake news in the news: An analysis of partisan coverage of the fake news phenomenon. In G. Fitzpatrick, J. Karahalios, A. Lampinen, and A. Monroy-Hernández (Eds.), *Companion of the 2018 ACM Conference on Computer Supported Cooperative Work and Social Computing* (pp. 289–292). Association for Computing Machinery.

Chemero, A. (2003). An outline of a theory of affordances. *Ecological Psychology, 15*(2), 181–195.

Chemero, A. (2011). *Radical embodied cognitive science*. MIT Press.

Chemero, A. (2016). Sensorimotor empathy. *Journal of Consciousness Studies, 23*(5–6), 138–152.

Chemero, A., & Turvey, M. T. (2008). Autonomy and hypersets. *Biosystems, 91*(2), 320–330.

Chomsky, N. (2017). *Who rules the world?* (reprint with new Afterword). Picador.

Chopra, D., & Kafatos, M. (2017). *You are the universe*. Harmony.

Chrysikou, E. G., Hamilton, R. H., Coslett, H. B., Datta, A., Bikson, M., & Thompson-Schill, S. L. (2013). Noninvasive transcranial direct current stimulation over the left prefrontal cortex facilitates cognitive flexibility in tool use. *Cognitive Neuroscience, 4*(2), 81–89.

Chudek, M., & Henrich, J. (2011). Culture–gene coevolution, norm-psychology and the emergence of human prosociality. *Trends in Cognitive Sciences, 15*(5), 218–226.

Churchland, P. M. (2013). *Matter and consciousness* (3rd ed.). MIT Press.

Churchland, P. M., & Churchland, P. S. (1998). *On the contrary: Critical essays, 1987–1997*. MIT Press.

Clark, A. (2003). *Natural-born cyborgs*. Oxford University Press.

Clark, A. (2008). *Supersizing the mind: Embodiment, action, and cognitive extension*. Oxford University Press.

Clark, A., & Chalmers, D. (1998). The extended mind. *Analysis, 58*(1), 7–19.

Clark, H. H. (1996). *Using language*. Cambridge University Press.

Clauset, A., Shalizi, C. R., & Newman, M. E. (2009). Power-law distributions in empirical data. *SIAM Review, 51*(4), 661–703.

Clavagnier, S., Falchier, A., & Kennedy, H. (2004). Long-distance feedback projections to area V1: Implications for multisensory integration, spatial awareness, and visual consciousness. *Cognitive, Affective, & Behavioral Neuroscience, 4*(2), 117–126.

Cluff, T., Boulet, J., & Balasubramaniam, R. (2011). Learning a stick-balancing task involves task-specific coupling between posture and hand displacements. *Experimental Brain Research, 213*(1), 15–25.

Cluff, T., Riley, M. A., & Balasubramaniam, R. (2009). Dynamical structure of hand trajectories during pole balancing. *Neuroscience Letters, 464*(2), 88–92.

Cottrell, G. W. (1993). Approaches to the inverse dogmatics problem: Time for a return to localist networks? *Connection Science, 5*(1), 95–97.

Cottrell, G. W., & Tsung, F. S. (1993). Learning simple arithmetic procedures. *Connection Science, 5*(1), 37–58.

Cowey, A. (2010). The blindsight saga. *Experimental Brain Research, 200*(1), 3–24.

Cowley, S., & Vallée-Tourangeau, F. (Eds.) (2017). *Cognition beyond the brain* (2nd ed.). Springer.

Crick, F. C., & Koch, C. (2005). What is the function of the claustrum? *Philosophical Transactions of the Royal Society of London B: Biological Sciences, 360*(1458), 1271–1279.

Crutchfield, J. P. (1994). The calculi of emergence: Computation, dynamics and induction. *Physica D: Nonlinear Phenomena, 75*(1), 11–54.

Crutchfield, J. P. (2012). Between order and chaos. *Nature Physics, 9*, 17–24.

Cutting, J. E. (1993). Perceptual artifacts and phenomena: Gibson's role in the 20th century. *Advances in Psychology, 99*, 231–260.

Dale, R., & Duran, N. D. (2011). The cognitive dynamics of negated sentence verification. *Cognitive Science, 35*(5), 983–996.

Dale, R., Fusaroli, R., Duran, N., & Richardson, D. C. (2013). The self-organization of human interaction. *Psychology of Learning and Motivation, 59*, 43–95.

Dale, R., & Spivey, M. J. (2005). From apples and oranges to symbolic dynamics: A framework for conciliating notions of cognitive representation. *Journal of Experimental & Theoretical Artificial Intelligence, 17*(4), 317–342.

Dale, R., & Spivey, M. J. (2018). Weaving oneself into others: Coordination in conversational systems. In B. Oben & G. Brône (Eds.), *Eyetracking in interaction: Studies on the role of eye gaze in dialogue* (pp. 67–90). John Benjamins.

Damasio, A. (1994). *Descartes' error: Emotion, rationality, and the human brain.* Putnam.

Damasio, H., Grabowski, T., Frank, R., Galaburda, A. M., & Damasio, A. R. (1994). The return of Phineas Gage: Clues about the brain from the skull of a famous patient. *Science, 264*(5162), 1102–1105.

Darwin, F. (1908). The address of the president of the British Association for the Advancement of Science. *Science, 28*(716), 353–362.

David, S. V., Vinje, W. E., & Gallant, J. L. (2004). Natural stimulus statistics alter the receptive field structure of V1 neurons. *Journal of Neuroscience, 24*(31), 6991–7006.

Davis, T. J., Brooks, T. R., & Dixon, J. A. (2016). Multi-scale interactions in interpersonal coordination. *Journal of Sport and Health Science, 5*(1), 25–34.

Davis, T. J., Kay, B. A., Kondepudi, D., & Dixon, J. A. (2016). Spontaneous interentity coordination in a dissipative structure. *Ecological Psychology, 28*(1), 23–36.

Dawkins, R. (2012). *The magic of reality: How we know what's really true*. Simon and Schuster.

Dawkins, R., & Wong, Y. (2016). *The ancestor's tale: A pilgrimage to the dawn of evolution* (revised and expanded). Mariner.

Dehaene, S. (2011). *The number sense: How the mind creates mathematic* (revised edition). Oxford University Press.

Dell, G. (1986). A spreading-activation theory of retrieval in sentence production. *Psychological Review, 93*(3), 283–321.

Del Vicario, M., Bessi, A., Zollo, F., Petroni, F., Scala, A., Caldarelli, G.,...& Quattrociocchi, W. (2016). The spreading of misinformation online. *Proceedings of the National Academy of Sciences, 113*(3), 554–559.

Dennett, D. C. (1993). *Consciousness explained*. Penguin.

de Sa, V. R., & Ballard, D. H. (1998). Category learning through multimodality sensing. *Neural Computation, 10*(5), 1097–1117.

De Savigny, D., & Adam, T. (Eds.). (2009). *Systems thinking for health systems strengthening*. World Health Organization.

Desimone, R., & Duncan, J. (1995). Neural mechanisms of selective visual attention. *Annual Review of Neuroscience, 18*(1), 193–222.

de Waal, F. (2017). *Are we smart enough to know how smart animals are?* W. W. Norton.

Dietert, R. (2016). *Human superorganism: How the microbe is revolutionizing the pursuit of a healthy life*. Dutton.

Dietrich, E. (Ed.). (1994). *Thinking computers and virtual persons: Essays on the intentionality of machines*. Academic Press.

Dietrich, E. (2001). Homo sapiens 2.0: Why we should build the better robots of our nature. *Journal of Experimental & Theoretical Artificial Intelligence, 13*(4), 323–328.

Dietrich, E., & Markman, A. (2003). Discrete thoughts: Why cognition must use discrete representations. *Mind and Language, 18*, 95–119.

References

Dietrich, V., Nieschalk, M., Stoll, W., Rajan, R., & Pantev, C. (2001). Cortical reorganization in patients with high frequency cochlear hearing loss. *Hearing Research, 158*(1), 95–101.

Dilks, D. D., Cook, P., Weiller, S. K., Berns, H. P., Spivak, M., & Berns, G. S. (2015). Awake fMRI reveals a specialized region in dog temporal cortex for face processing. *PeerJ, 3,* e1115.

Di Paolo, E., Buhrmann, T., & Barandiaran, X. (2017). *Sensorimotor life.* Oxford University Press.

Dixon, J. A., Kay, B. A., Davis, T. J., & Kondepudi, D. (2016). End-directedness and context in nonliving dissipative systems. In E. Dzhafarov (Ed.), *Contextuality from quantum physics to psychology* (pp. 185–208). World Scientific Press.

Dobosz, K., & Duch, W. (2010). Understanding neurodynamical systems via fuzzy symbolic dynamics. *Neural Networks, 23*(4), 487–496.

Domingos, P., Prado, A. M., Wong, A., Gehring, C., & Feijo, J. A. (2015). Nitric oxide: A multitasked signaling gas in plants. *Molecular Plant, 8*(4), 506–520.

Donaldson, S., & Kymlicka, W. (2011). *Zoopolis: A political theory of animal rights.* Oxford University Press.

Dotov, D. G., Nie, L., & Chemero, A. (2010). A demonstration of the transition from ready-to-hand to unready-to-hand. *PLoS One, 5*(3), e9433.

Drake, F., & Sobel, D. (1992). *Is anyone out there? The scientific search for extraterrestrial intelligence.* Delacorte Press.

Drake, N. (2016). *Little book of wonders.* National Geographic.

Dreyfus, H. L. (1992). *What computers still can't do: A critique of artificial reason.* MIT Press.

Dugatkin, L. A. (1997). *Cooperation among animals: An evolutionary perspective.* Oxford University Press.

Dugatkin, L. A., & Wilson, D. S. (1991). Rover: A strategy for exploiting cooperators in a patchy environment. *American Naturalist, 138*(3), 687–701.

Duncker, K. (1945). On problem-solving. *Psychological Monographs, 58*(5, Whole No. 270).

Dunning, D., & Stern, L. B. (1994). Distinguishing accurate from inaccurate eyewitness identifications via inquiries about decision processes. *Journal of Personality and Social Psychology, 67*(5), 818–835.

Durgin, F. H., Evans, L., Dunphy, N., Klostermann, S., & Simmons, K. (2007). Rubber hands feel the touch of light. *Psychological Science, 18*(2), 152–157.

Eagleman, D. (2012). *Incognito: The secret lives of the brain*. Vintage.

Edelman, G. M. (1993). Neural Darwinism: Selection and reentrant signaling in higher brain function. *Neuron, 10*(2), 115–125.

Edelman, S. (2008). *Computing the mind: How the mind really works*. Oxford University Press.

Edelman, S. (2012). *The happiness of pursuit: What neuroscience can teach us about the good life*. Basic Books.

Edmiston, P., & Lupyan, G. (2017). Visual interference disrupts visual knowledge. *Journal of Memory and Language, 92*, 281–292.

Ellis, N. C. (2005). At the interface: Dynamic interactions of explicit and implicit language knowledge. *Studies in Second Language Acquisition, 27*(2), 305–352.

Elman, J., & McClelland, J. (1988). Cognitive penetration of the mechanisms of perception: Compensation for coarticulation of lexically restored phonemes. *Journal of Memory and Language, 27*(2), 143–165.

Elman, J. L. (1990). Finding structure in time. *Cognitive Science, 14*(2), 179–211.

Elman, J. L. (2009). On the meaning of words and dinosaur bones: Lexical knowledge without a lexicon. *Cognitive Science, 33*(4), 547–582.

Ely, R. J., & Thomas, D. A. (2001). Cultural diversity at work: The effects of diversity perspectives on work group processes and outcomes. *Administrative Science Quarterly, 46*(2), 229–273.

Emberson, L. L., Lupyan, G., Goldstein, M. H., & Spivey, M. J. (2010). Overheard cell-phone conversations: When less speech is more distracting. *Psychological Science, 21*(10), 1383–1388.

Estes, Z., & Barsalou, L. W. (2018). A comprehensive meta-analysis of spatial interference from linguistic cues: Beyond Petrova et al. (2018). *Psychological Science, 29*(9), 1558–1564.

Fadiga, L., Craighero, L., Buccino, G., & Rizzolatti, G. (2002). Speech listening specifically modulates the excitability of tongue muscles: A TMS study. *European Journal of Neuroscience, 15*(2), 399–402.

Falandays, J. B., Batzloff, B. J., Spevack, S. C., and Spivey, M. J. (in press). Interactionism in language: From neural networks to bodies to dyads. *Language, Cognition and Neuroscience*. doi:10.1080/23273798.2018.1501501

Falandays, J. B., & Spivey, M. J. (in press). Theory visualizations for bilingual models of lexical ambiguity resolution. In R. Heredia & A. Cieslicka (Eds.), *Bilingual lexical ambiguity resolution*. Cambridge University Press.

References

Farmer, J. D., & Lafond, F. (2016). How predictable is technological progress? *Research Policy, 45*(3), 647–665.

Farmer, T. A., Cargill, S. A., Hindy, N. C., Dale, R., & Spivey, M. J. (2007). Tracking the continuity of language comprehension: Computer mouse trajectories suggest parallel syntactic processing. *Cognitive Science, 31*(5), 889–909.

Farnè, A., Serino, A., & Làdavas, E. (2007). Dynamic size-change of peri-hand space following tool-use: Determinants and spatial characteristics revealed through cross-modal extinction. *Cortex, 43*(3), 436–443.

Farrow, R. (2018). *War on peace: The end of diplomacy and the decline of American influence*. W. W. Norton.

Faulkenberry, T. J. (2016). Testing a direct mapping versus competition account of response dynamics in number comparison. *Journal of Cognitive Psychology, 28*(7), 825–842.

Favela, L. H., & Martin, J. (2017). "Cognition" and dynamical cognitive science. *Minds and Machines, 27*(2), 331–355.

Fazio, R. H., Jackson, J. R., Dunton, B. C., & Williams, C. J. (1995). Variability in automatic activation as an unobtrusive measure of racial attitudes: A bona fide pipeline? *Journal of Personality and Social Psychology, 69*(6), 1013–1027.

Fechner, G. T. (1848). *Nanna oder über das Seelenleben der Pflanzen* [Nanna—or on the soul life of plants]. Leipzig: Leopold Voss.

Fedorenko, E., Nieto-Castanon, A., & Kanwisher, N. (2012). Lexical and syntactic representations in the brain: An fMRI investigation with multi-voxel pattern analyses. *Neuropsychologia, 50*(4), 499–513.

Fedorenko, E., & Thompson-Schill, S. L. (2014). Reworking the language network. *Trends in Cognitive Sciences, 18*(3), 120–126.

Fehr, E., & Fischbacher, U. (2003). The nature of human altruism. *Nature, 425*(6960), 785–791.

Fekete, T., van Leeuwen, C., & Edelman, S. (2016). System, subsystem, hive: Boundary problems in computational theories of consciousness. *Frontiers in Psychology, 7*, 1041.

Feldman, J. A., & Ballard, D. H. (1982). Connectionist models and their properties. *Cognitive Science, 6*(3), 205–254.

Felleman, D. J., & Van Essen, D. C. (1991). Distributed hierarchical processing in the primate cerebral cortex. *Cerebral Cortex, 1*(1), 1–47.

Ferreira, F., Apel, J., & Henderson, J. M. (2008). Taking a new look at looking at nothing. *Trends in Cognitive Sciences, 12*(11), 405–410.

Field, D. J. (1994). What is the goal of sensory coding? *Neural Computation, 6*(4), 559–601.

Filevich, E., Kühn, S., & Haggard, P. (2013). There is no free won't: Antecedent brain activity predicts decisions to inhibit. *PloS One, 8*(2), e53053.

Finlay, B. L., & Workman, A. D. (2013). Human exceptionalism. *Trends in Cognitive Sciences, 17*(5), 199–201.

Firestone, C., & Scholl, B. J. (2016). Cognition does not affect perception: Evaluating the evidence for "top-down" effects. *Behavioral and Brain Sciences, 39*, e229.

Flusberg, S., Matlock, T., & Thibodeau, P. (2017). Metaphors for the war (or race) against climate change. *Environmental Communication*, 1–15.

Flusberg, S., Matlock, T., & Thibodeau, P. (2018). War metaphors in public discourse. *Metaphor & Symbol, 33*, 1–18.

Fodor, J. A. (1983). *The modularity of mind: An essay on faculty psychology*. MIT Press.

Forgays, D. G., & Forgays, D. K. (1992). Creativity enhancement through flotation isolation. *Journal of Environmental Psychology, 12*(4), 329–335.

Fossey, D. (2000). *Gorillas in the mist*. Houghton Mifflin Harcourt.

Francken, J. C., Kok, P., Hagoort, P., & De Lange, F. P. (2014). The behavioral and neural effects of language on motion perception. *Journal of Cognitive Neuroscience, 27*(1), 175–184.

Franconeri, S. L., Lin, J. Y., Enns, J. T., Pylyshyn, Z. W., & Fisher, B. (2008). Evidence against a speed limit in multiple-object tracking. *Psychonomic Bulletin & Review, 15*(4), 802–808.

Frangeul, L., Pouchelon, G., Telley, L., Lefort, S., Luscher, C., & Jabaudon, D. (2016). A cross-modal genetic framework for the development and plasticity of sensory pathways. *Nature, 538*(7623), 96–98.

Frank, A., & Sullivan III, W. T. (2016). A new empirical constraint on the prevalence of technological species in the universe. *Astrobiology, 16*(5), 359–362.

Freeman, J. B., Dale, R., & Farmer, T. (2011). Hand in motion reveals mind in motion. *Frontiers in Psychology, 2*, 59.

Freeman, J. B., & Johnson, K. L. (2016). More than meets the eye: Split-second social perception. *Trends in Cognitive Sciences, 20*(5), 362–374.

Freeman, J. B., Pauker, K., & Sanchez, D. T. (2016). A perceptual pathway to bias: Interracial exposure reduces abrupt shifts in real-time race perception that predict mixed-race bias. *Psychological Science, 27*(4), 502–517.

References

Freeman, J. B., Stolier, R. M., Ingbretsen, Z. A., & Hehman, E. A. (2014). Amygdala responsivity to high-level social information from unseen faces. *Journal of Neuroscience*, *34*(32), 10573–10581.

Froese, T., Iizuka, H., & Ikegami, T. (2014). Embodied social interaction constitutes social cognition in pairs of humans: A minimalist virtual reality experiment. *Scientific Reports*, *4*, 3672.

Fröhlich, E. (2010). The neuron: The basis for processing and propagation of information in the nervous system. *NeuroQuantology*, *8*(3), 403–415.

Fusaroli, R., Bahrami, B., Olsen, K., Roepstorff, A., Rees, G., Frith, C., & Tylén, K. (2012). Coming to terms: Quantifying the benefits of linguistic coordination. *Psychological Science*, *23*(8), 931–939.

Fusaroli, R., & Tylén, K. (2016). Investigating conversational dynamics: Interactive alignment, interpersonal synergy, and collective task performance. *Cognitive Science*, *40*(1), 145–171.

Fuster, J. M. (2001). The prefrontal cortex—an update: Time is of the essence. *Neuron*, *30*(2), 319–333.

Galantucci, B., Fowler, C. A., & Turvey, M. T. (2006). The motor theory of speech perception reviewed. *Psychonomic Bulletin & Review*, *13*(3), 361–377.

Gallagher, S. (2017). *Enactivist interventions*. Oxford University Press.

Gallese, V., Fadiga, L., Fogassi, L., & Rizzolatti, G. (1996). Action recognition in the premotor cortex. *Brain*, *119*, 593–609.

Gallese, V., & Lakoff, G. (2005). The brain's concepts: The role of the sensory-motor system in conceptual knowledge. *Cognitive Neuropsychology*, *22*(3–4), 455–479.

Gandhi, S. P., Heeger, D. J., & Boynton, G. M. (1999). Spatial attention affects brain activity in human primary visual cortex. *Proceedings of the National Academy of Sciences*, *96*(6), 3314–3319.

Gardner, H. (2011). *Frames of mind: The theory of multiple intelligences* (3rd ed.). Basic Books.

Gardner, R. A., & Gardner, B. T. (1969). Teaching sign language to a chimpanzee. *Science*, *165*(3894), 664–672.

Gendron, M., Lindquist, K. A., Barsalou, L., & Barrett, L. F. (2012). Emotion words shape emotion percepts. *Emotion*, *12*(2), 314–325.

Gentilucci, M., Benuzzi, F., Gangitano, M., & Grimaldi, S. (2001). Grasp with hand and mouth: A kinematic study on healthy subjects. *Journal of Neurophysiology*, *86*(4), 1685–1699.

Gerkey, D. (2013). Cooperation in context: Public goods games and post-Soviet collectives in Kamchatka, Russia. *Current Anthropology, 54*(2), 144–176.

Getz, L. M., & Toscano, J. C. (2019). Electrophysiological evidence for top-down lexical influences on early speech perception. *Psychological Science, 30*(6), 830–841.

Gibbs, R. W. (1994). *The poetics of mind: Figurative thought, language, and understanding.* Cambridge University Press.

Gibbs, R. W. (2005). *Embodiment and cognitive science.* Cambridge University Press.

Gibbs, R. W. (2006). Metaphor interpretation as embodied simulation. *Mind and Language, 21*(3), 434–458.

Gibbs, R. W., Strom, L. K., & Spivey-Knowlton, M. J. (1997). Conceptual metaphors in mental imagery for proverbs. *Journal of Mental Imagery, 21*, 83–110.

Gibson, J. J. (1966). *The senses considered as perceptual systems.* Boston: Houghton Mifflin.

Gibson, J. J. (2014). *The ecological approach to visual perception* (Classic ed.). Psychology Press.

Gibson, J. J., & Bridgeman, B. (1987). The visual perception of surface texture in photographs. *Psychological Research, 49*(1), 1–5.

Gilden, D. L., Thornton, T., & Mallon, M. W. (1995). 1/f noise in human cognition. *Science, 267*(5205), 1837–1839.

Gilroy, S., & Trewavas, A. J. (2001). Signal processing and transduction in plant cells: The end of the beginning? *Nature Reviews Molecular Cell Biology, 2*, 307–314.

Giuliani, N. R., Mann, T., Tomiyama, A. J., & Berkman, E. T. (2014). Neural systems underlying the reappraisal of personally craved foods. *Journal of Cognitive Neuroscience, 26*(7), 1390–1402.

Giummarra, M. J., & Moseley, G. L. (2011). Phantom limb pain and bodily awareness: Current concepts and future directions. *Current Opinion in Anesthesiology, 24*(5), 524–531.

Gladwell, M. (2007). *Blink.* Back Bay Books.

Glenberg, A. M., & Kaschak, M. P. (2002). Grounding language in action. *Psychonomic Bulletin & Review, 9*(3), 558–565.

Glucksberg, S. (1964). Functional fixedness: Problem solution as a function of observing responses. *Psychonomic Science, 1*, 117–118.

Glushko, R. J. (1979). The organization and activation of orthographic knowledge in reading aloud. *Journal of Experimental Psychology: Human Perception and Performance, 5*(4), 674–691.

References

Godfrey-Smith, P. (2016). Mind, matter, and metabolism. *Journal of Philosophy, 113*(10), 481–506.

Goel, N. S., Knox, L. B., & Norman, J. M. (1991). From artificial life to real life: Computer simulation of plant growth. *International Journal of General Systems, 18*(4), 291–319.

Goldberg, G., & Bloom, K. K. (1990). The alien hand sign: Localization, lateralization and recovery. *American Journal of Physical Medicine & Rehabilitation, 69*(5), 228–238.

Goldman-Rakic, P. S. (1995). Architecture of the prefrontal cortex and the central executive. *Annals of the New York Academy of Sciences, 769*(1), 71–84.

Goldstein, M. H., King, A. P., & West, M. J. (2003). Social interaction shapes babbling: Testing parallels between birdsong and speech. *Proceedings of the National Academy of Sciences, 100*(13), 8030–8035.

Goodall, J. (2010). *Through a window: My thirty years with the chimpanzees of Gombe*. Houghton Mifflin Harcourt.

Goodman, G., Poznanski, R. R., Cacha, L., & Bercovich, D. (2015). The two-brains hypothesis: Towards a guide for brain–brain and brain–machine interfaces. *Journal of Integrative Neuroscience, 14*(3), 281–293.

Gordon, C., Anderson, S., & Spivey, M. (2014). Is a *diamond* more elegant than a diamond? The role of sensory-grounding in conceptual content. In M. Bello, P. Guarini, M. McShane, & B. Scassellati (Eds.), *Proceedings of the 36th Annual Conference of the Cognitive Science Society* (pp. 2293–2297). Cognitive Science Society.

Gordon, C. L., Spivey, M. J., & Balasubramaniam, R. (2017). Corticospinal excitability during the processing of handwritten and typed words and non-words. *Neuroscience Letters, 651*, 232–236.

Gosling, S. (2009). *Snoop: What your stuff says about you*. Basic Books.

Gosling, S. D., Augustine, A. A., Vazire, S., Holtzman, N., & Gaddis, S. (2011). Manifestations of personality in online social networks: Self-reported Facebook-related behaviors and observable profile information. *Cyberpsychology, Behavior, and Social Networking, 14*(9), 483–488.

Gosling, S. D., Craik, K. H., Martin, N. R., & Pryor, M. R. (2005). Material attributes of personal living spaces. *Home Cultures, 2*(1), 51–87.

Gosling, S. D., Ko, S. J., Mannarelli, T., & Morris, M. E. (2002). A room with a cue: Personality judgments based on offices and bedrooms. *Journal of Personality and Social Psychology, 82*(3), 379–398.

Gow, D. W., & Olson, B. B. (2016). Sentential influences on acoustic-phonetic processing: A Granger causality analysis of multimodal imaging data. *Language, Cognition and Neuroscience, 31*(7), 841–855.

Grant, E. R., & Spivey, M. J. (2003). Eye movements and problem solving: Guiding attention guides thought. *Psychological Science, 14*(5), 462–466.

Green, C. S., Pouget, A., & Bavelier, D. (2010). Improved probabilistic inference as a general learning mechanism with action video games. *Current Biology, 20*(17), 1573–1579.

Greenberg, D. L. (2004). President Bush's false [flashbulb] memory of 9/11/01. *Applied Cognitive Psychology, 18*(3), 363–370.

Greenwald, A. G., McGhee, D. E., & Schwartz, J. L. (1998). Measuring individual differences in implicit cognition: The implicit association test. *Journal of Personality and Social Psychology, 74*(6), 1464–1480.

Greenwald, A. G., Nosek, B. A., & Banaji, M. R. (2003). Understanding and using the implicit association test: I. An improved scoring algorithm. *Journal of Personality and Social Psychology, 85*(2), 197–216.

Grèzes, J., Tucker, M., Armony, J., Ellis, R., & Passingham, R. E. (2003). Objects automatically potentiate action: An fMRI study of implicit processing. *European Journal of Neuroscience, 17*(12), 2735–2740.

Grinspoon, D. (2016). *Earth in human hands: Shaping our planet's future.* Grand Central Publishing.

Grosch, E. G., & Hazen, R. M. (2015). Microbes, mineral evolution, and the rise of microcontinents—origin and coevolution of life with early Earth. *Astrobiology, 15*(10), 922–939.

Gross, C. G. (1992). Representation of visual stimuli in inferior temporal cortex. *Philosophical Transactions of the Royal Society of London B: Biological Sciences, 335*(1273), 3–10.

Gross, C. G. (2002). Genealogy of the "grandmother cell." *The Neuroscientist, 8*(5), 512–518.

Gross, S. R. (2017). "What we think, what we know and what we think we know about false convictions." *Ohio Journal of Criminal Law, 14*(2), 753–786.

Grossberg, S. (1980). How does a brain build a cognitive code? *Psychological Review, 87*(1), 1–51.

Guastello, S. (2001). *Management of emergent phenomena.* Psychology Press.

Gudjonsson, G. H. (1992). *The psychology of interrogations, confessions and testimony.* Wiley.

Güntürkün, O., & Bugnyar, T. (2016). Cognition without cortex. *Trends in Cognitive Sciences, 20*(4), 291–303.

Haken, H., Kelso, J. S., & Bunz, H. (1985). A theoretical model of phase transitions in human hand movements. *Biological Cybernetics, 51*(5), 347–356.

Hall, L., Strandberg, T., Pärnamets, P., Lind, A., Tärning, B., & Johansson, P. (2013). How the polls can be both spot on and dead wrong: Using choice blindness to shift political attitudes and voter intentions. *PloS One, 8*(4), e60554.

Hameroff, S., & Penrose, R. (1996). Conscious events as orchestrated space-time selections. *Journal of Consciousness Studies, 3*(1), 36–53.

Hameroff, S., & Penrose, R. (2014). Consciousness in the universe: A review of the "Orch OR" theory. *Physics of Life Reviews, 11*(1), 39–78.

Hammer, M., & Menzel, R. (1995). Learning and memory in the honeybee. *Journal of Neuroscience, 15*(3), 1617–1630.

Hanczyc, M. M., & Ikegami, T. (2010). Chemical basis for minimal cognition. *Artificial Life, 16*(3), 233–243.

Haney, C. (2006). *Reforming punishment: Psychological limits to the pains of imprisonment*. American Psychological Association.

Haney, C., & Zimbardo, P. (1998). The past and future of US prison policy: Twenty-five years after the Stanford Prison Experiment. *American Psychologist, 53*(7), 709–727.

Hanna, J. E., Tanenhaus, M. K., & Trueswell, J. C. (2003). The effects of common ground and perspective on domains of referential interpretation. *Journal of Memory and Language, 49*(1), 43–61.

Hanning, N. M., Jonikaitis, D., Deubel, H., & Szinte, M. (2015). Oculomotor selection underlies feature retention in visual working memory. *Journal of Neurophysiology, 115*(2), 1071–1076.

Harari, Y. N. (2014). *Sapiens: A brief history of humankind*. Harvill Secker.

Harari, Y. N. (2016). *Homo deus: A brief history of tomorrow*. Random House.

Harris, S. (2012). *Free will*. Simon and Schuster.

Harris-Warrick, R. M., & Marder, E. (1991). Modulation of neural networks for behavior. *Annual Review of Neuroscience, 14*(1), 39–57.

Hassan, A., & Josephs, K. A. (2016). Alien hand syndrome. *Current Neurology and Neuroscience Reports, 16*(8), 1–10.

Hasson, U., Andric, M., Atilgan, H., & Collignon, O. (2016). Congenital blindness is associated with large-scale reorganization of anatomical networks. *NeuroImage, 128*, 362–372.

Hasson, U., Ghazanfar, A. A., Galantucci, B., Garrod, S., & Keysers, C. (2012). Brain-to-brain coupling: A mechanism for creating and sharing a social world. *Trends in Cognitive Sciences, 16*(2), 114–121.

Hasson, U., Nir, Y., Levy, I., Fuhrmann, G., & Malach, R. (2004). Intersubject synchronization of cortical activity during natural vision. *Science, 303*(5664), 1634–1640.

Hauk, O., Johnsrude, I., & Pulvermüller, F. (2004). Somatotopic representation of action words in human motor and premotor cortex. *Neuron, 41*(2), 301–307.

Havlin, S., Buldyrev, S. V., Bunde, A., Goldberger, A. L., Ivanov, P. C., Peng, C. K., & Stanley, H. E. (1999). Scaling in nature: From DNA through heartbeats to weather. *Physica A: Statistical Mechanics and Its Applications, 273*(1), 46–69.

Hawkley, L. C., Masi, C. M., Berry, J. D., & Cacioppo, J. T. (2006). Loneliness is a unique predictor of age-related differences in systolic blood pressure. *Psychology and Aging, 21*(1), 152–164.

Haxby, J. V., Gobbini, M. I., Furey, M. L., Ishai, A., Schouten, J. L., & Pietrini, P. (2001). Distributed and overlapping representations of faces and objects in ventral temporal cortex. *Science, 293*(5539), 2425–2430.

Hayhoe, M. (2000). Vision using routines: A functional account of vision. *Visual Cognition, 7*(1–3), 43–64.

Hazen, R. M. (2013). *The story of Earth: The first 4.5 billion years from stardust to living planet*. Penguin.

Hazen, R. M., & Sverjensky, D. A. (2010). Mineral surfaces, geochemical complexities, and the origins of life. *Cold Spring Harbor Perspectives in Biology, 2*(5), a002162.

He, B. J. (2014). Scale-free brain activity: Past, present, and future. *Trends in Cognitive Sciences, 18*(9), 480–487.

Hebb, D. O. (1958). The motivating effects of exteroceptive stimulation. *American Psychologist, 13*(3), 109–113.

Heersmink, R. (2018). The narrative self, distributed memory, and evocative objects. *Philosophical Studies, 175*(8), 1829–1849.

Hehman, E., Stolier, R. M., & Freeman, J. B. (2015). Advanced mouse-tracking analytic techniques for enhancing psychological science. *Group Processes & Intergroup Relations, 18*(3), 384–401.

Henderson, J. M., & Hollingworth, A. (1999). The role of fixation position in detecting scene changes across saccades. *Psychological Science, 10*(5), 438–443.

Henrich, J. (2015). *The secret of our success: How culture is driving human evolution, domesticating our species, and making us smarter*. Princeton University Press.

Henrich, J., Ensminger, J., McElreath, R., Barr, A., Barrett, C., Bolyanatz, A.,...& Lesorogol, C. (2010). Markets, religion, community size, and the evolution of fairness and punishment. *Science, 327*(5972), 1480–1484.

Herrmann, C. S., Pauen, M., Min, B. K., Busch, N. A., & Rieger, J. W. (2008). Analysis of a choice-reaction task yields a new interpretation of Libet's experiments. *International Journal of Psychophysiology, 67*(2), 151–157.

Herzog, H. (2011). The impact of pets on human health and psychological well-being: Fact, fiction, or hypothesis? *Current Directions in Psychological Science, 20*(4), 236–239.

Hickok, G. (2009). Eight problems for the mirror neuron theory of action understanding in monkeys and humans. *Journal of Cognitive Neuroscience, 21*(7), 1229–1243.

Hill, K., Barton, M., & Hurtado, A. M. (2009). The emergence of human uniqueness: Characters underlying behavioral modernity. *Evolutionary Anthropology: Issues, News, and Reviews, 18*(5), 187–200.

Hills, T. T., Jones, M. N., & Todd, P. M. (2012). Optimal foraging in semantic memory. *Psychological Review, 119*(2), 431–440.

Hindy, N. C., Ng, F. Y., & Turk-Browne, N. B. (2016). Linking pattern completion in the hippocampus to predictive coding in visual cortex. *Nature Neuroscience, 19*(5), 665–667.

Hirst, W., Phelps, E. A., Buckner, R. L., Budson, A. E., Cuc, A., Gabrieli, J. D., ... & Meksin, R. (2009). Long-term memory for the terrorist attack of September 11: Flashbulb memories, event memories, and the factors that influence their retention. *Journal of Experimental Psychology: General, 138*(2), 161–176.

Hoffmeyer, J. (2012). The natural history of intentionality: A biosemiotic approach. In T. Deacon, T. Schilhab, & F. Stjernfelt (Eds.), *The symbolic species evolved* (pp. 97–116). Springer.

Hofman, P. M., Van Riswick, J. G., & Van Opstal, A. J. (1998). Relearning sound localization with new ears. *Nature Neuroscience, 1*(5), 417–421.

Hofstadter, D. R., & Dennett, D. C. (1981). *The mind's I: Fantasies and reflections on self & soul*. Basic Books.

Holbrook, C., Izuma, K., Deblieck, C., Fessler, D. M. T., & Iacoboni, M. (2015). Neuromodulation of group prejudice and religious belief. *Social Cognitive & Affective Neuroscience, 11*(3), 387–394.

Holbrook, C., Pollack, J., Zerbe, J. G., & Hahn-Holbrook, J. (2018). Perceived supernatural support heightens battle confidence: A knife combat field study. *Religion, Brain & Behavior*. doi:10.1080/2153599X.2018.1464502

Holland, J. H. (2000). *Emergence: From chaos to order*. Oxford University Press.

Hommel, B., Müsseler, J., Aschersleben, G., & Prinz, W. (2001). The theory of event coding: A framework for perception and action planning. *Behavioral and Brain Sciences, 24*(5), 849–937.

Hood, L., & Tian, Q. (2012). Systems approaches to biology and disease enable translational systems medicine. *Genomics, Proteomics & Bioinformatics, 10*(4), 181–185.

Hotton, S., & Yoshimi, J. (2011). Extending dynamical systems theory to model embodied cognition. *Cognitive Science, 35*(3), 444–479.

Hove, M. J., & Keller, P. E. (2015). Impaired movement timing in neurological disorders: Rehabilitation and treatment strategies. *Annals of the New York Academy of Sciences, 1337*(1), 111–117.

Hove, M. J., & Risen, J. L. (2009). It's all in the timing: Interpersonal synchrony increases affiliation. *Social Cognition, 27*(6), 949–960.

Hove, M. J., Suzuki, K., Uchitomi, H., Orimo, S., & Miyake, Y. (2012). Interactive rhythmic auditory stimulation reinstates natural 1/f timing in gait of Parkinson's patients. *PloS One, 7*(3), e32600.

Huette, S., & McMurray, B. (2010). Continuous dynamics of color categorization. *Psychonomic Bulletin & Review, 17*(3), 348–354.

Huettig, F., Quinlan, P. T., McDonald, S. A., & Altmann, G. T. (2006). Models of high-dimensional semantic space predict language-mediated eye movements in the visual world. *Acta Psychologica, 121*(1), 65–80.

Huijing, P. A. (2009). Epimuscular myofascial force transmission: A historical review and implications for new research. *Journal of Biomechanics, 42*(1), 9–21.

Hurley, S. L. (1998). Vehicles, contents, conceptual structure, and externalism. *Analysis, 58*(1), 1–6.

Hurst, H. E. (1951). Long-term storage capacity of reservoirs. *Transactions of the American Society of Civil Engineers, 116*, 770–808.

Hutcherson, C. A., Seppala, E. M., & Gross, J. J. (2008). Loving-kindness meditation increases social connectedness. *Emotion, 8*(5), 720–724.

Hutchins, E. (1995). *Cognition in the wild.* MIT Press.

Hutchins, E. (2010). Cognitive ecology. *Topics in Cognitive Science, 2*(4), 705–715.

Ikegami, T. (2013). A design for living technology: Experiments with the mind time machine. *Artificial Life, 19*(3–4), 387–400.

Ingold, T. (2000). Evolving skills. In H. Rose & S. Rose, *Alas, poor Darwin: Arguments against evolutionary psychology* (pp. 273–297). Harmony Books.

Iriki, A. (2006). The neural origins and implications of imitation, mirror neurons and tool use. *Current Opinion in Neurobiology, 16*(6), 660–667.

Iriki, A., Tanaka, M., & Iwamura, Y. (1996). Coding of modified body schema during tool use by macaque postcentral neurones. *Neuroreport, 7*(14), 2325–2330.

References

Irwin, D. E., Yantis, S., & Jonides, J. (1983). Evidence against visual integration across saccadic eye movements. *Perception & Psychophysics, 34*(1), 49–57.

Ivanov, P. C., Nunes Amaral, L. A., Goldberger, A. L., Havlin, S., Rosenblum, M. G., Stanley, H. E., & Struzik, Z. R. (2001). From 1/f noise to multifractal cascades in heartbeat dynamics. *Chaos: An Interdisciplinary Journal of Nonlinear Science, 11*(3), 641–652.

Järvilehto, T. (1999). The theory of the organism-environment system: III. Role of efferent influences on receptors in the formation of knowledge. *Integrative Physiological and Behavioral Science, 34*(2), 90–100.

Järvilehto, T. (2009). The theory of the organism-environment system as a basis of experimental work in psychology. *Ecological Psychology, 21*(2), 112–120.

Järvilehto, T., & Lickliter, R. (2006). Behavior: Role of genes. In *Encyclopedia of Life Sciences*. Wiley.

Jensen, D. (2016). *The myth of human supremacy*. Seven Stories Press.

Jillette, P. (2012). *Every day is an atheist holiday*. Plume Press.

Johansson, P., Hall, L., & Sikström, S. (2008). From change blindness to choice blindness. *Psychologia, 51*(2), 142–155.

Johansson, R., & Johansson, M. (2014). Look here, eye movements play a functional role in memory retrieval. *Psychological Science, 25*(1), 236–242.

Johnson, J. B. (1925). The Schottky effect in low frequency circuits. *Physical Review, 26*(1), 71–85.

Johnson, S. (2002). *Emergence: The connected lives of ants, brains, cities, and software*. Scribner.

Johnson, W. (2016). *Eyes wide open: Buddhist instructions on merging body and vision*. Inner Traditions.

Joly-Mascheroni, R. M., Senju, A., & Shepherd, A. J. (2008). Dogs catch human yawns. *Biology Letters, 4*(5), 446–448.

Jordan, J. S. (2013). The wild ways of conscious will: What we do, how we do it, and why it has meaning. *Frontiers in Psychology, 4*, 574.

Juarrero, A. (1999). *Dynamics in action: Intentional behavior as a complex system*. MIT Press.

Jurafsky, D. (2014). *The language of food: A linguist reads the menu*. W. W. Norton.

Kaas, J. (2000). The reorganization of sensory and motor maps after injury in adult mammals. In M. Gazzaniga (Ed.), *The new cognitive neurosciences* (pp. 223–236). MIT Press.

Kagan, Y. Y. (2010). Earthquake size distribution: Power-law with exponent ß = ½? *Tectonophysics, 490*(1), 103–114.

Kahn, H. (1983). *Some comments on multipolarity and stability* (Discussion Paper No. HI-3662-DP). Hudson Institute.

Kahneman, D. (2013). *Thinking fast and slow.* Farrar, Straus and Giroux.

Kaiser, A., Snezhko, A., & Aranson, I. S. (2017). Flocking ferromagnetic colloids. *Science Advances, 3*(2), e1601469.

Kanan, C., & Cottrell, G. (2010). Robust classification of objects, faces, and flowers using natural image statistics. In *2010 IEEE Conference on Computer Vision and Pattern Recognition* (pp. 2472–2479). IEEE.

Kandel, E. (2016). *Reductionism in art and brain science: Bridging the two cultures.* Columbia University Press.

Kaschak, M. P., & Borreggine, K. L. (2008). Temporal dynamics of the action–sentence compatibility effect. *Quarterly Journal of Experimental Psychology, 61*(6), 883–895.

Kassin, S. M. (2005). On the psychology of confessions: Does innocence put innocents at risk? *American Psychologist, 60*(3), 215–228.

Kassin, S. M., & Kiechel, K. L. (1996). The social psychology of false confessions: Compliance, internalization, and confabulation. *Psychological Science, 7*(3), 125–128.

Kassin, S. M., Meissner, C. A., & Norwick, R. J. (2005). "I'd know a false confession if I saw one": A comparative study of college students and police investigators. *Law and Human Behavior, 29*(2), 211–227.

Kauffman, S. (1996). *At home in the universe: The search for the laws of self-organization and complexity.* Oxford University Press.

Kawakami, K., Dovidio, J. F., Moll, J., Hermsen, S., & Russin, A. (2000). Just say no (to stereotyping): Effects of training in the negation of stereotypic associations on stereotype activation. *Journal of Personality and Social Psychology, 78*(5), 871–888.

Kawasaki, M., Yamada, Y., Ushiku, Y., Miyauchi, E., & Yamaguchi, Y. (2013). Inter-brain synchronization during coordination of speech rhythm in human-to-human social interaction. *Scientific Reports, 3*, 1692.

Keil, P. G. (2015). Human-sheepdog distributed cognitive systems: An analysis of interspecies cognitive scaffolding in a sheepdog trial. *Journal of Cognition and Culture, 15*(5), 508–529.

Kello, C. T. (2013). Critical branching neural networks. *Psychological Review, 120*(1), 230–254.

Kello, C. T., Anderson, G. G., Holden, J. G., & Van Orden, G. C. (2008). The pervasiveness of 1/f scaling in speech reflects the metastable basis of cognition. *Cognitive Science, 32*(7), 1217–1231.

Kello, C. T., Beltz, B. C., Holden, J. G., & Van Orden, G. C. (2007). The emergent coordination of cognitive function. *Journal of Experimental Psychology: General, 136*(4), 551–568.

Kello, C. T., Brown, G. D., Ferrer-i-Cancho, R., Holden, J. G., Linkenkaer-Hansen, K., Rhodes, T., & Van Orden, G. C. (2010). Scaling laws in cognitive sciences. *Trends in Cognitive Sciences, 14*(5), 223–232.

Kelso, J. S. (1997). *Dynamic patterns: The self-organization of brain and behavior*. MIT Press.

Keltner, D., & Haidt, J. (2003). Approaching awe, a moral, spiritual, and aesthetic emotion. *Cognition and Emotion, 17*(2), 297–314.

Kieslich, P. J., & Hilbig, B. E. (2014). Cognitive conflict in social dilemmas: An analysis of response dynamics. *Judgment and Decision Making, 9*(6), 510–522.

Kirchhoff, M., Parr, T., Palacios, E., Friston, K., & Kiverstein, J. (2018). The Markov blankets of life: Autonomy, active inference and the free energy principle. *Journal of the Royal Society Interface, 15*, 20170792.

Kirsh, D. (1995). The intelligent use of space. *Artificial Intelligence, 73*(1–2), 31–68.

Kirsh, D., & Maglio, P. (1994). On distinguishing epistemic from pragmatic action. *Cognitive Science, 18*(4), 513–549.

Kleckner, D., Scheeler, M. W., & Irvine, W. T. (2014). The life of a vortex knot. *Physics of Fluids, 26*(9), 091105.

Klosterman, C. (2017). *But what if we're wrong? Thinking about the present as if it were the past*. Penguin.

Knoblich, G., & Jordan, J. S. (2003). Action coordination in groups and individuals: Learning anticipatory control. *Journal of Experimental Psychology: Learning, Memory, and Cognition, 29*(5), 1006–1016.

Knoeferle, P., & Crocker, M. W. (2007). The influence of recent scene events on spoken comprehension: Evidence from eye movements. *Journal of Memory and Language, 57*(4), 519–543.

Koike, T., Tanabe, H. C., & Sadato, N. (2015). Hyperscanning neuroimaging technique to reveal the "two-in-one" system in social interactions. *Neuroscience Research, 90*, 25–32.

Kokot, G., & Snezhko, A. (2018). Manipulation of emergent vortices in swarms of magnetic rollers. *Nature ommunications, 9*(1), 2344.

Kolbert, E. (2014). *The sixth extinction: An unnatural history*. A&C Black.

Konvalinka, I., & Roepstorff, A. (2012). The two-brain approach: How can mutually interacting brains teach us something about social interaction? *Frontiers in Human Neuroscience, 6*, 215.

Konvalinka, I., Xygalatas, D., Bulbulia, J., Schjødt, U., Jegindø, E. M., Wallot, S., ... Roepstorff, A. (2011). Synchronized arousal between performers and related spectators in a fire-walking ritual. *Proceedings of the National Academy of Sciences, 108*(20), 8514–8519.

Koop, G. J. (2013). An assessment of the temporal dynamics of moral decisions. *Judgment and Decision Making, 8*(5), 527–539.

Kosinski, M., Stillwell, D., & Graepel, T. (2013). Private traits and attributes are predictable from digital records of human behavior. *Proceedings of the National Academy of Sciences, 110*(15), 5802–5805.

Kosslyn, S. M., Ganis, G., & Thompson, W. L. (2001). Neural foundations of imagery. *Nature Reviews Neuroscience, 2*(9), 635–642.

Kosslyn, S. M., Thompson, W. L., & Ganis, G. (2006). *The case for mental imagery.* Oxford University Press.

Koubeissi, M. Z., Bartolomei, F., Beltagy, A., & Picard, F. (2014). Electrical stimulation of a small brain area reversibly disrupts consciousness. *Epilepsy & Behavior, 37*, 32–35.

Kourtzi, Z., & Kanwisher, N. (2000). Activation in human MT/MST by static images with implied motion. *Journal of Cognitive Neuroscience, 12*(1), 48–55.

Kövecses, Z. (2003). *Metaphor and emotion: Language, culture, and body in human feeling.* Cambridge University Press.

Krajbich, I., Armel, C., & Rangel, A. (2010). Visual fixations and the computation and comparison of value in simple choice. *Nature Neuroscience, 13*(10), 1292–1298.

Krajbich, I., & Smith, S. M. (2015). Modeling eye movements and response times in consumer choice. *Journal of Agricultural & Food Industrial Organization, 13*(1), 55–72.

Kriete, T., & Noelle, D. C. (2015). Dopamine and the development of executive dysfunction in autism spectrum disorders. *PloS One, 10*(3), e0121605.

Kriete, T., Noelle, D. C., Cohen, J. D., & O'Reilly, R. C. (2013). Indirection and symbol-like processing in the prefrontal cortex and basal ganglia. *Proceedings of the National Academy of Sciences, 110*(41), 16390–16395.

Kroll, J. F., & Bialystok, E. (2013). Understanding the consequences of bilingualism for language processing and cognition. *Journal of Cognitive Psychology, 25*(5), 497–514.

Krubitzer, L. (1995). The organization of neocortex in mammals: Are species differences really so different? *Trends in Neurosciences, 18*(9), 408–417.

Kruglanski, A. W., Bélanger, J. J., Gelfand, M., Gunaratna, R., Hettiarachchi, M., Reinares, F., Orehek, E., ... Sharvit, K. (2013). Terrorism: A (self) love story. *American Psychologist, 68*(7), 559–575.

References

Kuhlen, A. K., Allefeld, C., & Haynes, J. D. (2012). Content-specific coordination of listeners' to speakers' EEG during communication. *Frontiers in Human Neuroscience, 6,* 266.

Kveraga, K., Ghuman, A. S., & Bar, M. (2007). Top-down predictions in the cognitive brain. *Brain and Cognition, 65*(2), 145–168.

Laeng, B., & Teodorescu, D. S. (2002). Eye scanpaths during visual imagery reenact those of perception of the same visual scene. *Cognitive Science, 26*(2), 207–231.

Lahav, A., Saltzman, E., & Schlaug, G. (2007). Action representation of sound: Audiomotor recognition network while listening to newly acquired actions. *Journal of Neuroscience, 27*(2), 308–314.

Lakoff, G., & Johnson, M. (1980). *Metaphors we live by.* University of Chicago Press.

Lakoff, G., & Johnson, M. (1999). *Philosophy in the flesh: The embodied mind and its challenge to Western thought.* Basic Books.

Lakoff, G., & Núñez, R. E. (2000). *Where mathematics comes from: How the embodied mind brings mathematics into being.* Basic Books.

Laschi, C., Mazzolai, B., & Cianchetti, M. (2016). Soft robotics: Technologies and systems pushing the boundaries of robot abilities. *Science Robotics, 1,* eaah3690.

Laughlin, R. (2006). *A different universe: Reinventing physics from the bottom down.* Basic Books.

Laureys, S., Pellas, F., Van Eeckhout, P., Ghorbel, S., Schnakers, C., Perrin, F., ... Goldman, S. (2005). The locked-in syndrome: What is it like to be conscious but paralyzed and voiceless? *Progress in Brain Research, 150,* 495–611.

Lauwereyns, J. (2012). *Brain and the gaze: On the active boundaries of vision.* MIT Press.

Leary, M. R., & Tangney, J. P. (2012). *The handbook of self and identity* (2nd ed.). Guilford Press.

LeCun, Y., Bengio, Y., & Hinton, G. (2015). Deep learning. *Nature, 521*(7553), 436–444.

Lenggenhager, B., Tadi, T., Metzinger, T., & Blanke, O. (2007). Video ergo sum: Manipulating bodily self-consciousness. *Science, 317*(5841), 1096–1099.

Lents, N. (2018). *Human errors: A panorama of our glitches, from pointless bones to broken genes.* Houghton Mifflin Harcourt.

Lestel, D. (1998). How chimpanzees have domesticated humans: Towards an anthropology of human-animal communication. *Anthropology Today, 14*(3), 12–15.

Lestel, D. (2016). *Eat this book: A carnivore's manifesto.* Columbia University Press.

Lestel, D., Brunois, F., & Gaunet, F. (2006). Etho-ethnology and ethno-ethology. *Social Science Information, 45*(2), 155–177.

Lettvin, J. Y., Maturana, H. R., McCulloch, W. S., & Pitts, W. H. (1959). What the frog's eye tells the frog's brain. *Proceedings of the IRE, 47*(11), 1940–1951.

Levitsky, S., & Ziblatt, D. (2018). *How democracies die*. Crown.

Lewandowsky, S., & Oberauer, K. (2016). Motivated rejection of science. *Current Directions in Psychological Science, 25*(4), 217–222.

Lewis, M. (2016). *The undoing project: A friendship that changed our minds*. W. W. Norton.

Libby, L. K., & Eibach, R. P. (2011). Self-enhancement or self-coherence? Why people shift visual perspective in mental images of the personal past and future. *Personality and Social Psychology Bulletin, 37*(5), 714–726.

Liberman, A. M., Cooper, F. S., Shankweiler, D. P., & Studdert-Kennedy, M. (1967). Perception of the speech code. *Psychological Review, 74*(6), 431–461.

Liberman, A. M., & Whalen, D. H. (2000). On the relation of speech to language. *Trends in Cognitive Sciences, 4*(5), 187–196.

Libet, B. (1985). Unconscious cerebral initiative and the role of conscious will in voluntary action. *Behavioral and Brain Sciences, 8*(4), 529–566.

Libet, B. (1999). Do we have free will? *Journal of Consciousness Studies, 6*(8–9), 47–57.

Lilly, J. C. (1977). *The deep self: Profound relaxation and the tank isolation technique*. Simon and Schuster.

Lin, Y. C., & Lin, P. Y. (2016). Mouse tracking traces the "Camrbidge Unievrsity" effects in monolingual and bilingual minds. *Acta Psychologica, 167*, 52–62.

Linkenkaer-Hansen, K., Nikouline, V. V., Palva, J. M., & Ilmoniemi, R. J. (2001). Long-range temporal correlations and scaling behavior in human brain oscillations. *Journal of Neuroscience, 21*(4), 1370–1377.

Linster, B. G. (1992). Evolutionary stability in the infinitely repeated prisoners' dilemma played by two-state Moore machines. *Southern Economic Journal, 58*(4), 880–903.

Lipson, H. (2014). Challenges and opportunities for design, simulation, and fabrication of soft robots. *Soft Robotics, 1*(1), 21–27.

Lipson, H., & Pollack, J. B. (2000). Automatic design and manufacture of robotic lifeforms. *Nature, 406*(6799), 974–978.

Liu, R. T. (2017). The microbiome as a novel paradigm in studying stress and mental health. *American Psychologist, 72*(7), 655–667.

Lobben, M., & Bochynska, A. (2018). Grounding by attention simulation in peripersonal space: Pupils dilate to pinch grip but not big size nominal classifiers. *Cognitive Science, 42*(2), 576–599.

Loftus, E. F., & Pickrell, J. E. (1995). The formation of false memories. *Psychiatric Annals, 25*(12), 720–725.

Long, G. M., & Toppino, T. C. (2004). Enduring interest in perceptual ambiguity: Alternating views of reversible figures. *Psychological Bulletin, 130*(5), 748–768.

Lopez, R. B., Stillman, P. E., Heatherton, T. F., & Freeman, J. B. (2018). Minding one's reach (to eat): The promise of computer mouse-tracking to study self-regulation of eating. *Frontiers in Nutrition, 5*, 43.

Lorenz, E. (2000). The butterfly effect. *World Scientific Series on Nonlinear Science A, 39*, 91–94.

Louwerse, M. M. (2011). Symbol interdependency in symbolic and embodied cognition. *Topics in Cognitive Science, 3*(2), 273–302.

Louwerse, M. M., Dale, R., Bard, E. G., & Jeuniaux, P. (2012). Behavior matching in multimodal communication is synchronized. *Cognitive Science, 36*(8), 1404–1426.

Louwerse, M. M., & Zwaan, R. A. (2009). Language encodes geographical information. *Cognitive Science, 33*(1), 51–73.

Lovelock, J. E., & Margulis, L. (1974). Atmospheric homeostasis by and for the biosphere: The Gaia hypothesis. *Tellus, 26*(1–2), 2–10.

Luce, E. (2017). *The retreat of Western liberalism*. Atlantic Monthly Press.

Lupyan, G., & Clark, A. (2015). Words and the world: Predictive coding and the language-perception-cognition interface. *Current Directions in Psychological Science, 24*(4), 279–284.

Lupyan, G., Mirman, D., Hamilton, R., & Thompson-Schill, S. L. (2012). Categorization is modulated by transcranial direct current stimulation over left prefrontal cortex. *Cognition, 124*(1), 36–49.

Lupyan, G., & Spivey, M. J. (2008). Perceptual processing is facilitated by ascribing meaning to novel stimuli. *Current Biology, 18*(10), R410–R412.

Lupyan, G., & Spivey, M. J. (2010). Making the invisible visible: Verbal but not visual cues enhance visual detection. *PLoS ONE, 5*(7), e11452.

Maass, A., & Russo, A. (2003). Directional bias in the mental representation of spatial events: Nature or culture? *Psychological Science, 14*(4), 296–301.

Macknik, S. L., King, M., Randi, J., Robbins, A., Teller, Thompson, J., & Martinez-Conde, S. (2008). Attention and awareness in stage magic: Turning tricks into research. *Nature Reviews Neuroscience, 9*(11), 871–879.

Macknik, S. L., Martinez-Conde, S., & Blakeslee, S. (2010). *Sleights of mind: What the neuroscience of magic reveals about our everyday deceptions*. Macmillan.

Macmillan, M. (2002). *An odd kind of fame: Stories of Phineas Gage*. MIT Press.

Maglio, P. P., Kwan, S. K., & Spohrer, J. (2015). Commentary—Toward a research agenda for human-centered service system innovation. *Service Science, 7*(1), 1–10.

Maglio, P. P., & Spohrer, J. (2013). A service science perspective on business model innovation. *Industrial Marketing Management, 42*(5), 665–670.

Magnuson, J. S. (2005). Moving hand reveals dynamics of thought. *Proceedings of the National Academy of Sciences of the United States of America, 102*(29), 9995–9996.

Magnuson, J. S., McMurray, B., Tanenhaus, M. K., & Aslin, R. N. (2003). Lexical effects on compensation for coarticulation: The ghost of Christmas past. *Cognitive Science, 27*(2), 285–298.

Magnuson, J. S., Tanenhaus, M. K., Aslin, R. N., & Dahan, D. (2003). The time course of spoken word learning and recognition: Studies with artificial lexicons. *Journal of Experimental Psychology: General, 132*(2), 202–227.

Mahon, B. Z., & Caramazza, A. (2008). A critical look at the embodied cognition hypothesis and a new proposal for grounding conceptual content. *Journal of Physiology-Paris, 102*(1), 59–70.

Maier, N. R. F. (1931). Reasoning in humans: II. The solution of a problem and its appearance in consciousness. *Journal of Comparative Psychology, 12*(2), 181–194.

Malafouris, L. (2010). Metaplasticity and the human becoming: Principles of neuroarchaeology. *Journal of Anthropological Sciences, 88*(4), 49–72.

Mancardi, D., Varetto, G., Bucci, E., Maniero, F., & Guiot, C. (2008). Fractal parameters and vascular networks: Facts & artifacts. *Theoretical Biology and Medical Modelling, 5*(1), 12.

Mandelbrot, B. B. (2013). *Fractals and scaling in finance: Discontinuity, concentration, risk*. Springer.

Mandler, J. M. (1992). How to build a baby: II. Conceptual primitives. *Psychological Review, 99*(4), 587–604.

Mandler, J. M. (2004). *The foundations of mind: Origins of conceptual thought*. Oxford University Press.

Maravita, A., & Iriki, A. (2004). Tools for the body (schema). *Trends in Cognitive Sciences, 8*(2), 79–86.

Maravita, A., Spence, C., & Driver, J. (2003). Multisensory integration and the body schema: Close to hand and within reach. *Current Biology, 13*(13), R531–R539.

Marian, V., & Spivey, M. (2003). Bilingual and monolingual processing of competing lexical items. *Applied Psycholinguistics, 24*(2), 173–193.

Marslen-Wilson, W., & Tyler, L. (1980). The temporal structure of spoken language understanding. *Cognition, 8*(1), 1–71.

Martin, M., & Augustine, K. (Eds.) (2015). *The myth of an afterlife: The case against life after death.* Rowman & Littlefield.

Maruna, S., & Immarigeon, R. (Eds.). (2013). *After crime and punishment.* Routledge.

Massen, J. J., Church, A. M., & Gallup, A. C. (2015). Auditory contagious yawning in humans: An investigation into affiliation and status effects. *Frontiers in Psychology, 6,* 1735.

Matlock, T. (2010). Abstract motion is no longer abstract. *Language and Cognition, 2*(2), 243–260.

Matlock, T., Coe, C. M., & Westerling, A. L. (2017). Monster wildfires and metaphor in risk communication. *Metaphor & Symbol, 32*(4), 250–261.

Matsumoto, R., Nair, D. R., LaPresto, E., Najm, I., Bingaman, W., Shibasaki, H., & Lüders, H. O. (2004). Functional connectivity in the human language system: A cortico-cortical evoked potential study. *Brain, 127*(10), 2316–2330.

Matsuzawa, T. (Ed.). (2008). *Primate origins of human cognition and behavior* (corrected ed.). Springer.

Maturana, H. R., & Varela, F. J. (1987). *The tree of knowledge: The biological roots of human understanding.* New Science Library/Shambhala Publications.

Maturana, H. R., & Varela, F. J. (1991). *Autopoiesis and cognition: The realization of the living* (1st ed.). D. Reidel Publishing.

Maznevski, M. L. (1994). Understanding our differences: Performance in decision-making groups with diverse members. *Human Relations, 47*(5), 531–552.

McAuliffe, K. (2016). *This is your brain on parasites: How tiny creatures manipulate our behavior and shape society.* Houghton Mifflin Harcourt.

McClelland, J., & Elman, J. (1986). The TRACE model of speech perception. *Cognitive Psychology, 18*(1), 1–86.

McCubbins, M. D., and Turner, M. (2020). Collective action in the wild. In A. Pennisi & A. Falzone (Eds.), *The Extended Theory of Cognitive Creativity.* Springer.

McCulloch, W. S., & Pitts, W. (1943). A logical calculus of the ideas immanent in nervous activity. *Bulletin of Mathematical Biophysics, 5*(4), 115–133.

McGurk, H., & MacDonald, J. (1976). Hearing lips and seeing voices. *Nature, 264*(5588), 746–748.

McKinstry, C., Dale, R., & Spivey, M. J. (2008). Action dynamics reveal parallel competition in decision making. *Psychological Science, 19*(1), 22–24.

McLaughlin, J., Osterhout, L., & Kim, A. (2004). Neural correlates of second-language word learning: Minimal instruction produces rapid change. *Nature Neuroscience, 7*(7), 703–704.

McMurray, B. (2007). Defusing the childhood vocabulary explosion. *Science, 317*(5838), 631.

McMurray, B., Tanenhaus, M. K., Aslin, R. N., & Spivey, M. J. (2003). Probabilistic constraint satisfaction at the lexical/phonetic interface: Evidence for gradient effects of within-category VOT on lexical access. *Journal of Psycholinguistic Research, 32*(1), 77–97.

McRae, K., Brown, & Elman, J. E. (in press). Prediction-based learning and processing of event knowledge. *Topics in Cognitive Science*.

McRae, K., de Sa, V. R., & Seidenberg, M. S. (1997). On the nature and scope of featural representations of word meaning. *Journal of Experimental Psychology: General, 126*(2), 99–130.

McRae, K., Spivey-Knowlton, M., & Tanenhaus, M. (1998). Modeling the effects of thematic fit (and other constraints) in on-line sentence comprehension. *Journal of Memory and Language, 38*(3), 283–312.

Meadows, D. (2008). *Thinking in systems: A primer*. Chelsea Green.

Mechsner, F., Kerzel, D., Knoblich, G., & Prinz, W. (2001). Perceptual basis of bimanual coordination. *Nature, 414*(6859), 69–73.

Meijer, D. K., & Korf, J. (2013). Quantum modeling of the mental state: The concept of a cyclic mental workspace. *Syntropy, 1*, 1–41.

Mele, A. R. (Ed.). (2014). *Surrounding free will: Philosophy, psychology, neuroscience*. Oxford University Press.

Merker, B. (2007). Consciousness without a cerebral cortex: A challenge for neuroscience and medicine. *Behavioral and Brain Sciences, 30*(1), 63–81.

Mermin, N. D. (1998). What is quantum mechanics trying to tell us? *American Journal of Physics, 66*(9), 753–767.

Metcalfe, A., & Game, A. (2012). "In the beginning is relation": Martin Buber's alternative to binary oppositions. *Sophia, 51*(3), 351–363.

Meteyard, L., Cuadrado, S. R., Bahrami, B., & Vigliocco, G. (2012). Coming of age: A review of embodiment and the neuroscience of semantics. *Cortex, 48*(7), 788–804.

Meteyard, L., Zokaei, N., Bahrami, B., & Vigliocco, G. (2008). Visual motion interferes with lexical decision on motion words. *Current Biology, 18*(17), R732–R733.

Michaels, C. F. (2003). Affordances: Four points of debate. *Ecological Psychology, 15*(2), 135–148.

Michard, E., Lima, P. T., Borges, F., Silva, A. C., Portes, M. T., Carvalho, J. E., … Feijó, J. A. (2011). Glutamate receptor-like genes form Ca2+ channels in pollen tubes and are regulated by pistil D-serine. *Science, 332*(6028), 434–437.

Mikhailov, A. S., & Ertl, G. (2017). *Chemical complexity: Self-organization processes in molecular systems.* Springer.

Miklósi, Á. (2014). *Dog behaviour, evolution, and cognition.* Oxford University Press.

Milardi, D., Bramanti, P., Milazzo, C., Finocchio, G., Arrigo, A., Santoro, G., … Gaeta, M. (2015). Cortical and subcortical connections of the human claustrum revealed in vivo by constrained spherical deconvolution tractography. *Cerebral Cortex, 25*(2), 406–414.

Miller, E. K., & Cohen, J. D. (2001). An integrative theory of prefrontal cortex function. *Annual Review of Neuroscience, 24*(1), 167–202.

Miller, J. G. (1978). *Living systems.* McGraw-Hill.

Millikan, R. G. (2004). Existence proof for a viable externalism. In R. Schantz (Ed.), *The Externalist Challenge* (pp. 227–238). De Gruyter.

Misselhorn, C. (Ed.). (2015). *Collective agency and cooperation in natural and artificial systems.* Springer.

Mitchell, J. P., Macrae, C. N., & Banaji, M. R. (2006). Dissociable medial prefrontal contributions to judgments of similar and dissimilar others. *Neuron, 50*(4), 655–663.

Mitchell, M. (2009). *Complexity: A guided tour.* Oxford University Press.

Molotch, H. (2017). Objects in sociology. In A. Clarke (Ed.), *Design anthropology: Object cultures in transition* (pp.19–35). Bloomsbury.

Monsell, S., & Driver, J. (2000). Banishing the control homunculus. In S. Monsell & J. Driver (Eds.), *Attention and Performance: Vol. 18. Control of cognitive processes* (pp. 1–32). MIT Press.

Montague, P. R., Dayan, P., Person, C., & Sejnowski, T. J. (1995). Bee foraging in uncertain environments using predictive Hebbian learning. *Nature, 377*(6551), 725–728.

Moore, A., & Malinowski, P. (2009). Meditation, mindfulness and cognitive flexibility. *Consciousness and Cognition, 18*(1), 176–186.

Morrow, G. (2018). *Awestruck: A journal for finding awe year-round.* William Morrow of Harper Collins.

Morsella, E., Godwin, C. A., Jantz, T. K., Krieger, S. C., & Gazzaley, A. (2016). Homing in on consciousness: An action-based synthesis. *Behavioral and Brain Sciences, 39*, e168.

Motter, B. C. (1993). Focal attention produces spatially selective processing in visual cortical areas V1, V2, and V4 in the presence of competing stimuli. *Journal of Neurophysiology, 70*(3), 909–919.

Moyes, H. (Ed.). (2012). *Sacred darkness: A global perspective on the ritual use of caves.* University Press of Colorado.

Moyes, H., Rigoli, L., Huette, S., Montello, D., Matlock, T., & Spivey, M. J. (2017). Darkness and the imagination: The role of environment in the development of spiritual beliefs. In C. Papadopoulos & H. Moyes (Eds.), *The Oxford handbook of light in archaeology.* Oxford University Press.

Moyo, D. (2018). *Edge of chaos: Why democracy is failing to deliver economic growth and how to fix it.* Basic Books.

Mozer, M. C., Pashler, H., & Homaei, H. (2008). Optimal predictions in everyday cognition: The wisdom of individuals or crowds? *Cognitive Science, 32*(7), 1133–1147.

Mukamel, R., Ekstrom, A. D., Kaplan, J., Iacoboni, M., & Fried, I. (2010). Single-neuron responses in humans during execution and observation of actions. *Current Biology, 20*(8), 750–756.

Mumford, D. (1992). On the computational architecture of the neocortex. *Biological Cybernetics, 66*(3), 241–251.

Murphy, N., & Brown, W. S. (2007). *Did my neurons make me do it? Philosophical and neurobiological perspectives on moral responsibility and free will.* Oxford University Press.

Nakagaki, T., Yamada, H., & Tóth, Á. (2000). Intelligence: Maze-solving by an amoeboid organism. *Nature, 407*(6803), 470.

Nazir, T. A., Boulenger, V., Roy, A., Silber, B., Jeannerod, M., & Paulignan, Y. (2008). Language-induced motor perturbations during the execution of a reaching movement. *Quarterly Journal of Experimental Psychology, 61*(6), 933–943.

Neisser, U. (1976). *Cognition and reality: Principles and implications of cognitive psychology.* W. H. Freeman.

Neisser, U. (1991). Two perceptually given aspects of the self and their development. *Developmental Review, 11*(3), 197–209.

Neisser, U., & Harsch, N. (1992). Phantom flashbulbs: False recollections of hearing the news about Challenger. In E. Winograd & U. Neisser (Eds.), *Affect and accuracy in recall: Studies of "flashbulb" memories* (pp. 9–31). Cambridge University Press.

Neisser, U., Winograd E., Bergman, E. T., Schreiber, C. A., Palmer, S. E., & Weldon, M. S. (1996). Remembering the earthquake: Direct experience vs. hearing the news. *Memory, 4*(4), 337–358.

Nekovee, M., Moreno, Y., Bianconi, G., & Marsili, M. (2007). Theory of rumour spreading in complex social networks. *Physica A: Statistical Mechanics and Its Applications*, *374*(1), 457–470.

Nelson, R. K. (1982). *Make prayers to the Raven: A Koyukon view of the northern forest*. University of Chicago Press.

Noë, A. (2005). *Action in perception*. MIT Press.

Noë, A. (2009). *Out of our heads: Why you are not your brain, and other lessons from the biology of consciousness*. Macmillan.

Noelle, D. C. (2012). On the neural basis of rule-guided behavior. *Journal of Integrative Neuroscience*, *11*(4), 453–475.

Northoff, G., & Panksepp, J. (2008). The trans-species concept of self and the subcortical–cortical midline system. *Trends in Cognitive Sciences*, *12*(7), 259–264.

O'Connell, M. (2017). *To be a machine: Adventures among cyborgs, utopians, and the futurists solving the modest problem of death*. Doubleday.

Odum, H. T. (1988). Self-organization, transformity, and information. *Science*, *242*(4882), 1132–1139.

Ohl, S., & Rolfs, M. (2017). Saccadic eye movements impose a natural bottleneck on visual short-term memory. *Journal of Experimental Psychology: Learning, Memory, and Cognition*, *43*(5), 736–748.

O'Hora, D., Dale, R., Piiroinen, P. T., & Connolly, F. (2013). Local dynamics in decision making: The evolution of preference within and across decisions. *Scientific Reports*, *3*, 2210.

Olsen, R. K., Chiew, M., Buchsbaum, B. R., & Ryan, J. D. (2014). The relationship between delay period eye movements and visuospatial memory. *Journal of Vision*, *14*(1), 8.

Olshausen, B. A., & Field, D. J. (2004). Sparse coding of sensory inputs. *Current Opinion in Neurobiology*, *14*(4), 481–487.

Onnis, L., & Spivey, M. J. (2012). Toward a new scientific visualization for the language sciences. *Information*, *3*(1), 124–150.

Oppenheimer, D. M., & Frank, M. C. (2008). A rose in any other font would not smell as sweet: Effects of perceptual fluency on categorization. *Cognition*, *106*(3), 1178–1194.

O'Regan, J. K. (1992). Solving the "real" mysteries of visual perception: The world as an outside memory. *Canadian Journal of Psychology*, *46*(3), 461–488.

O'Regan, J. K., & Lévy-Schoen, A. (1983). Integrating visual information from successive fixations: Does trans-saccadic fusion exist? *Vision Research*, *23*, 765–769.

O'Regan, J. K., & Noë, A. (2001). A sensorimotor account of vision and visual consciousness. *Behavioral and Brain Sciences, 24*(5), 939–973.

O'Regan, J. K., Rensink, R. A., & Clark, J. J. (1999). Change-blindness as a result of "mudsplashes." *Nature, 398*(6722), 34.

Orehek, E., Sasota, J. A., Kruglanski, A. W., Dechesne, M., & Ridgeway, L. (2014). Interdependent self-construals mitigate the fear of death and augment the willingness to become a martyr. *Journal of Personality and Social Psychology, 107*(2), 265–275.

O'Reilly, R. C. (2006). Biologically based computational models of high-level cognition. *Science, 314*(5796), 91–94.

Oreskes, N., & Conway, E. M. (2011). *Merchants of doubt: How a handful of scientists obscured the truth on issues from tobacco smoke to global warming*. Bloomsbury.

Ortner, C. N., Kilner, S. J., & Zelazo, P. D. (2007). Mindfulness meditation and reduced emotional interference on a cognitive task. *Motivation and Emotion, 31*(4), 271–283.

Ostarek, M., Ishag, I., Joosen, D., & Huettig, F. (2018). Saccade trajectories reveal dynamic interactions of semantic and spatial information during the processing of implicitly spatial words. *Journal of Experimental Psychology: Learning, Memory, and Cognition, 44*(10), 1658–1670.

Page, S. E. (2007). *The difference: How the power of diversity creates better groups, firms, schools and societies*. Princeton University Press.

Palagi, E., Leone, A., Mancini, G., & Ferrari, P. F. (2009). Contagious yawning in gelada baboons as a possible expression of empathy. *Proceedings of the National Academy of Sciences, 106*(46), 19262–19267.

Pallas, S. L. (2001). Intrinsic and extrinsic factors that shape neocortical specification. *Trends in Neurosciences, 24*(7), 417–423.

Pallas, S. L., Roe, A. W., & Sur, M. (1990). Visual projections induced into the auditory pathway of ferrets. I. Novel inputs to primary auditory cortex (AI) from the LP/pulvinar complex and the topography of the MGN-AI projection. *Journal of Comparative Neurology, 298*(1), 50–68.

Papies, E. K. (2016). Health goal priming as a situated intervention tool: How to benefit from nonconscious motivational routes to health behaviour. *Health Psychology Review, 10*(4), 408–424.

Papies, E. K., Barsalou, L. W., & Custers, R. (2012). Mindful attention prevents mindless impulses. *Social Psychological and Personality Science, 3*(3), 291–299.

Pärnamets, P., Johansson, P., Hall, L., Balkenius, C., Spivey, M. J., & Richardson, D. C. (2015). Biasing moral decisions by exploiting the dynamics of eye gaze. *Proceedings of the National Academy of Sciences, 112*(13), 4170–4175.

Parnia, S., Spearpoint, K., de Vos, G., Fenwick, P., Goldberg, D., Yang, J., ... Wood, M. (2014). AWARE—AWAreness during REsuscitation—a prospective study. *Resuscitation, 85*(12), 1799–1805.

Patterson, F., & Linden, E. (1981). *The education of Koko*. Andre Deutsch Limited.

Patterson, F. G., & Cohn, R. H. (1990). Language acquisition by a lowland gorilla: Koko's first ten years of vocabulary development. *Word, 41*(2), 97–143.

Paukner, A., & Anderson, J. R. (2006). Video-induced yawning in stumptail macaques (*Macaca arctoides*). *Biology Letters, 2*(1), 36–38.

Paul, C., Valero-Cuevas, F. J., & Lipson, H. (2006). Design and control of tensegrity robots for locomotion. *IEEE Transactions on Robotics, 22*(5), 944–957.

Penrose, R. (1994). *Shadows of the mind*. Oxford University Press.

Pepperberg, I. M. (2009). *The Alex studies: Cognitive and communicative abilities of grey parrots*. Harvard University Press.

Pereboom, D. (2006). *Living without free will*. Cambridge University Press.

Perlman, M., Patterson, F. G., & Cohn, R. H. (2012). The human-fostered gorilla Koko shows breath control in play with wind instruments. *Biolinguistics, 6*(3–4), 433–444.

Perri, A., Widga, C., Lawler, D., Martin, T., Loebel, T., Farnsworth, K.... Buenger, B. (2019). New evidence of the earliest domestic dogs in the Americas. *American Antiquity, 84*(1), 68–87.

Petrova, A., Navarrete, E., Suitner, C., Sulpizio, S., Reynolds, M., Job, R., & Peressotti, F. (2018). Spatial congruency effects exist, just not for words: Looking into Estes, Verges, and Barsalou (2008). *Psychological Science, 29*(7), 1195–1199.

Pezzulo, G., Barsalou, L. W., Cangelosi, A., Fischer, M. H., McRae, K., & Spivey, M. J. (2011). The mechanics of embodiment: A dialog on embodiment and computational modeling. *Frontiers in Psychology, 2*(5), 1–21.

Pezzulo, G., Barsalou, L. W., Cangelosi, A., Fischer, M. H., McRae, K., & Spivey, M. J. (2013). Computationally grounded cognition: A new alliance between grounded cognition and computational modeling. *Frontiers in Psychology, 3*, 612.

Pezzulo, G., Verschure, P. F., Balkenius, C., & Pennartz, C. M. (2014). The principles of goal-directed decision-making: From neural mechanisms to computation and robotics. *Philosophical Transactions of the Royal Society B, 369*(1655), 20130470.

Pfeifer, R., Lungarella, M., & Iida, F. (2012). The challenges ahead for bio-inspired "soft" robotics. *Communications of the Association for Computing Machinery, 55*(11), 76–87.

Pick, A. D., Pick Jr, H. L., Jones, R. K., & Reed, E. S. (1982). James Jerome Gibson: 1904–1979. *American Journal of Psychology, 95*(4), 692–700.

Pickering, M. J., & Garrod, S. (2004). Toward a mechanistic psychology of dialogue. *Behavioral and Brain Sciences, 27*(2), 169–190.

Piff, P. K., Dietze, P., Feinberg, M., Stancato, D. M., & Keltner, D. (2015). Awe, the small self, and prosocial behavior. *Journal of Personality and Social Psychology, 108*(6), 883–899.

Pinker, S. (2018). *Enlightenment now: The case for reason, science, humanism, and progress*. Viking Press.

Piore, A. (2017). *The body builders: Inside the science of the engineered human*. HarperCollins.

Platek, S. M., Critton, S. R., Myers, T. E., & Gallup, G. G. (2003). Contagious yawning: The role of self-awareness and mental state attribution. *Cognitive Brain Research, 17*(2), 223–227.

Podobnik, B., Jusup, M., Kovac, D., & Stanley, H. E. (2017). Predicting the rise of EU right-wing populism in response to unbalanced immigration. *Complexity, 2017*, 1580526.

Prigogine, I., & Stengers, I. (1984). *Order out of chaos: Man's new dialogue with nature*. Bantam Books.

Proffitt, D. R. (2006). Embodied perception and the economy of action. *Perspectives on Psychological Science, 1*(2), 110–122.

Provine, R. R. (2005). Yawning: The yawn is primal, unstoppable and contagious, revealing the evolutionary and neural basis of empathy and unconscious behavior. *American Scientist, 93*(6), 532–539.

Pulvermüller, F. (1999). Words in the brain's language. *Behavioral and Brain Sciences, 22*(2), 253–279.

Pulvermüller, F., Hauk, O., Nikulin, V. V., & Ilmoniemi, R. J. (2005). Functional links between motor and language systems. *European Journal of Neuroscience, 21*(3), 793–797.

Putnam, H. (1973). Meaning and reference. *Journal of Philosophy, 70*(19), 699–711.

Putnam, H. (1981). *Reason, truth and history* (Vol. 3). Cambridge University Press.

Pylyshyn, Z. (1980). The "causal power" of machines. *Behavioral and Brain Sciences, 3*(3), 442–444.

Pylyshyn, Z. W. (2001). Visual indexes, preconceptual objects, and situated vision. *Cognition, 80*(1), 127–158.

Pylyshyn, Z. W. (2007). *Things and places: How the mind connects with the world*. MIT Press.

Pylyshyn, Z. W., & Storm, R. W. (1988). Tracking multiple independent targets: Evidence for a parallel tracking mechanism. *Spatial Vision, 3*(3), 179–197.

Quiroga, R. Q., Kreiman, G., Koch, C., & Fried, I. (2008). Sparse but not "grandmother-cell" coding in the medial temporal lobe. *Trends in Cognitive Sciences, 12*(3), 87–91.

Rachman, G. (2017). *Easternization: Asia's rise and America's decline from Obama to Trump and beyond.* Other Press.

Ramanathan, V., Han, H., & Matlock, T. (2017). Educating children to bend the curve: For a stable climate, sustainable nature and sustainable humanity. In A. M. Battro, P. Lena, M. S. Sorondo, & J. von Braun (Eds.), *Children and sustainable development: Ecological education in a globalized world* (pp. 3–16). Springer.

Ramachandran, V. S., & Blakeslee, S. (1998). *Phantoms in the brain: Probing the mysteries of the human mind.* William Morrow.

Randall, L. (2017). *Dark matter and the dinosaurs: The astounding interconnectedness of the universe.* Random House.

Rao, R. P., & Ballard, D. H. (1999). Predictive coding in the visual cortex: A functional interpretation of some extra-classical receptive-field effects. *Nature Neuroscience, 2*(1), 79–87.

Rayner, K., & Pollatsek, A. (1983). Is visual information integrated across saccades? *Perception & Psychophysics, 34*(1), 39–48.

Reed, E., & Jones, R. (Eds.) (1982). *Reasons for realism: Selected essays of James J. Gibson.* Erlbaum.

Reed, E. S. (2014). The intention to use a specific affordance: A conceptual framework for psychology. In R. Wozniak & K. Fischer (Eds.), *Development in context* (pp. 61–92). Psychology Press.

Reese, R. A. (2013). *Sustainable or bust.* CreateSpace.

Reese, R. A. (2016). *Understanding sustainability.* CreateSpace.

Reid, C. R., Latty, T., Dussutour, A., & Beekman, M. (2012). Slime mold uses an externalized spatial "memory" to navigate in complex environments. *Proceedings of the National Academy of Sciences, 109*(43), 17490–17494.

Reimers, J. R., McKemmish, L. K., McKenzie, R. H., Mark, A. E., & Hush, N. S. (2014). The revised Penrose–Hameroff orchestrated objective-reduction proposal for human consciousness is not scientifically justified. *Physics of Life Reviews, 11*(1), 101–103.

Rhodes, T., & Turvey, M. T. (2007). Human memory retrieval as Lévy foraging. *Physica A: Statistical Mechanics and Its Applications, 385*(1), 255–260.

Richardson, D. C. (2017). *Man vs. mind.* Quarto.

Richardson, D. C., Altmann, G. T., Spivey, M. J., & Hoover, M. A. (2009). Much ado about eye movements to nothing. *Trends in Cognitive Sciences, 13*(6), 235–236.

Richardson, D. C., & Dale, R. (2005). Looking to understand: The coupling between speakers' and listeners' eye movements and its relationship to discourse comprehension. *Cognitive Science, 29*(6), 1045–1060.

Richardson, D. C., Dale, R., & Kirkham, N. Z. (2007). The art of conversation is coordination. *Psychological Science, 18*(5), 407–413.

Richardson, D. C., & Kirkham, N. Z. (2004). Multimodal events and moving locations: Eye movements of adults and 6-month-olds reveal dynamic spatial indexing. *Journal of Experimental Psychology: General, 133*(1), 46–62.

Richardson, D. C., & Matlock, T. (2007). The integration of figurative language and static depictions: An eye movement study of fictive motion. *Cognition, 102*(1), 129–138.

Richardson, D. C., & Spivey, M. J. (2000). Representation, space and Hollywood Squares: Looking at things that aren't there anymore. *Cognition, 76*(3), 269–295.

Richardson, D. C., Spivey, M. J., Barsalou, L. W., & McRae, K. (2003). Spatial representations activated during real-time comprehension of verbs. *Cognitive Science, 27*(5), 767–780.

Richardson, D. C., Spivey, M. J., & Cheung, J. (2001). Motor representations in memory and mental models: Embodiment in cognition. In J. D. Moore & K. Stenning (Eds.), *Proceedings of the 23rd Annual Meeting of the Cognitive Science Society* (pp. 867–872). Erlbaum.

Richardson, D. C., Spivey, M. J., Edelman, S., & Naples, A. D. (2001). "Language is spatial": Experimental evidence for image schemas of concrete and abstract verbs. In J. D. Moore & K. Stenning (Eds.), *Proceedings of the 23rd Annual Meeting of the Cognitive Science Society* (pp. 873–878). Erlbaum.

Richardson, M. J., & Kallen, R. W. (2016). Symmetry-breaking and the contextual emergence of human multiagent coordination and social activity. In E. Dzhafarov (Ed.), *Contextuality from quantum physics to psychology* (pp. 229–286). World Scientific.

Riley, M. A., Richardson, M., Shockley, K., & Ramenzoni, V. C. (2011). Interpersonal synergies. *Frontiers in Psychology, 2*, 38.

Risko, E. F., & Gilbert, S. J. (2016). Cognitive offloading. *Trends in Cognitive Sciences, 20*(9), 676–688.

Rizzolatti, G., & Arbib, M. A. (1998). Language within our grasp. *Trends in Neurosciences, 21*(5), 188–194.

Rizzolatti, G., Fogassi, L., & Gallese, V. (2006). Mirrors in the mind. *Scientific American, 295*(5), 54–61.

Rodny, J. J., Shea, T. M., & Kello, C. T. (2017). Transient localist representations in critical branching networks. *Language, Cognition and Neuroscience, 32*(3), 330–341.

Rohwer, F. (2010). *Coral reefs in the microbial seas*. Plaid Press.

Rolls, E. T. (2017). Cortical coding. *Language, Cognition and Neuroscience, 32*(3), 316–329.

Rosch, E. (1975). Cognitive representations of semantic categories. *Journal of Experimental Psychology: General, 104*(3), 192–233.

Rosen, R. (1991). *Life itself: A comprehensive inquiry into the nature, origin, and fabrication of life*. Columbia University Press.

Rosenblatt, F. (1958). The perceptron: A probabilistic model for information storage and organization in the brain. *Psychological Review, 65*(6), 386–408.

Rosenblum, L. D. (2008). Speech perception as a multimodal phenomenon. *Current Directions in Psychological Science, 17*(6), 405–409.

Roy, D. (2005). Grounding words in perception and action: Computational insights. *Trends in Cognitive Sciences, 9*(8), 389–396.

Rozenblit, L., Spivey, M., & Wojslawowicz, J. (2002). Mechanical reasoning about gear-and-belt diagrams: Do eye-movements predict performance? In M. Anderson, B. Meyer, & P. Olivier (Eds.), *Diagrammatic representation and reasoning* (pp. 223–240). Springer.

Rozin, P., & Fallon, A. E. (1987). A perspective on disgust. *Psychological Review, 94*(1), 23–41.

Rumelhart, D., Hinton, G., & Williams, R. (1986). Learning representations by back-propagating errors. *Nature, 323*(9), 533–536.

Rumelhart, D., & McClelland, J. (1982). An interactive activation model of context effects in letter perception: II. The contextual enhancement effect and some tests and extensions of the model. *Psychological Review, 89*(1), 60–94.

Rumelhart, D., McClelland, J., & the PDP Research Group. (1986). *Parallel distributed processing: Explorations in the microstructure of cognition* (Vols. 1–2). MIT Press.

Runyon, J. B., Mescher, M. C., & De Moraes, C. M. (2006). Volatile chemical cues guide host location and host selection by parasitic plants. *Science, 313*(5795), 1964–1967.

Rupert, R. D. (2004). Challenges to the hypothesis of extended cognition. *Journal of Philosophy, 101*(8), 389–428.

Rupert, R. D. (2009). *Cognitive systems and the extended mind*. Oxford University Press.

Rynecki, E. (2016). *Chasing portraits: A great-granddaughter's quest for her lost art legacy*. Berkley Press.

Ryskin, R. A., Wang, R. F., & Brown-Schmidt, S. (2016). Listeners use speaker identity to access representations of spatial perspective during online language comprehension. *Cognition, 147*, 75–84.

Saffran, J. R., Aslin, R. N., & Newport, E. L. (1996). Statistical learning by 8-month-old infants. *Science, 274*(5294), 1926–1928.

Safina, C. (2016). *Beyond words: What animals think and feel.* Picador.

Sagan, C., & Drake, F. (1975). The search for extraterrestrial intelligence. *Scientific American, 232*(5), 80–89.

Samuel, A. G. (1981). Phonemic restoration: Insights from a new methodology. *Journal of Experimental Psychology: General, 110*(4), 474–494.

Santiago, J., Román, A., & Ouellet, M. (2011). Flexible foundations of abstract thought: A review and a theory. In A. Maas & T. Schubert (Eds.), *Spatial dimensions of social thought* (pp. 31–108). Mouton de Gruyter.

Santos, L. R., Flombaum, J. I., & Phillips, W. (2007). The evolution of human mindreading: How non-human primates can inform social cognitive neuroscience. In S. Platek, J. Keenan, & T. Shackelford (Eds.), *Evolutionary cognitive neuroscience* (pp. 433–456). MIT Press.

Sapolsky, R. (2019). This is your brain on nationalism. *Foreign Affairs, 98*(2), 42–47.

Savage-Rumbaugh, E. S. (1986). *Ape language: From conditioned response to symbol.* Columbia University Press.

Sawyer, R. K. (2005). *Social emergence: Societies as complex systems.* Cambridge University Press.

Saygin, A. P., McCullough, S., Alac, M., & Emmorey, K. (2010). Modulation of BOLD response in motion-sensitive lateral temporal cortex by real and fictive motion sentences. *Journal of Cognitive Neuroscience, 22*(11), 2480–2490.

Scalia, F., Grant, A. C., Reyes, M., & Lettvin, J. Y. (1995). Functional properties of regenerated optic axons terminating in the primary olfactory cortex. *Brain Research, 685*(1), 187–197.

Schacter, D. L. (1992). Implicit knowledge: New perspectives on unconscious processes. *Proceedings of the National Academy of Sciences, 89*(23), 11113–11117.

Schmidhuber, J. (2015). Deep learning in neural networks: An overview. *Neural Networks, 61*, 85–117.

Schmidt, R. C., Carello, C., & Turvey, M. T. (1990). Phase transitions and critical fluctuations in the visual coordination of rhythmic movements between people. *Journal of Experimental Psychology: Human Perception and Performance, 16*(2), 227–247.

References

Schmolck, H., Buffalo, E. A., & Squire, L. R. (2000). Memory distortions develop over time: Recollections of the OJ Simpson trial verdict after 15 and 32 months. *Psychological Science*, *11*(1), 39–45.

Scholl, B. J., & Pylyshyn, Z. W. (1999). Tracking multiple items through occlusion: Clues to visual objecthood. *Cognitive Psychology*, *38*(2), 259–290.

Schoot, L., Hagoort, P., & Segaert, K. (2016). What can we learn from a two-brain approach to verbal interaction? *Neuroscience & Biobehavioral Reviews*, *68*, 454–459.

Schulte-Mecklenbeck, M., Kühberger, A., & Ranyard, R. (2011). The role of process data in the development and testing of process models of judgment and decision making. *Judgment and Decision Making*, *6*(8), 733–739.

Schwaber, M. K., Garraghty, P. E., & Kaas, J. H. (1993). Neuroplasticity of the adult primate auditory cortex following cochlear hearing loss. *Otology & Neurotology*, *14*(3), 252–258.

Schwartz, C. E., & Sendor, R. M. (1999). Helping others helps oneself: Response shift effects in peer support. *Social Science & Medicine*, *48*(11), 1563–1575.

Scott, P. (2013). *Physiology and behaviour of plants*. Wiley.

Searle, J. R. (1990). Is the brain's mind a computer program? *Scientific American*, *262*(1), 26–31.

Sebanz, N., Bekkering, H., & Knoblich, G. (2006). Joint action: Bodies and minds moving together. *Trends in Cognitive Sciences*, *10*(2), 70–76.

Sebanz, N., Knoblich, G., Prinz, W., & Wascher, E. (2006). Twin peaks: An ERP study of action planning and control in coacting individuals. *Journal of Cognitive Neuroscience*, *18*(5), 859–870.

Sedivy, J., & Carlson, G. N. (2011). *Sold on language*. Wiley.

Segal, G. (2000). *A slim book about narrow content*. MIT Press.

Seidenberg, M., & McClelland, J. (1989). A distributed, developmental model of word recognition and naming. *Psychological Review*, *96*(4), 523–568.

Sekuler, R., Sekuler, A. B., & Lau, R. (1997). Sound alters visual motion perception. *Nature*, *385*(6614), 308.

Sengupta, B., Laughlin, S. B., & Niven, J. E. (2014). Consequences of converting graded to action potentials upon neural information coding and energy efficiency. *PLoS Computational Biology*, *10*(1), e1003439.

Senju, A., Kikuchi, Y., Akechi, H., Hasegawa, T., Tojo, Y., & Osanai, H. (2009). Brief report: Does eye contact induce contagious yawning in children with autism spectrum disorder? *Journal of Autism and Developmental Disorders*, *39*(11), 1598–1602.

Senju, A., Maeda, M., Kikuchi, Y., Hasegawa, T., Tojo, Y., & Osanai, H. (2007). Absence of contagious yawning in children with autism spectrum disorder. *Biology Letters, 3*(6), 706–708.

Serra, I., & Corral, Á. (2017). Deviation from power law of the global seismic moment distribution. *Scientific Reports, 7*, 40045.

Seung, S. (2013). *Connectome: How the brain's wiring makes us who we are*. Mariner Books.

Shahin, A. J., Backer, K. C., Rosenblum, L. D., & Kerlin, J. R. (2018). Neural mechanisms underlying cross-modal phonetic encoding. *Journal of Neuroscience, 38*(7), 1835–1849.

Shams, L., Kamitani, Y., & Shimojo, S. (2000). Illusions: What you see is what you hear. *Nature, 408*(6814), 788.

Shapiro, L. (2011). *Embodied cognition*. Routledge.

Shatner, W. (2017). *Spirit of the horse: A celebration in fact and fable*. Thomas Dunne Books.

Shaw, R., & Turvey, M. (1999). Ecological foundations of cognition: II. Degrees of freedom and conserved quantities in animal-environment systems. *Journal of Consciousness Studies, 6*(11-12), 111–123.

Shebani, Z., & Pulvermüller, F. (2013). Moving the hands and feet specifically impairs working memory for arm- and leg-related action words. *Cortex, 49*(1), 222–231.

Shockley, K., Santana, M. V., & Fowler, C. A. (2003). Mutual interpersonal postural constraints are involved in cooperative conversation. *Journal of Experimental Psychology: Human Perception and Performance, 29*(2), 326–332.

Shostak, S. (1998). *Sharing the universe: Perspectives on extraterrestrial life*. Berkeley Hills Books.

Silva, K., Bessa, J., & de Sousa, L. (2012). Auditory contagious yawning in domestic dogs (*Canis familiaris*): First evidence for social modulation. *Animal Cognition, 15*(4), 721–724.

Simmons, W. K., Martin, A., & Barsalou, L. W. (2005). Pictures of appetizing foods activate gustatory cortices for taste and reward. *Cerebral Cortex, 15*(10), 1602–1608.

Simons, D. J., & Levin, D. T. (1997). Change blindness. *Trends in Cognitive Sciences, 1*(7), 261–267.

Simons, D. J., & Levin, D. T. (1998). Failure to detect changes to people during a real-world interaction. *Psychonomic Bulletin & Review, 5*(4), 644–649.

Skarda, C. A., & Freeman, W. J. (1987). How brains make chaos in order to make sense of the world. *Behavioral and Brain Sciences, 10*(2), 161–173.

References

Skrbina, D. (2005). *Panpsychism in the West*. MIT Press.

Smalarz, L., & Wells, G. L. (2015). Contamination of eyewitness self-reports and the mistaken-identification problem. *Current Directions in Psychological Science, 24*(2), 120–124.

Smaldino, P. E. (2016). Not even wrong: Imprecision perpetuates the illusion of understanding at the cost of actual understanding. *Behavioral and Brain Sciences, 39*, e163.

Smaldino, P. E., Schank, J. C., & McElreath, R. (2013). Increased costs of cooperation help cooperators in the long run. *American Naturalist, 181*(4), 451–463.

Smart, P. (2012). The web-extended mind. *Metaphilosophy, 43*(4), 446–463.

Smeding, A., Quinton, J. C., Lauer, K., Barca, L., & Pezzulo, G. (2016). Tracking and simulating dynamics of implicit stereotypes: A situated social cognition perspective. *Journal of Personality and Social Psychology, 111*(6), 817–834.

Smith, A. V., Proops, L., Grounds, K., Wathan, J., & McComb, K. (2016). Functionally relevant responses to human facial expressions of emotion in the domestic horse (*Equus caballus*). *Biology Letters, 12*(2), 20150907.

Smith, L. B. (2005). Action alters shape categories. *Cognitive Science, 29*(4), 665–679.

Smith, L. B., & Breazeal, C. (2007). The dynamic lift of developmental process. *Developmental Science, 10*(1), 61–68.

Smith, L. B., & Gasser, M. (2005). The development of embodied cognition: Six lessons from babies. *Artificial Life, 11*(1–2), 13–29.

Smith, R., Rathcke, T., Cummins, F., Overy, K., & Scott, S. (2014). Communicative rhythms in brain and behaviour. *Philosophical Transactions of the Royal Society B, 369*, 20130389.

Snezhko, A., & Aranson, I. S. (2011). Magnetic manipulation of self-assembled colloidal asters. *Nature Materials, 10*(9), 698–703.

Solman, G. J. F., & Kingstone, A. (2017). Arranging objects in space: Measuring task-relevant organizational behaviors during goal pursuit. *Cognitive Science, 41*(4), 1042–1070.

Song, J. H., & Nakayama, K. (2009). Hidden cognitive states revealed in choice reaching tasks. *Trends in Cognitive Science, 13*(8), 360–366.

Soon, C. S., Brass, M., Heinze, H. J., & Haynes, J. D. (2008). Unconscious determinants of free decisions in the human brain. *Nature Neuroscience, 11*(5), 543–545.

Soon, C. S., He, A. H., Bode, S., & Haynes, J. D. (2013). Predicting free choices for abstract intentions. *Proceedings of the National Academy of Sciences, 110*(15), 6217–6222.

Spevack, S. C., Falandays, J. B., Batzloff, B. J., and Spivey, M. J. (2018). Interactivity of language. *Language and Linguistics Compass, 12*(7), e12282.

Spiegelhalder, K., Ohlendorf, S., Regen, W., Feige, B., van Elst, L. T., Weiller, C., …Tüscher, O. (2014). Interindividual synchronization of brain activity during live verbal communication. *Behavioural Brain Research*, *258*, 75–79.

Spivey, M. J. (2000). Turning the tables on the Turing test: The Spivey test. *Connection Science*, *12*(1), 91–94.

Spivey, M. J. (2007). *The continuity of mind*. Oxford University Press.

Spivey, M. J. (2012). The spatial intersection of minds. *Cognitive Processing*, *13*(1), 343–346.

Spivey, M. J. (2013). The emergence of intentionality. *Ecological Psychology*, *25*(3), 233–239.

Spivey, M. J. (2017). Fake news and false corroboration: Interactivity in rumor networks. In G. Gunzelmann, A. Howes, T. Tenbrink, & E. J. Davelaar (Eds.), *Proceedings of the 39th Annual Conference of the Cognitive Science Society* (pp. 3229–3234). Cognitive Science Society.

Spivey, M. J. (2018). Discovery in complex adaptive systems. *Cognitive Systems Research*, *51*, 40–55.

Spivey, M. J., & Batzloff, B. J. (2018). Bridgemanian space constancy as a precursor to extended cognition. *Consciousness and Cognition*, *64*, 164–175.

Spivey, M. J., & Cardon, C. D. (2015). Methods for studying adult bilingualism. In J. Schwieter (Ed.), *The Cambridge handbook of bilingual language processing* (pp. 108–132). Cambridge University Press.

Spivey, M. J., & Geng, J. J. (2001). Oculomotor mechanisms activated by imagery and memory: Eye movements to absent objects. *Psychological Research*, *65*(4), 235–241.

Spivey, M. J., Grosjean, M., & Knoblich, G. (2005). Continuous attraction toward phonological competitors. *Proceedings of the National Academy of Sciences of the United States of America*, *102*(29), 10393–10398.

Spivey, M. J., & Richardson, D. (2009). Language processing embodied and embedded. In P. Robbins & M. Aydede (Eds.), *The Cambridge handbook of situated cognition* (pp. 382–400). Cambridge University Press.

Spivey, M. J., Richardson, D. C., & Fitneva, S. A. (2004). Thinking outside the brain: Spatial indices to visual and linguistic information. In J. Henderson & F. Ferreira (Eds.), *The interface of language, vision, and action: Eye movements and the visual world* (pp. 161–189). Psychology Press.

Spivey, M. J., & Spevack, S. C. (2017). An inclusive account of mind across spatiotemporal scales of cognition. *Journal of Cultural Cognitive Science*, *1*(1), 25–38.

Spivey, M. J., & Spirn, M. J. (2000). Selective visual attention modulates the direct tilt aftereffect. *Perception & Psychophysics*, *62*(8), 1525–1533.

References

Spivey-Knowlton, M., & Saffran, J. (1995). Inducing a grammar without an explicit teacher: Incremental distributed prediction feedback. In J. D. Moore & J. F. Lehman (Eds.), *Proceedings of the 17th Annual Conference of the Cognitive Science Society* (pp. 230–235). Erlbaum.

Spivey-Knowlton, M. J. (1996). *Integration of visual and linguistic information: Human data and model simulations* (Unpublished PhD dissertation). University of Rochester.

Sporns, O., & Kötter, R. (2004). Motifs in brain networks. *PLoS Biology, 2*(11), e369.

Spratling, M. W. (2012). Predictive coding accounts for V1 response properties recorded using reverse correlation. *Biological Cybernetics, 106*(1), 37–49.

Stager, C. (2014). *Your atomic self: The invisible elements that connect you to everything else in the universe.* Thomas Dunne Books.

Stanfield, R. A., & Zwaan, R. A. (2001). The effect of implied orientation derived from verbal context on picture recognition. *Psychological Science, 12*(2), 153–156.

Stanley, J. (2018). *How fascism works: The politics of us and them.* Random House.

Steegen, S., Dewitte, L., Tuerlinckx, F., & Vanpaemel, W. (2014). Measuring the crowd within again: A pre-registered replication study. *Frontiers in Psychology, 5*, 786.

Steels, L. (2003). Evolving grounded communication for robots. *Trends in Cognitive Sciences, 7*(7), 308–312.

Steinbock, O., Tóth, Á., & Showalter, K. (1995). Navigating complex labyrinths: Optimal paths from chemical waves. *Science, 267*(5199), 868–871.

Stellar, J. E., Gordon, A. M., Piff, P. K., Cordaro, D., Anderson, C. L., Bai, Y., … Keltner, D. (2017). Self-transcendent emotions and their social functions: Compassion, gratitude, and awe bind us to others through prosociality. *Emotion Review, 9*(3), 200–207.

Stellar, J. E., John-Henderson, N., Anderson, C. L., Gordon, A. M., McNeil, G. D., & Keltner, D. (2015). Positive affect and markers of inflammation: Discrete positive emotions predict lower levels of inflammatory cytokines. *Emotion, 15*(2), 129–133.

Stenger, V. (2008). *God: The failed hypothesis.* Prometheus.

Stenger, V. (2011). *The fallacy of fine-tuning: Why the universe is not designed for us.* Prometheus.

Stephen, D. G., Boncoddo, R. A., Magnuson, J. S., & Dixon, J. A. (2009). The dynamics of insight: Mathematical discovery as a phase transition. *Memory & Cognition, 37*(8), 1132–1149.

Stevens, J. A., Fonlupt, P., Shiffrar, M., & Decety, J. (2000). New aspects of motion perception: Selective neural encoding of apparent human movements. *Neuroreport, 11*(1), 109–115.

Stiefel, K. M., Merrifield, A., & Holcombe, A. O. (2014). The claustrum's proposed role in consciousness is supported by the effect and target localization of *Salvia divinorum*. *Frontiers in Integrative Neuroscience, 8*, 20.

Stoffregen, T. A. (2000). Affordances and events. *Ecological Psychology, 12*(1), 1–29.

Stolier, R. M., & Freeman, J. B. (2017). A neural mechanism of social categorization. *Journal of Neuroscience, 37*(23), 5711–5721.

Strandberg, T., Sivén, D., Hall, L., Johansson, P., & Pärnamets, P. (2018). False beliefs and confabulation can lead to lasting changes in political attitudes. *Journal of Experimental Psychology: General, 147*(9), 1382–1399.

Strassman, R. (2000). *DMT: The spirit molecule: A doctor's revolutionary research into the biology of near-death and mystical experiences*. Inner Traditions/Bear & Company.

Strawson, G. (2006). Realistic monism: Why physicalism entails panpsychism. *Journal of Consciousness Studies, 13*(10–11), 3–31.

Strogatz, S. (2004). *Sync: The emerging science of spontaneous order*. Penguin.

Suedfeld, P., Metcalfe, J., & Bluck, S. (1987). Enhancement of scientific creativity by flotation REST (restricted environmental stimulation technique). *Journal of Environmental Psychology, 7*(3), 219–231.

Sugden, R. (2004). *The economics of rights, co-operation and welfare* (2nd ed.). Springer.

Sutton, J. (2008). Material agency, skills, and history: Distributed cognition and the archaeology of memory. In L. Malafouris & Carl Knappett (Eds.), *Material agency: Towards a non-anthropocentric approach* (pp. 37–55). Springer.

Sutton, J. (2010). Exograms and interdisciplinarity: History, the extended mind, and the civilizing process. In R. Menary (Ed.), *The extended mind* (pp. 189–225). MIT Press.

Sutton, J., & Keene, N. (2016). Cognitive history and material culture. In D. Gaimster, T. Hamling, & C. Richardson (Eds.), *The Routledge handbook of material culture in early modern Europe* (pp. 44–56). Routledge.

Swenson, R. (1989). Emergent attractors and the law of maximum entropy production: Foundations to a theory of general evolution. *Systems Research and Behavioral Science, 6*(3), 187–197.

Swenson, R., & Turvey, M. T. (1991). Thermodynamic reasons for perception—action cycles. *Ecological Psychology, 3*(4), 317–348.

Szary, J., Dale, R., Kello, C. T., & Rhodes, T. (2015). Patterns of interaction-dominant dynamics in individual versus collaborative memory foraging. *Cognitive Processing, 16*(4), 389–399.

Talaifar, S., & Swann, W. B. (2016). Differentiated selves can surely be good for the group, but let's get clear about why. *Behavioral and Brain Sciences*, *39*, e165.

Tallal, P., Merzenich, M. M., Miller, S., & Jenkins, W. (1998). Language learning impairments: Integrating basic science, technology, and remediation. *Experimental Brain Research*, *123*(1–2), 210–219.

Tanenhaus, M. K., Spivey-Knowlton, M. J., Eberhard, K. M., & Sedivy, J. C. (1995). Integration of visual and linguistic information in spoken language comprehension. *Science*, *268*(5217), 1632–1634.

Tauber, A. I. (2017). *Immunity: The evolution of an idea*. Oxford University Press.

Tegmark, M. (2000). Importance of quantum decoherence in brain processes. *Physical Review E*, *61*(4), 4194–4206.

Tero, A., Takagi, S., Saigusa, T., Ito, K., Bebber, D. P., Fricker, M. D., ... Nakagaki, T. (2010). Rules for biologically inspired adaptive network design. *Science*, *327*(5964), 439–442.

Thagard, P. (2010). *The brain and the meaning of life*. Princeton University Press.

Thagard, P. (2019). *Mind-society: From brains to social sciences and professions*. Oxford University Press.

Theiner, G. (2014). A beginner's guide to group minds. In M. Sprevack & J. Kallestrup (Eds.), *New waves in philosophy of mind* (pp. 301–322). Palgrave Macmillan.

Theiner, G. (2017). Collaboration, exploitation, and distributed animal cognition. *Comparative Cognition & Behavior Reviews*, *13*, 41–47.

Theiner, G., Allen, C., & Goldstone, R. L. (2010). Recognizing group cognition. *Cognitive Systems Research*, *11*(4), 378–395.

Thellier, M. (2012). A half-century adventure in the dynamics of living systems. In U. Lüttge, W. Beyschlag, B. Büdel, and D. Francis (Eds.), *Progress in Botany 73* (pp. 3–53). Springer.

Thellier, M. (2017). *Plant responses to environmental stimuli: The role of specific forms of plant memory*. Springer.

Thellier, M., & Lüttge, U. (2013). Plant memory: A tentative model. *Plant Biology*, *15*(1), 1–12.

Theobald, D. L. (2010). A formal test of the theory of universal common ancestry. *Nature*, *465*(7295), 219–222.

Thiam, P., & Sit, J. (2015). *Senegal: Modern Senegalese recipes from the source to the bowl*. Lake Isle Press.

Thomas, L. E. (2017). Action experience drives visual-processing biases near the hands. *Psychological Science, 28*(1), 124–131.

Thomas, L. E., & Lleras, A. (2007). Moving eyes and moving thought: On the spatial compatibility between eye movements and cognition. *Psychonomic Bulletin & Review, 14*(4), 663–668.

Thompkins, A. M., Deshpande, G., Waggoner, P., & Katz, J. S. (2016). Functional magnetic resonance imaging of the domestic dog: Research, methodology, and conceptual issues. *Comparative Cognition & Behavior Reviews, 11*, 63–82.

Thompson, E. (2007). *Mind in life: Biology, phenomenology, and the sciences of mind.* Harvard University Press.

Tipler, F. J. (1980). Extraterrestrial intelligent beings do not exist. *Quarterly Journal of the Royal Astronomical Society, 21*, 267–281.

Tollefsen, D. P., Dale, R., & Paxton, A. (2013). Alignment, transactive memory, and collective cognitive systems. *Review of Philosophy and Psychology, 4*(1), 49–64.

Tomasello, M. (2008). *Origins of human communication.* Bradford Books.

Tomiyama, A. J., Hunger, J. M., Nguyen-Cuu, J., & Wells, C. (2016). Misclassification of cardiometabolic health when using body mass index categories in NHANES 2005–2012. *International Journal of Obesity, 40*(5), 883–886.

Tononi, G., Sporns, O., & Edelman, G. M. (1994). A measure for brain complexity: Relating functional segregation and integration in the nervous system. *Proceedings of the National Academy of Sciences, 91*(11), 5033–5037.

Topolinski, S., Boecker, L., Erle, T. M., Bakhtiari, G., & Pecher, D. (2017). Matching between oral inward–outward movements of object names and oral movements associated with denoted objects. *Cognition and Emotion, 31*(1), 3–18.

Topolinski, S., Maschmann, I. T., Pecher, D., & Winkielman, P. (2014). Oral approach–avoidance: Affective consequences of muscular articulation dynamics. *Journal of Personality and Social Psychology, 106*(6), 885–896.

Torres, P. (2016). *The End: What science and religion tell us about the apocalypse.* Pitchstone.

Trewavas, A. (2003). Aspects of plant intelligence. *Annals of Botany, 92*(1), 1–20.

Trueswell, J., & Tanenhaus, M. K. (Eds.). (2005). *Approaches to studying world-situated language use: Bridging the language-as-product and language-as-action traditions.* MIT Press.

Trueswell, J. C., Sekerina, I., Hill, N. M., & Logrip, M. L. (1999). The kindergarten-path effect: Studying on-line sentence processing in young children. *Cognition, 73*(2), 89–134.

Tse, P. (2013). *The neural basis of free will: Criterial causation.* MIT Press.

Tucker, M., & Ellis, R. (1998). On the relations between seen objects and components of potential actions. *Journal of Experimental Psychology: Human Perception and Performance, 24*(3), 830–846.

Turchin, P. (2016). *Ages of discord: A structural-demographic analysis of American history*. Beresta Press.

Turnbaugh, P. J., Ley, R. E., Hamady, M., Fraser-Liggett, C., Knight, R., & Gordon, J. I. (2007). The human microbiome project: Exploring the microbial part of ourselves in a changing world. *Nature, 449*(7164), 804–810.

Turner, M. (2014). *The origin of ideas: Blending, creativity, and the human spark*. Oxford University Press.

Turney, J. (2015). *I, superorganism: Learning to love your inner ecosystem*. Icon Books.

Turvey, M. T. (2004). Impredicativity, dynamics, and the perception-action divide. In V. K. Jirsa & J. A. S. Kelso (Eds.), *Coordination dynamics: Issues and trends* (pp. 1–20). Springer.

Turvey, M. T. (2013). Ecological perspective on perception-action: What kind of science does it entail? In W. Prinz, M. Beisert, & A. Herwig (Eds.), *Action science: Foundations of an emerging discipline* (pp. 139–170). MIT Press.

Turvey, M. T. (2018). *Lectures on perception: An ecological perspective*. Routledge.

Turvey, M. T., & Carello, C. (1986). The ecological approach to perceiving-acting: A pictorial essay. *Acta Psychologica, 63*(1-3), 133–155.

Turvey, M. T., & Carello, C. (2012). On intelligence from first principles: Guidelines for inquiry into the hypothesis of physical intelligence (PI). *Ecological Psychology, 24*(1), 3–32.

Turvey, M. T., & Fonseca, S. T. (2014). The medium of haptic perception: A tensegrity hypothesis. *Journal of Motor Behavior, 46*(3), 143–187.

Turvey, M. T., & Shaw, R. (1999). Ecological foundations of cognition: I. Symmetry and specificity of animal-environment systems. *Journal of Consciousness Studies, 6*(11-12), 111–123.

Tuszynski, J. A. (2014). The need for a physical basis of cognitive process: A review of the "Orch OR" theory by Hameroff and Penrose. *Physics of Life Reviews, 11*(1), 79–80.

Tversky, B. (2019). *Mind in motion: How action shapes thought*. Basic Books.

Tyson, N. d. (2017). *Astrophysics for people in a hurry*. W. W. Norton.

Usher, M., Stemmler, M., & Olami, Z. (1995). Dynamic pattern formation leads to 1/f noise in neural populations. *Physical Review Letters, 74*(2), 326.

Vakoch, D., & Dowd, M. (Eds.). (2015). *The Drake equation: Estimating the prevalence of extraterrestrial life through the ages* (Cambridge Astrobiology, Vol. 8). Cambridge University Press.

van den Heuvel, M. P., Bullmore, E. T., & Sporns, O. (2016). Comparative connectomics. *Trends in Cognitive Sciences, 20*(5), 345–361.

Vanderschraaf, P. (2006). War or peace? A dynamical analysis of anarchy. *Economics and Philosophy, 22*(2), 243–279.

Vanderschraaf, P. (2018). *Strategic justice: Conventions and problems of balancing divergent interests*. Oxford University Press.

van der Wel, R. P., Sebanz, N., & Knoblich, G. (2014). Do people automatically track others' beliefs? Evidence from a continuous measure. *Cognition, 130*(1), 128–133.

van der Wel, R. P. R. D., Knoblich, G., & Sebanz, N. (2011). Let the force be with us: Dyads exploit haptic coupling for coordination. *Journal of Experimental Psychology: Human Perception and Performance, 37*(5), 1420–1431.

Van Orden, G. C., & Holden, J. G. (2002). Intentional contents and self-control. *Ecological Psychology, 14*(1–2), 87–109.

Van Orden, G. C., Holden, J. G., & Turvey, M. T. (2003). Self-organization of cognitive performance. *Journal of Experimental Psychology: General, 132*(3), 331–350.

Van Orden, G. C., Holden, J. G., & Turvey, M. T. (2005). Human cognition and 1/f scaling. *Journal of Experimental Psychology: General, 134*(1), 117–123.

Van Orden, G. C., Kloos, H., & Wallot, S. (2011). Living in the pink: Intentionality, wellbeing, and complexity. *Philosophy of Science (Complex Systems), 10*, 629–672.

Van Steveninck, R. D. R., & Laughlin, S. B. (1996). The rate of information transfer at graded-potential synapses. *Nature, 379*(6566), 642–645.

Varela, F. J. (1997). Patterns of life: Intertwining identity and cognition. *Brain and Cognition, 34*(1), 72–87.

Verga, L., & Kotz, S. A. (2019). Putting language back into ecological communication contexts. *Language, Cognition and Neuroscience, 34*(4), 536–544.

Vinson, D. W., Abney, D. H., Amso, D., Anderson, M. L., Chemero, T., Cutting, J. E., Dale, R., … Spivey, M. (2016). Perception, as you make it. Commentary on Firestone & Scholl's "Cognition does not affect perception: Evaluating the evidence for 'top-down' effects." *Behavioral and Brain Sciences, 39*, e260.

Von Melchner, L., Pallas, S. L., & Sur, M. (2000). Visual behaviour mediated by retinal projections directed to the auditory pathway. *Nature, 404*(6780), 871–876.

von Zimmerman, J., Vicary, S., Sperling, M., Orgs, G., & Richardson, D. C. (2018). The choreography of group affiliation. *Trends in Cognitive Sciences, 10*(1), 80–94.

References

Vosoughi, S., Roy, D., & Aral, S. (2018). The spread of true and false news online. *Science, 359*(6380), 1146–1151.

Vukovic, N., Fuerra, M., Shpektor, M., Myachykov, A., & Shtyrov, Y. (2017). Primary motor cortex functionally contributes to language comprehension: An online rTMS study. *Neuropsychologia, 96*, 222–229.

Vul, E., & Pashler, H. (2008). Measuring the crowd within: Probabilistic representations within individuals. *Psychological Science, 19*(7), 645–647.

Wagenmakers, E. J., Farrell, S., & Ratcliff, R. (2004). Estimation and interpretation of 1/f α noise in human cognition. *Psychonomic Bulletin & Review, 11*(4), 579–615.

Wagman, J. B., Stoffregen, T. A., Bai, J., & Schloesser, D. S. (2017). Perceiving nested affordances for another person's actions. *Quarterly Journal of Experimental Psychology, 71*(3), 1–24.

Walker, S. I., Packard, N., & Cody, G. D. (Eds.). (2017). Special issue on "Reconceptualizing the origins of life." *Philosophical Transactions of the Royal Society A, 375*(2109).

Wallsten, T. S., Budescu, D. V., Erev, I., & Diederich, A. (1997). Evaluating and combining subjective probability estimates. *Journal of Behavioral Decision Making, 10*(3), 243–268.

Ward, L. M. (2002). *Dynamical cognitive science*. MIT Press.

Warlaumont, A. S., Richards, J. A., Gilkerson, J., & Oller, D. K. (2014). A social feedback loop for speech development and its reduction in autism. *Psychological Science, 25*(7), 1314–1324.

Warlaumont, A. S., Westermann, G., Buder, E. H., & Oller, D. K. (2013). Prespeech motor learning in a neural network using reinforcement. *Neural Networks, 38*, 64–75.

Warren, W. H. (1984). Perceiving affordances: Visual guidance of stair climbing. *Journal of Experimental Psychology: Human Perception and Performance, 10*(5), 683–703.

Watson, W. E., Kumar, K., & Michaelsen, L. K. (1993). Cultural diversity's impact on interaction process and performance: Comparing homogeneous and diverse task groups. *Academy of Management Journal, 36*(3), 590–602.

Watts, A. W. (1966). *The book: On the taboo against knowing who you are*. Pantheon Books.

Webb, B. (1996). A cricket robot. *Scientific American, 275*(6), 94–99.

Webb, S. (2015). *If the universe is teeming with aliens…where is everybody? Seventy-five solutions to the Fermi paradox and the problem of extraterrestrial life* (2nd ed.). Springer.

Wegner, D. (2002). *The illusion of conscious will*. Bradford Books/MIT Press.

Weil, A. (2001). *Breathing: The master key to self healing*. Sounds True Audiobooks.

Weiskrantz, L., Warrington, E. K., Sanders, M. D., & Marshall, J. (1974). Visual capacity in the hemianopic field following a restricted occipital ablation. *Brain*, *97*(1), 709–728.

Weizenbaum, J. (1966). ELIZA—a computer program for the study of natural language communication between man and machine. *Communications of the ACM*, *9*(1), 36–45.

Westerling, A. L., Hidalgo, H. G., Cayan, D. R., & Swetnam, T. W. (2006). Warming and earlier spring increase western US forest wildfire activity. *Science*, *313*(5789), 940–943.

Whishaw, I. Q. (1990). The decorticate rat. In B. Kolb & R. Tees (Eds.), *The cerebral cortex of the rat* (pp. 239–267). MIT Press.

Whitelaw, M. (2004). *Metacreation: Art and artificial life*. MIT Press.

Whitwell, R. L., Striemer, C. L., Nicolle, D. A., & Goodale, M. A. (2011). Grasping the non-conscious: Preserved grip scaling to unseen objects for immediate but not delayed grasping following a unilateral lesion to primary visual cortex. *Vision Research*, *51*(8), 908–924.

Wightman, F. L., & Kistler, D. J. (1989). Headphone simulation of free-field listening. I: Stimulus synthesis. *Journal of the Acoustical Society of America*, *85*(2), 858–867.

Wilber, K. (2001). *No boundary: Eastern and Western approaches to personal growth* (with new Preface). Shambhala Press.

Wilber, K. (2017). *A brief history of everything* (20th anniversary ed., with new Afterword). Shambhala Press.

Wilkinson, A., Sebanz, N., Mandl, I., & Huber, L. (2011). No evidence of contagious yawning in the red-footed tortoise *Geochelone carbonaria*. *Current Zoology*, *57*(4), 477–484.

Wilson, D. S. (2002). *Darwin's cathedral: Evolution, religion, and the nature of society*. University of Chicago Press.

Wilson, M. (2002). Six views of embodied cognition. *Psychonomic Bulletin & Review*, *9*(4), 625–636.

Wilson, R. A. (1994). Wide computationalism. *Mind*, *103*(411), 351–372.

Wilson-Mendenhall, C. D., Simmons, W. K., Martin, A., & Barsalou, L. W. (2013). Contextual processing of abstract concepts reveals neural representations of nonlinguistic semantic content. *Journal of Cognitive Neuroscience*, *25*(6), 920–935.

Winawer, J., Huk, A. C., & Boroditsky, L. (2008). A motion aftereffect from still photographs depicting motion. *Psychological Science*, *19*(3), 276–283.

Winfree, A. T. (1984). The prehistory of the Belousov-Zhabotinsky oscillator. *Journal of Chemical Education, 61*(8), 661–663.

Winograd, T. (1972). *Understanding natural language*. Academic Press.

Winter, B., Marghetis, T., & Matlock, T. (2015). Of magnitudes and metaphors: Explaining cognitive interactions between space, time, and number. *Cortex, 64*, 209–224.

Wise, T. (2012). *Dear white America: Letter to a new minority*. City Lights.

Witt, J. K. (2011). Action's effect on perception. *Current Directions in Psychological Science, 20*(3), 201–206.

Witt, J. K., & Proffitt, D. R. (2008). Action-specific influences on distance perception: A role for motor simulation. *Journal of Experimental Psychology: Human Perception and Performance, 34*(6), 1479–1492.

Witt, J. K., Proffitt, D. R., & Epstein, W. (2005). Tool use affects perceived distance, but only when you intend to use it. *Journal of Experimental Psychology: Human Perception and Performance, 31*(5), 880–888.

Wojnowicz, M. T., Ferguson, M. J., Dale, R., & Spivey, M. J. (2009). The self-organization of explicit attitudes. *Psychological Science, 20*(11), 1428–1435.

Woodley, D. (2017). *Globalization and capitalist geopolitics*. Routledge.

Wright, R. (2017). *Why Buddhism is true: The science and philosophy of meditation and enlightenment*. Simon and Schuster.

Wu, L. L., & Barsalou, L. W. (2009). Perceptual simulation in conceptual combination: Evidence from property generation. *Acta Psychologica, 132*(2), 173–189.

Yang, M., Chan, H., Zhao, G., Bahng, J. H., Zhang, P., Král, P., & Kotov, N. A. (2017). Self-assembly of nanoparticles into biomimetic capsid-like nanoshells. *Nature Chemistry, 9*(3), 287–294.

Yee, E., Huffstetler, S., & Thompson-Schill, S. L. (2011). Function follows form: Activation of shape and function features during object identification. *Journal of Experimental Psychology: General, 140*(3), 348–363.

Yee, E., & Sedivy, J. C. (2006). Eye movements to pictures reveal transient semantic activation during spoken word recognition. *Journal of Experimental Psychology: Learning, Memory, and Cognition, 32*(1), 1–14.

Yee, E., & Thompson-Schill, S. L. (2016). Putting concepts into context. *Psychonomic Bulletin & Review, 23*(4), 1015–1027.

Yoshimi, J. (2012). Active internalism and open dynamical systems. *Philosophical Psychology, 25*(1), 1–24.

Yoshimi, J., & Vinson, D. W. (2015). Extending Gurwitsch's field theory of consciousness. *Consciousness and Cognition, 34,* 104–123.

Young, J. (2012). *What the robin knows: How birds reveal the secrets of the natural world.* Houghton Mifflin Harcourt.

Young, M. P., & Yamane, S. (1992). Sparse population coding of faces in the inferotemporal cortex. *Science, 29,* 1327–1331.

Zakaria, F. (2012). *The post-American world, release 2.0.* W. W. Norton.

Zatorre, R. J., Chen, J. L., & Penhune, V. B. (2007). When the brain plays music: Auditory–motor interactions in music perception and production. *Nature Reviews Neuroscience, 8*(7), 547–558.

Zayas, V., & Hazan, C. (2014). *Bases of adult attachment.* Springer.

Zeidan, F., Johnson, S. K., Diamond, B. J., David, Z., & Goolkasian, P. (2010). Mindfulness meditation improves cognition: Evidence of brief mental training. *Consciousness and Cognition, 19*(2), 597–605.

Zhabotinsky, A. (1964). Periodical process of oxidation of malonic acid solution. *Biofizika, 9,* 306–311.

Zubiaga, A., Liakata, M., Procter, R., Hoi, G. W. S., & Tolmie, P. (2016). Analysing how people orient to and spread rumours in social media by looking at conversational threads. *PloS One, 11*(3), e0150989.

Zwaan, R. A., & Pecher, D. (2012). Revisiting mental simulation in language comprehension: Six replication attempts. *PloS One, 7*(12), e51382.

Zwaan, R. A., & Taylor, L. J. (2006). Seeing, acting, understanding: Motor resonance in language comprehension. *Journal of Experimental Psychology: General, 135*(1), 1–11.

Index

Ackerman, Diane, 201, 288
Action-perception cycle, 126–133, 147, 153, 184–188, 276–277
Active matter, 218–219, 221, 223, 289–290
Adamatsky, Andy, 193, 283, 287
Affordances, 127–129, 132–133, 276
Agent-based modeling, 165–168, 283–284
Aldenderfer, Mark, 231–232, 291
Alien hand syndrome, 19, 263–264
Allen, James, 103–104, 274
Anderson, Michael, 69–70, 270
Animal (nonhuman) cognition, 66–68, 177, 182–185, 270, 286
Anthony, David W., 187, 287
Artificial Intelligence, 11, 75–77, 101–110, 221–223, 271–274
Artificial life, 96–97, 218–223, 229, 272, 289–290
Atmanspacher, Harald, 142–143, 261, 278
Auditory cortex, 56–59, 116, 268–269
Augustine, Keith, 28, 262–263
Autocatakinetics, 217–218, 223, 289–290
Autocatalysis, 203, 216–217, 223, 227, 228, 289–290
Autopoiesis, 95–97, 203, 216, 272
Axelrod, Robert, 167, 284

Backer, Kristina, 59, 269
Bajcsy, Ruzena, 103, 274
Bak, Per, 226–228, 290
Balasubramaniam, Ramesh, 131–132, 273, 276, 291
Ballard, Dana, 138–139, 267, 268, 269, 277
Banaji, Mahzarin, 162–163, 283
Bar, Moshe, 52–53, 268
Barad, Karen, 293
Barlow, Horace, 23, 52, 261–262
Barnes, Rory, 210, 289
Barrett, Lisa Feldman, 94–95, 272
Barrett, Robert, 160, 282
Barsalou, Lawrence, 62–63, 82–83, 90, 269, 271–274, 294
Bateson, Gregory, 202–203, 288
Bauby, Jean-Dominique, 79–80, 271
Baumeister, Roy, 162, 174, 283
Beer, Randall, 96–97, 265, 272
beim Graben, Peter, 142–143, 265, 278
Belousov, Boris, 216–217, 228, 289, 294
Bénard convection cells, 215–216, 220
Bergen, Benjamin, 85, 271–272
Berns, Gregory, 183, 286
Blakeslee, Sandra, 257
Blindsight, 34–35, 264
Boroditsky, Lera, 65–66, 269, 272
Bosbach, Simone, 80, 271
Botvinick, Matthew, 134–135, 277
Bourdain, Anthony, 150–151, 279

Boynton, Geoffrey, 54, 268
"Brain in a vat," 74–82, 103
Breazeal, Cynthia, 105–106, 274
Bridgeman, Bruce, 123–126, 131, 137, 252, 275–276
Brooks, Rodney, 104, 274
Buber, Martin, 253, 294
Butterfly effect, 227, 247, 255
BZ reaction, 216–217, 228, 289

Calvo-Merino, Beatriz, 98, 273
Cangelosi, Angelo, 105, 274
Capra, Fritjof, 202, 288
Carello, Claudia, 219–221, 223, 229, 275, 280, 287, 290, 294
Carriveau, Rupp, 217, 289, 294
Chamovitz, Daniel, 191–192, 287
Change blindness, 4–5, 17, 34, 258
Chatterjee, Anjan, 88–90, 264, 272
Chemero, Anthony, 272, 275, 276, 282
Choice blindness, 9–10, 258
Chopra, Deepak, 20, 293
Clark, Andy, 102–104, 233, 253, 274–275, 282, 292, 294
Climate change, 198–201, 247–248, 288
Cluff, Tyler, 131–132, 276
Cognitive linguistics, 86–92, 272–274
Cohen, Jonathan, 134–135, 263, 277
Complex systems, 41–44, 140–143, 159, 168–169, 173–174, 201, 219, 244–245, 278, 282, 293
Computer vision, 103, 119, 274
Concepts and categories, 61–66, 269
Conceptual metaphor, 86–87, 272–273
Consciousness, 21, 28, 68–69, 79, 190, 196, 241–245, 261, 270, 293–294
Cooperation
  between brain regions, 30, 64, 263, 267–270
  between brain and body, 110, 267, 277
  between neurons, 20, 22, 261–262
  between nations, 169–172, 197–198, 284–285, 288

  between other animals, 184–185, 283, 286–287
  between people, 151, 161, 165–175, 283–284, 293
Cottrell, Gary, 262, 267, 286
Creativity, 30, 231, 251, 293
Crick, Francis, 69, 270
Criminogenic circumstances, 40, 264–265
Cultural diversity, 170–172, 283–285

Dale, Rick, 11, 154–156, 259, 268, 278, 281–282
Damasio, Antonio, 92, 258, 263, 272
Darwin, Charles, 23, 190–191, 219, 283, 287
Darwin, Francis, 190–191, 287
Davis, Tehran, 218, 282, 290, 294
Decety, Jean, 98, 273
Descartes, René, 26–28, 293–294
Desimone, Robert, 54, 268
de Waal, Frans, 185, 286
Dietert, Rodney, 180–181, 285
Dietrich, Eric, 222, 267, 274, 290
Distributed representation, 24–25, 82–84, 90, 261–262
Dixon, J., 217–218, 220, 277, 282, 290, 294
Drake, Frank, 211–213, 227, 289
Drake, Nadia, 227, 290, 293
Dreams, 77–78, 271
Dreyfus, Hubert, 76–77, 271, 274
Duncan, John, 54, 268
Dunning, David, 8, 258
Durgin, Frank, 134–135, 277
Dynamical systems, 41–44, 140–143, 151–153, 159, 168–169, 173–174, 201, 244–245, 278, 282, 293

Eagleman, David, 258–259
Ecological perception, 120–133, 219–220, 275–276
Edelman, Shimon, 261, 270, 272, 278, 292

Index

Electroencephalography (EEG), 31–35, 60–61, 154, 264
ELIZA, 75–76
Ellis, Rob, 127, 276
Elman, Jeffrey, 59–60, 252, 264, 269
Ely, Robin, 171–172, 284
Embodied cognition, 81–86, 95–101, 109–110, 271–274
   in concepts, 63–66, 262
   in emotion, 92–95, 272
   in language, 86–92, 272–274
   in robotics, 101–110, 274
Emergence, 203–204, 215–227, 267, 291, 294
   in decision-making, 20–22, 41–44, 69–70, 265
   in dynamical simulations, 141–143, 278
   in robotics, 104, 274
Emotion, 92–95, 272
Entropy, 203, 219–220, 229
Evolution
   biological, 67–68, 95, 121–122, 173–174, 182, 185, 191, 194–197, 202–204, 212, 218–219, 228–230, 262, 283–284, 288, 291
   cultural, 173–174, 187–188, 202, 231–233, 262, 283–284
Extended cognition, 121, 135–137, 184
   in joint action, 153–159, 280–282
   in problem solving, 128–131, 171–172, 276–277, 284
   in visual memory, 138–140, 277–278
Extraterrestrial life, 209–214, 288–289
Eye movements, 10, 22, 79, 122–126, 260
   in change blindness, 34
   in conversation, 154–156
   in decision-making, 13–14, 22
   in figurative language, 91–92
   in problem solving, 129–130
   in visual imagery, 82–83
   in visual memory, 138–140

Fadiga, Luciano, 100, 273
False confessions, 8–9, 258
False rumors, 170, 247, 266, 287
Fascia (connective tissue), 108, 115–117, 120, 243, 254
Fazio, Russell, 163–164, 283
Fechner, Gustav, 190, 223, 287
Ferguson, Melissa, 164, 283
Fermi paradox, 213, 289
Flashbulb memories, 6–8, 258
Floatation tank, 78–79, 271
Fodor, Jerry, 48–51, 267, 275
Fossey, Dian, 188, 287
Frank, Adam, 211, 289
Frank, Michael, 64, 269
Freeman, Jonathan, 93, 259, 272, 283
Free will, 20–22, 26, 31–44, 264–266
Functional magnetic resonance imaging (fMRI), 33, 53–54, 60, 93, 98, 281
Fusaroli, Riccardo, 157, 282
Fuster, Joaquin, 30, 263

Gage, Phineas, 29–30
Gaia hypothesis, 202, 234
Gallese, Vittorio, 97–98, 273
Game theory, 165–168, 283–284
Gardner, Howard, 184–185, 286–287
Garrod, Simon, 153–154, 282
Geng, Joy, 82–83, 271
Gentilucci, Maurizio, 100–101, 273
Gibbs, Raymond, 87, 258, 272, 276
Gibson, James, 120–131, 219–220, 275–276
Gladwell, Malcolm, 8, 259
Glenberg, Art, 84–85, 271
Glucksberg, Samuel, 128–130, 276
Goodale, Melvyn, 35, 264
Goodall, Jane, 188, 287
Gosling, Samuel, 145, 278–279
Gow, David, 60–61, 269
Grant, Elizabeth, 129, 276
Greenwald, Anthony, 162–163, 283

Gross, Charles, 261–262
Group cognition, 157–161, 184, 282–283

Haggard, Patrick, 32, 263, 273
Haken-Kelso-Bunz model, 152, 280
Hall, Lars, 9–10, 258
Hameroff, Stuart, 21, 261
Haney, Craig, 40, 265
Harari, Yuval, 222, 285, 290
Hasson, Uri, 269, 281
Hayhoe, Mary, 34, 139, 258, 264, 277
Haynes, John-Dylan, 33, 36, 242, 263, 282
Hazen, Robert, 229–230, 291
Hebb, Donald, 78, 271
Henrich, Joseph, 173–174, 285
Holobiont Theory, 181–182, 285
Homo sapiens, 37, 173–174, 222, 232–233, 236, 243, 291
Homunculus, 48–49, 267
Hotton, Scott, 141–142, 278
Hove, Michael, 291
Human-computer interaction, 159–160, 282
Hurst, Harold, 225, 290
Hutchins, Edwin, 159, 282, 288, 292

Ikegami, Takashi, 221, 223, 282, 290
Image schema, 87–92, 272–273
Inaccurate testimony, 8, 258
Ingold, Tim, 204, 288
Innocence Project, 8
Intelligence, 75–77, 103–110, 184–185, 211–212, 220–223, 271, 273, 285–286, 289
Interactionism, 43–44, 49–51, 56–70, 95–96, 267–270
International politics, 169–173, 284–285
Iriki, Atsushi, 133, 277
Irwin, David, 123, 275–276

Järvilehto, Timo, 140, 142, 278
Johansson, Roger, 278

Johnson, Bert, 225, 290
Johnson, Mark, 86, 272
Johnson, Will, 146, 279
Josephson junction, 280
Juarrero, Alicia, 41–42, 265
Jurafsky Daniel, 279

Kaas, Jon, 57, 269
Kahneman, Daniel, 259
Kanwisher, Nancy, 91, 272
Kaschak, Michael, 85, 271
Kassin, Saul, 9, 258
Kauffman, Stuart, 214, 290, 291, 293
Kay, Bruce, 218, 290, 294
Keenan, Maynard James, 190
Kello, Chris, 44, 225, 262, 265, 282, 291
Kelso, J. Scott, 151–153, 156, 225, 280
Kim, Albert, 35, 264
Kirkham, Natasha, 155, 272, 282
Kirsh, David, 130–131, 137, 277, 278
Klosterman, Chuck, 2, 246–248
Knoblich, Günther, 158, 259, 271, 280, 282
Koch, Christof, 69, 262, 270
Kondepudi, Dilip, 218, 290, 294
Konvalinka, Ivana, 280
Kosinski, Michal, 279
Kosslyn, Stephen, 82–83, 271
Kourtzi, Zoe, 91, 272
Kruglanski, Arie, 162, 283
Kuhlen, Anna, 154, 282

Laeng, Bruno, 278
Lakoff, George, 86–89, 272–273
Langacker, Ron, 87–89
Language processing, 83–92, 98–101, 153–158, 259–260, 264, 268–269, 272, 281–282
Lestel, Dominique, 188–189, 287
Lettvin, Jerome, 261, 268
Levin, Dan, 5, 258
Lewis, Michael, 259

Liberman, Alvin, 100, 273
Libet, Benjamin, 31–33, 36–37, 44, 242, 263–265
Living technology, 221–223, 289–290
Locked-in syndrome, 79–80, 271
Louwerse, Max, 156–157, 269, 272, 282
Lovelock, James, 202, 234, 292
Lupyan, Gary, 55–56, 262–263, 268–271, 282

Macknik, Steve, 4, 257
Macmillan, Malcolm, 29, 263
Maglio, Paul, 130–131, 159–161, 277, 282
Magnetoencephalography (MEG), 53, 60–61
Mandler, Jean, 89–90, 272
Margulis, Lynn, 202, 234, 292
Martin, Alex, 63, 269, 271, 273
Martinez-Conde, Susana, 257
Matlock, Teenie, 91, 198–199, 272, 288, 291
Matsumoto, Riki, 59, 269
Maturana, Humberto, 203, 261, 272, 288
McClelland, James, 50–51, 59–60, 267, 269
McComb, Karen, 187–188, 287
McGurk, Harry, 58–60, 269
McKinstry, Chris, 11, 259
McLaughlin, Judith, 35, 264
McMurray, Bob, 259, 260, 264, 269, 280–281
McRae, Ken, 90, 269, 272
Meditation, 2, 78–79, 250–251, 294
Mental representations, 62–63, 121–128, 275–277
Mental rotation, 126, 130–131
Merker, Björn, 67–69, 270
Mermin, David, 253, 294
Meteyard, Lotte, 81, 271–272
Microbes, 178–182, 229–230, 285, 291
Miklósi, Adam, 183, 286

Modularity, 48–51, 62–70, 93–94, 242, 267–268
Moore's law, 201–202
Morphological computation, 106–108, 117–120, 129, 136, 274
Motor coordination, 35, 104, 131–132, 151–158, 185, 188, 281–282
Motor cortex, 31–32, 97–100, 120–121, 152–153
Motor theory of speech perception, 100, 273
Mousetracking, 11–12, 164, 259
Moyes, Holley, 231, 291

Nazir, Tatiana, 99–100, 273
Near-death experiences, 27–28, 262
Neisser, Ulric, 6–8, 128, 130, 252, 258
Neuronal population code, 24–25, 82–84, 90, 261–262
Neurons, 1, 20–26, 33, 42, 48–52, 56–57, 70–71, 101–102, 115–117, 253–254, 260–261
 "grandmother" cells, 23–25, 261–262
 mirror neurons, 97–98, 273
 multi-sensory parietal neurons, 133, 277
 sensory reorganization of neurons, 135, 268–269
Noelle, David, 263

Odum, Howard, 233–234, 292
Olson, Bruna, 60–61, 269
1/f noise, 43–44, 225–228, 290–291
Oppenheimer, Daniel, 64, 269
Osterhout, Lee, 35, 264

Pallas, Sarah, 56–57, 268
Parietal cortex, 35, 133, 263
Pärnamets, Philip, 13–15, 37, 258, 260
Parnia, Sam, 28, 262
Penrose, Roger, 21, 261
Perceptual simulation, 63–64, 83–85, 90–92

Pereboom, Derk, 37–38, 265
Pets, 66–67, 177, 182–183, 186–189, 283, 286, 291–292
Pezzulo, Giovanni, 105, 274, 283
Physical intelligence, 220, 221, 223, 229, 289–290
Pickering, Martin, 153–154, 282
Pineal gland, 26–28, 293–294
Pink noise, 43–44, 225–228, 290–291
Plant cognition, 190–194, 287
Power law, 226–227, 290–291
Prefrontal cortex, 29–33, 41–45, 48–53, 56, 61, 69, 263
Prejudice, 162–165, 170–172
Prigogine, Ilya, 203, 218, 227, 265, 288–289
Prisoner's dilemma, 165–168, 283–284
Problem solving, 128–131, 171–172, 276, 284
Proffitt, Dennis, 133–134, 277
Proprioception, 120–124, 135
Pulvermüller, Friedmann, 98–99, 105, 273
Putnam, Hilary, 75–81, 271, 274
Pylyshyn, Zenon, 101–102, 119, 137–138, 273–274, 276–277

Quantum mechanics, 20–23, 45, 49, 244–245, 253, 261, 280, 293, 294

Ramachandran, V. S., 110–111, 135, 274, 277
Randi, James, 3–4, 257
Receptive fields, 23–24, 52, 57, 133, 267–268
Reese, Richard Adrian, 199, 288
Richardson, Daniel, 13, 90–91, 130, 139–140, 154–156, 259, 272, 277–278
Rizzolatti, Giacomo, 97–98, 100, 273
Robots, 76–77, 101–110, 116–120, 212–215, 220–223, 274, 290
Roepstorff, Andreas, 280

Rosen, Robert, 204, 288, 291
Roy, Deb, 105, 274, 287
Rozenblit, Leon, 5
Rumelhart, David, 50–51, 267
Rynecki, Elizabeth, 252, 293

Sandberg, Anders, 212
Schmidt, Richard, 153, 156, 225, 280
Sebanz, Natalie, 158, 259, 282, 286
Sekuler, Allison, 54, 268
Self-organization, 42–44, 95–96, 104, 202–204, 215–221, 223, 227–229, 234–235, 289–291
Senju, Atsushi, 186–187, 286
Sensory deprivation, 78–79, 271
Sensory transduction, 115–120, 274–275
"Sequoia," 280–281
Service science, 159–161, 171, 282
Shahin, Antoine, 59, 269
Shams, Ladan, 54–55, 58, 268
Shebani, Zubaida, 99, 273
Shockley Kevin, 156, 282
Simmons, Kyle, 63, 269, 273, 277
Simons, Dan, 5, 258
Smaldino, Paul, 167–168, 283–284
Smith, Linda, 85–86, 271, 274
Snezhko, Alexey, 218, 220, 221, 223, 290
Social cognition, 161–165, 282–283
Social media, 95, 233, 266, 287
Sparse coding, 24–26, 31, 261–262
Speech perception, 58–61, 269
Stager, Curt, 1, 257
Stanley, Jason, 164, 285
Stark, Lawrence, 124, 276
Stereotypes, 162–165, 170–172
Strogatz, Steven, 153, 280, 290
Subcortex, 66–69, 77, 93–94, 263, 270
Sullivan, Woodruff, 211, 289
Sutton, John, 232–233, 275, 291
Swenson, Rod, 218–219, 290
Symmetry breaking, 215–216

Synchrony, 31, 151–158, 218–219, 225, 246, 252, 280–282
Synesthesia, 47
Systems theory, 41–44, 131–132, 140–143, 152–159, 180–181, 186–187, 197, 201–204, 233–234

Talmy, Leonard, 88–91, 234
Tanenhaus, Michael, 258, 260, 269, 282, 295
Tauber, Alfred, 181–182, 285
Tensegrity, 108
Tero, Atsushi, 193, 287
Theiner, Georg, 184, 275, 286
Thellier, Michel, 192, 287
Thiam, Pierre, 279
Thompson-Schill, Sharon, 30, 262–263, 269, 276
Tit-for-tat strategy, 165–168, 283–284
Torres, Phil, 210, 285
Transcranial direct current stimulation (tDCS), 30, 263
Transcranial magnetic stimulation (TMS), 99–100, 273
Tucker, Mike, 127, 276
Turner, Mark, 294
Turvey, Michael, 219–221, 223, 229, 265, 272, 273, 274, 275, 280, 287, 288, 290, 291, 294

United Nations, 172, 197–198

Vanderschraaf, Peter, 166–167, 284
Van Orden, Guy, 42–44, 70, 225, 252, 265, 291, 294
Varela, Francisco, 95–97, 203, 272, 288
Visual cortex, 34, 47, 52–57, 82–83, 91, 267–268
Visual imagery, 82–84, 87, 90–92, 271
Visual indexing, 137–140, 277
Visual perception, 51–56, 122–126, 268, 275–276

Watts, Alan, 247–249
Webb, Barbara, 106–107, 274
Wegner, Daniel, 36, 265
Weil, Andrew, 1, 257
Weizenbaum, Joseph, 75, 271
Wilber, Ken, 240–241, 292
Winograd, Terry, 76, 271
Wisdom of the crowd, 184, 282–283
Witt, Jessica, 133–134, 277
Wojnowicz, Michael, 164, 283
Wright, Robert, 251, 293–294
Wu, Ling-Ling, 63, 269

Yawn contagion, 186–187, 286
Yee, Eiling, 127, 260, 262, 269, 276
Yoshimi, Jeff, 141–142, 261, 278

Zhabotinsky, Anatol, 216–217, 228, 289
Zwaan, Rolf, 83–84, 90, 269, 271